T0258544

Oxidative Stress and Related Diseases

Volume I

Oxidative Stress and Related Diseases Volume I

Edited by **Nick Gilmour**

New York

Published by Callisto Reference,
106 Park Avenue, Suite 200,
New York, NY 10016, USA
www.callistoreference.com

Oxidative Stress and Related Diseases
Volume I
Edited by Nick Gilmour

International Standard Book Number: 978-1-63239-503-0 (Hardback)

Contents

Permissions

List of Contributors

Preface

This book has been an outcome of determined endeavour from a group of educationists in the field. The primary objective was to involve a broad spectrum of professionals from diverse cultural background involved in the field for developing new researches. The book not only targets students but also scholars pursuing higher research for further enhancement of the theoretical and practical applications of the subject.

This book responds to the urge to find, in a sole document, the impact of oxidative stress at distinct levels as well as treatment with antioxidants to react and decrease the damage. The contents of the book are grouped under two sections: cell biology, chemical free radicals & antioxidants and cell damage by oxidative stress. The book is a collaborative effort for progressing towards the prevention and treatment of chronic degenerative diseases. It is intended for health professionals, as researchers from across the world have contributed in it.

It was an honour to edit such a profound book and also a challenging task to compile and examine all the relevant data for accuracy and originality. I wish to acknowledge the efforts of the contributors for submitting such brilliant and diverse chapters in the field and for endlessly working for the completion of the book. Last, but not the least; I thank my family for being a constant source of support in all my research endeavours.

Editor

Cell Biology, Chemical Free Radicals and Antioxidant Defenses

The Exogenous Antioxidants

Alejandro Chehue Romero,
Elena G. Olvera Hernández, Telma Flores Cerón and
Angelina Álvarez Chávez

Additional information is available at the end of the chapter

1. Introduction

One theory that was initially questioned was the proposal of Dr. Denhan Harman (1956) of the University of Nebraska. He was the first researcher to propose Free radicals (FR) as an important cause of cellular aging. Today this theory enjoys wide approval. FR are "disequilibrated" molecules that travel through our organism attempting to capture an electron of the stable molecules to obtain its electrochemical stability [1,2].

FR perform many useful functions in the organism (in fact, our own bodies manufacture these in moderate amounts to combat, for example, infections). When the increase of the intracellular contents of FR exceeds the cells' antioxidant defenses and are not efficient for inhibiting them, this causes organic damage known as Oxidative stress (OS), which leads to a variety of physiological and biochemical changes that induce damage to biological molecules such as nucleic acids, proteins, lipids, etc., which consequently cause deterioration and cell death. An FR comprises any atom or group that possesses one or more unpaired electrons; thus, FR are very reactive[3].

OS traditionally has been considered a static cell-damage process that derives from the aerobic metabolism, and its clinical importance has been recognized to the point of currently being considered a central component of any pathological process. OS in diverse pathological states affects a wide variety of physiological functions, contributing to or providing biofeedback on the development of a great number of human degenerative diseases, such as atherosclerosis, diabetes, cardiomyopathies, chronic inflammatory diseases (rheumatoid arthritis, intestinal inflammatory intestinal disease, and pancreatitis), neurological diseases, high blood pressure, ocular diseases, and pulmonary and hematological disease, cancer, and immunodepression, asthma, among others [4].

This implication does not mean that Reactive oxygen species (ROS) always play a direct role in the development of the disease. In fact, reactive species predispose the organism to diseases caused by other agents. In many cases, oxidative damage is to a greater degree the consequence of the tissue damage that the disease produces than a cause of the disease itself and therefore can contribute to worsening of the tissue damage generated [3].

While our own body produces FR in moderate amounts, amounts that decrease when we age, we must also bear in mind ROS-generated exogenous sources in organisms, such as antibiotics, drugs, alcohol, tobacco, stress, contaminants, chemotherapy, and exposure to Ultraviolet (UV) and ionizing radiation.

On the other hand, numerous epidemiological studies suggest that more persons could avoid the appearance of pathological processes if they consumed antioxidant-rich diets (fruits and vegetables). Thus, it would be possible to protect the organism more efficiently against OS, with the presentation of lesser risk of developing human degenerative diseases.

This has led to conducting experiments to identify the specific components responsible for the positive effects on health by the consumption of foods of plant origin. One explanation that has found great acceptance is that this is due to the presence of antioxidant nutrients such as vitamins C and E, carotenoids, flavonoids, selenium, etc., which would interfere with oxidative damage to the DNA, proteins, and lipids [3].

Antioxidants are synthetic or natural substances that present in low concentrations compared with the biomolecules that they should protect. Antioxidants protect by retarding or inhibiting the harmful effects of FR. They are classified as follows: endogens (glutathione, co-enzyme Q, etc.), which are manufactured by the cell itself; exogens, which enter the organism through the diet (existing in determined foods) or through supplements with antioxidant formulations, and co-factors (copper, zinc, manganese, iron, and selenium). The consumption of antioxidant exogens can increase protection of the body and aid antioxidant endogens in combating diseases [5].

Fortunately, numerous foods and supplements that we ingest are rich in the antioxidants that protect against damage to the cells. Vitamin C, which is found in abundance in citrics and vegetables, is perhaps the best known antioxidant. Vitamin E, which is liposoluble, can be found in nuts, unrefined vegetable oils such as corn, cotton seed, and wheat germ, and in whole grains. Beta carotene, which is converted into vitamin A in the organism, can be found in dark-leafed vegetables, carrots, and sweet potatoes.

In recent years, plant-derived natural antioxidants have been used frequently, given that they present activity that is comparable with the most frequently employed synthetic antioxidants. Antioxidants are also found in a variety of herbs and foods that are to a great extent unknown in and not easily available in our environment, such as green tea, cardo mariano, ginkgo biloba, pine bark, and red wine; however, we do have dulcamara, dragon's blood, cat's claw, anamu/guinea hen weed), garlic, onion, aloe vera, and others that are very rich in antioxidants.

Many benefits are conferred on antioxidants against diverse pathological states; in adition to this, an unequaled richness in natural foods is exhibited as well as our obligation to take advantage of and assess these.

In the present work, the description is performed of the characteristics of the exogenous antioxidants with regard to their employment in human health [6].

2. Vitamins

Vitamins are organic micronutrients that possess no energetic value, are biologically active, and with diverse molecular structure, which are necessary for humans in very small quantities (micronutrients) and which should be supplied by the diet because humans are unable to synthesize and which are essential for maintaining health [7].

The majority of vitamins are not synthesized by the organism, some can be formed in variable amounts in the organism (vitamin D and niacin are synthesized endogenously; the former forms in the skin by exposure to the sun, niacin can be obtained from tryptophan, and vitamins K2, B1, B2, and biotin are synthesized by bacteria). However, this synthesis is generally not sufficient to cover the organism's needs. [8,9].

The functions of the vitamins and the need of the organism for these are highly varied. Persons always need vitamins and at all life stages. However, during specific periods such as growth, pregnancy, lactancy, and disease, the needs are increased [8].

The majority of vitamins have a basic function in the maintenance of health (doing honor to their name: "vita" means life. The term vitamin, proposed for the first time (in 1912) by Polish Chemist Casimir Funk, is demonstrated by the appearance of deficiency or deficiency-related diseases that were caused by the lack of vitamins in the diet; for example, lack of vitamin A can produce blindness and the lack of vitamin D can retard bone growth; vitamins also facilitate the metabolic reactions necessary for utilization of proteins, fats, and carbohydrates.

In addition, today we know that their nutritional role extends beyond that of the prevention of deficiency or deficiency-associated diseases. They can also aid in preventing some of the most prevalent chronic diseases in developed societies. Vitamin C, for example, prevents scurvy and also appears to prevent certain types of cancer. Vitamin E, a potent antioxidant, is a protector factor in cardiovascular disease and folates help in preventing fetal neural tube defects [9].

Traditionally, vitamins have been classified into two large groups in terms of their solubility as follows:

Liposoluble vitamins: A (retinol); D (ergocalciferol); E (tocopherol), and K (filoquinone and menadione), which are soluble in lipids but not in water; thus, they are generally vehiculized in the fat found in foods. These vitamins can accumulate and cause toxicity when ingested in large amounts [9].

These are fat-soluble compounds and are found associated in foods with fats, mainly of animal origin, and are absorbed with them. Therefore, any problem with respect to the absorption of fats will be an obstacle to the absorption of liposoluble vitamins. The latter are stored in moderate amounts in the vital organs, especially in the liver [8].

Hydrosoluble vitamins: The following are vitamins of the B group: [B_1 (thiamin); B_2 (riboflavin; B3 (niacin); pantothenic acid; B_6 (pyridoxine); biotin; folic acid, and B_{12} (cyanocobalamin)], and vitamin C (ascorbic acid), contained in the aqueous compartments of foods. [9].

These are water-soluble compounds that are found in foods of animal and plant origin. Different from liposoluble vitamins, water-soluble vitamins are not stored in the body; thus, they should be ingested daily with food to avoid their supply becoming exhausted [8]. The hydrosoluble vitamins participate as co-enzymes in processes linked with the metabolism of organic foods: carbohydrates; lipids, and proteins.

One important difference between these two vitamin groups lies in their final destiny in the organism. An excess of water-soluble vitamins is rapidly excreted in the urine; on the other hand, liposoluble vitamins cannot be eliminated in this manner; they accumulate in tissues and organs. This characteristic is associated with a greater risk of toxicity, which means the ingestion of excessive amounts of liposoluble vitamins, especially vitamins A and E. Vitamin B12 constitutes an exception because it is stored in the liver in important quantities.

2.1. Vitamin C

Vitamin C, also known as ascorbic acid (enantiomer, L-ascorbic acid) is an antioxidant hydrosoluble vitamin, this due to that it is an electron donor, which explains its being a reducer that directly neutralizes or reduces the damage exercised by electronically disequilibrated and instable reactive species, denominated Free radicals (FR).

Action: The presence of this vitamin is required for a certain number of metabolic reactions in all animals and plants and is created internally by nearly all organisms, humans comprising a notable exception [10].Vitamin C is essential for the biosynthesis of collagen proteins, carnitine (which is a pro-catabolic transporter of fatty acids in the mitochondria), neurotransmitters (mediators of cell communications, primarily of nerve expression), neuroendocrine peptides, and in the control of angiogenesis; it aids in the development of teeth and gums, bone, cartilage, iron absorption, the growth and repair of normal connective tissue, the metabolism of fats, and the scarring of wounds; it promotes resistance to infections by means of the immunological activity of the leukocytes [11].

In addition to the biological functions mentioned, there are an infinite number of scientific and pseudoscientific reports that qualify this vitamin as an immunomodulator, an antiviral influenza protector, an antiatherogenic, an antiangiogenic, and as an anti-inflammatory, and debate continues on its activity in cancer and its antioxidant properties, given that there is information that lends support to its procancerigenenous and to its role as a pro-oxidant. Currently, this vitamin is the most widely employed vitamin in drugs, premedication, and nutritional supplements worldwide [11]. Various lines of experimental and epidemiological evidence suggest that vitamin C is a powerful antioxidant in biological systems, both *in vitro* as well as *in vivo*. Health benefits have been attributed to vitamin C, such as the anticancerigenous, immunoregulator, antiinflammatory, and neuroprotector effect. Vitamin C rapidly eliminates Reactive oxygen species (ROS), Reactive nitrogen species (ROS), or both, and reduces the transitional metallic ions of specific biosynthetic enzymes; thus, it can prevent biological oxidation (García G.A., et al., 2006). The damage exercised by electronically disequilibrated and instable Reactive oxygen-derived species (ROS) (Free radicals, FR), nitrogen-derived FR, NOS), and sulfa-derived or mixed FR harm through oxidation any of the cellular macromolecular components. If these are not neutralized, so-called "propagation" or "amplification" is produced and, in the case of oxidation, the peroxides are again oxidized into peroxyls [12].

Clinical Uses: Vascular diseases, cancer, cataracts, High blood pressure, acute pancreatitis, the common cold, iron fixation in blood hemoglobin, dermatological uses (photochronoaging, photoprotection, prevention of contact dermatitis, non-scarring of wounds, and hyperpigmentation) [9].

Foods are substances or products of any nature that due to their characteristics and components are utilized for human nutrition. Ascorbic acid, commonly known as vitamin C, promotes resistance to infections by means of the immunological activity of leukocytes; it is useful for preventing and curing the common cold, as well as improving iron absorption in the human body and diminishing the incidence of anemia caused by lack of this mineral, which presents a high incidence in Mexican population.

Chemical structure: Ascorbic acid is a 6-carbon ketolactone that has a structural relationship with glucose; it is a white substance, stable in its dry form, but in solution it oxidizes easily, even more so if exposed to heat. An alkaline pH (>7), copper, and iron also accelerate its oxidation. Its chemical structure is reminiscent of that of glucose (in many mammals and plants, this vitamin is synthesized by glucose and galactose).

Vitamin C is found mainly in foods of plant origin and can present in two chemically interchangeable forms: ascorbic acid (the reduced form), and dihydroascorbic acid (the oxidated form) (See Figure 1), with both forms biologically functional and maintaining themselves in physiological equilibrium. If dihydroascorbic acid is hydrated, it is transformed into diketogluconic acid, which is not biologically active, and with this an irreversible transformation. This hydration occurs spontaneously on neutral or alkaline dissolution.

Deficit: It is well known that a deficiency of vitamin C causes scurvy in humans, thus the origin of the name "ascorbic" given to the acid [10].Scurvy was recognized for the first time in the XV and XVI Centuries as a serious disease contracted by sailors on long sea journeys (it appeared in adults after a nutritional need had existed for >6 months, because sailors had no access to fresh foods, including fruits and vegetables). Prior to the era of research on vitamins, the British Navy established the practice of supplying lemons and other citric fruits to their sailors to avoid scurvy [13]. Scurvy is related with defective collagen synthesis, which manifests itself as the lack of scarring, progressive asthenia, gum inflammation, falling out of the teeth, joint inflammation and pain, capillary fragility, and esquimosis, thus the importance of the ingestion of vitamin C in the diet [11].

Obtaining Vitamin C: This is a nutrient that is localized, above all, in citric fruits and vegetables. All fruits and vegetables contain a certain amount of vitamin C. Foods that tend to be greater sources of vitamin C are, among others, the following: citrics (oranges, limes, lemons, grapefruit); guavas; pineapple; strawberries; kiwis; mangoes; melon; watermelon, and cantaloupe and, as examples of vegetables, green peppers, tomatoes, broccoli, cabbage, cauliflower, green peas, asparagus, parsley, turnips, green tea, and other green-leafed vegetables (spinach), potatoes or sweet potatoes, and yams.

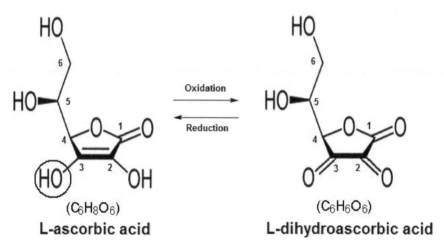

Figure 1. The oxidation-reduction (redox) reaction of vitamin C, molecular forms in equilibrium. L-dihydroascorbic acid also possesses biological activity, due to that in the body it is reduced to form ascorbic acid.

However, it is noteworthy that vitamin C diminishes on boiling, drying, or soaking foods; thus, it is convenient to consume these raw.

Daily recommended doses of ascorbic acid are 75 mg/day (for women) and 90 mg/day (for men). There are between 1.2 and 2 g (20 mg/kg body weight) of ascorbic acid available in the entire organism and its half-life ranges from 10–20 days [11–15].

Absorption: Vitamin C is easily absorbed in the small intestine, more precisely, in the duodenum. It enters the blood by active transport and perhaps also by diffusion. It would appear that the mechanism of absorption is saturable, due to that when large amounts of the vitamin are ingested, the percentage absorbed is much lower (Figure 2). In normal ingestions (30–180 mg), vitamin C is absorbed (bioavailability) at 70–90% vs. a 16% ingestion of 12 g. Its concentrations in plasma are 10–20 mcg/ml.

The vitamin C concentration in the leukocytes is in relation to the concentration of the vitamin in the tissues: therefore, by measuring the concentration of vitamin C in the leukocytes, we can know the real level of the vitamin in the tissues. The pool of vitamin C that humans possess under normal conditions is approximately 1,500 g. When this pool is full, vitamin C is eliminated at a high percentage by the urine in the form of oxalic acid (catabolite) or, if it is ingested in very high amounts, as ascorbic acid. If there are deficiencies, absorption is very high and there is no elimination by urine. Ascorbic acid is found at high concentrations in various tissues, for example, suprarenal, liver, spleen, and kidneys.Alcohol consumption diminishes absorption of the vitamin, and the smoking habit depletes the levels of the vitamin in the organism; thus, it is recommended that smokers and regular alcohol consumers supplement their diet with vitamin C.

The half-life of ascorbic acid in the organism is approximately16 days. Thus, the symptoms of scurvy do not appear for months in subjects with a diet deficient in vitamin C [7].

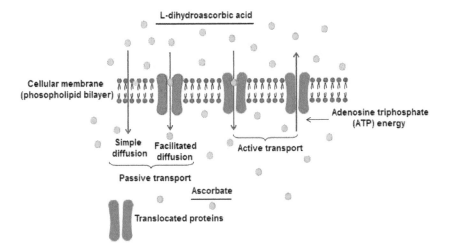

Figure 2. Mechanism of absorption of vitamin C. The L-dihydroascorbic acid molecule is better absorbed than that of L-ascorbic acid. Passive absorption is dependent on a glucose transporter and active absorption is dependent on Na^+.

Toxicity: It is scarcely probable for vitamin C intoxication (megadose) to occur because it is a hydrosoluble vitamin and excesses are eliminated through the urine. But if the daily dose of vitamin C exceeds 2,000 mg/day, the following can appear [16]:

• Diarrhea

• Smarting on urinating

• Prickling and irritation of the skin

• Important alterations of glucose in persons with diabetes

• Insomnia

• Excessive iron absorption

• Formation of oxalate and uric kidney stones.

Great controversy on the theme of Free radicals (FR) and antioxidants such as vitamin C continues, although there is conceptual dispute on whether these are the cause or consequence of the pathology. Biochemically, L-ascorbic acid donates two of its electrons from a double loop between the carbons in positions 2 and 3 (See Figure 1), and this donation is sequential, because the first molecular species generated after the loss of an electron is an FR denominated ascorbile acid. Similar to other FR with an unpaired electron, ascorbile is relatively stable and regularly non-reactive, with half-life of 10–15 seconds.

A great diversity of scientific works has allowed increasing the knowledge of the biological function of vitamin C, but this has also generated doubts, given that controversies have surfaced. One of these controversial points comprises the pro-oxidant activity of vitamin C [12,17]. In the meanwhile, there are starting points, such as that its pro-oxidant activity depends as much on the dose in the diet as on the presence of trace metals, such as iron and free copper, in order for these to produce Fenton-type reactions, and this is amplified by the additional presence of certain FR in the circulating medium [16,18]. This would also depend on the vitamin C-directed reaction.

2.2. Vitamin E

Discovered at the beginning of the 1920s in vegetable oils such as that of wheat germ by Herbert Evans and Katherine Bishop, vitamin E is also denominated tocopherol or the antisterile vitamin, due to its activity. Vitamin E is present in small amounts in all of the cells.

Vitamin E is a group of methylated phenolic compounds known as tocopherols and tocotrienols (a combination of the Greek words "τόκος" [birth] and "φέρειν" [possess or carry], which together mean "to carry a pregnancy"). Alpha-tocopherol is the most common of these and biologically that with the greatest vitaminic action. It is a lipophilic antioxidant that is localized in the cell membranes whose absorption and transport are found to be very highly linked with that of lipids. It is considered the most important lipid molecule protector because its action consists of protecting the polyunsaturated fatty acids of cell membrane phospholipids from cellular peroxidation, and also inhibiting the peroxidation of Low-density lipoproteins (LDL). It oxidizes the oxygen singlet, takes up hydroxyl FR, neutralizes peroxides, and captures the superoxide anion in order to convert it into less reactive forms [1].

Fortunately, the foods with the greatest amounts of Polyunsaturated fatty acids (PFA) also tend to have a high content of this vitamin. Sunflower seed oil, one of the foods richest in PFA, also has the highest content of vitamin E among all of the foods that we habitually consume. It is also found in other vegetable oils, in dry fruit, and in eggs. In the mean diet of Spaniards, vegetable oils furnish 79% of the vitamin E that they consume [9].

Ingestion that adequately covers the recommended allowance appears to behave as a factor of protection in cardiovascular disease, on protecting LDL from oxidation, one of the main risk factors of this pathology. Vitamin E acts jointly and synergically with the mineral selenium, another of the organism's antioxidants.

It can be easily destroyed by the action of heat and of oxygen in the air. Vitamin E is one of the least toxic liposoluble vitamins [9].

Action: It has been proposed that in addition to its antioxidant function, vitamin E can perform a specific physicochemical function in the ordering of the lipic membranes, especially of phospholipids rich in arachidonic acid (thus acting as a membrane stabilizer) [1].

In vivo, vitamin E acts when it breaks the chain of antioxidants, thus preventing the propagation of damage to the biological membranes that give rise to FR, something akin to a protective shield of the cells' membranes that allows them not to age or to deteriorate due to oxygen-containing FR, retarding cellular catabolism, impeding the chain reaction that can produce peroxides from ensuing. It participates in the hemo group and in vitamin E deficiency; hemolytic anemia appears as a result of damage by FR. It also exercises an antitoxic function, a protector in the face of various chemical agents, especially preventing the formation of peroxides from PFA, thus favoring the maintenance and stability of the biological membranes and of the lisosomes in erythrocytes, liver, and muscle [19].

Tocopherols act as intra- and extracellular liposoluble antioxidants within the body. In particular, the tocopherols protect the highly stored fatty acids (PFA) that are present in cellular and subcellular membranes, maintaining the integrity of the biological membranes, as well as other reactive compounds (e.g., vitamins A and C) from the oxidative damage that they could undergo on acting as FR traps. It has also been suggested that the tocopherols play an important role in cell respiration and in DNA and co-enzyme Q biosynthesis.

Tocopherols favor normal growth and development, act as an anticoagulant agent, stimulate the formation of red globules, stimulate the recycling of vitamin C, reduce the risk of the first mortal heart attack in males, protect against prostate cancer, improve immunity, and is a potent antioxidant against cancer in general, cardiac diseases, and FR, thus possessing a potent anti-aging function.

Vitamin E can reduce circulatory problems in the lower limbs, prevent coronary diseases, increase strength and muscular resistance (fostering achievement in sports), drive the sexual metabolism, and relieve menopausal symptoms. It can reduce the formation of scars (stimulating the curing of burns and wounds), could help in the treatment of acne, and is a potential treatment for diaper dermatitis and bee stings.

Chemical structure: The chemical formula for vitamin E ($C_{29}H_{50}O_2$) is utilized for designating a group of eight natural species (vitamers) of tocopherols and tocotrienols (α, β, γ, and δ). Together with vitamins A, D, and K, these constitute the group of liposoluble vitamins, characterized by deriving from the isoprenoid nucleus, soluble in lipids and organic solvents. They are essentials, given that the organism cannot synthesize them; therefore, their contribution is carried out through the diet in small amounts. For efficient absorption by the organism, these require the presence of fatty acids, bile, and lipolytic enzymes of the pancreas and intestinal mucosa [20].

Their structure comprises two primary parts: they contain a substitute aromatic ring denominated chromate and a long side chain (See Figure 3). These eight vitamers are divided into two basic groups: four tocopherols, and four tocotrienols, which are differentiated in the side-chain saturation; the tocopherols possess a saturated chain, and the tocotrienols, an unsaturated one with three double loops on carbons 3, 7, and 11 (Figure 4).

WIthin each group, the vitamers differ in the number and position of the methyl groups in the chromate ring, designating these as α, β, and δ (Figures 4 and 5) [19,20].

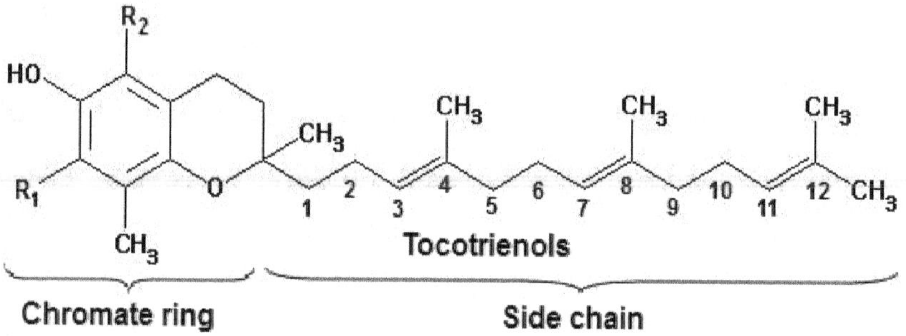

Figure 3. Components of the tocotrienol structure.

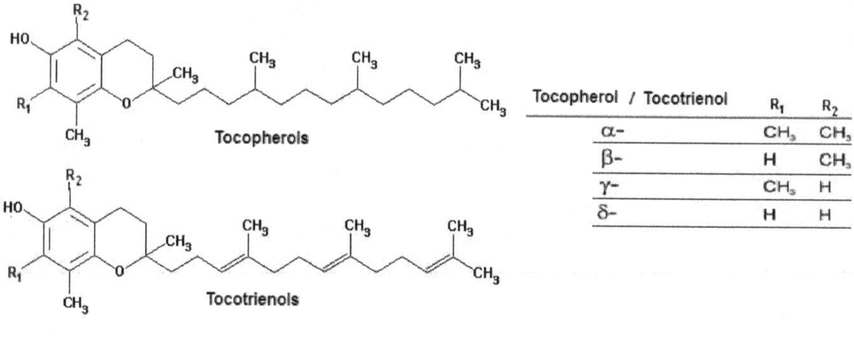

Tocopherol / Tocotrienol	R_1	R_2
α-	CH_3	CH_3
β-	H	CH_3
γ-	CH_3	H
δ-	H	H

Figure 4. Chemical structure of the possible stereoisomers of the tocopherols and tocotrienols that make up the natural vitamin E. The presence of the -CH_3 or -H groups in the chromate ring define that these substances as α, β, γ, and δ.

The presence of three chiral centers (position C2 of the chromate ring, positions C4 and C8 of the phytyl chain) allow there to be a total of eight configurations depending on the R or S orientation of the methyl group in each of the chiral centers (Figures 3 and 5) [19].

During vitamin E synthesis, equimolar amounts of these isomers (vitamers) are produced.

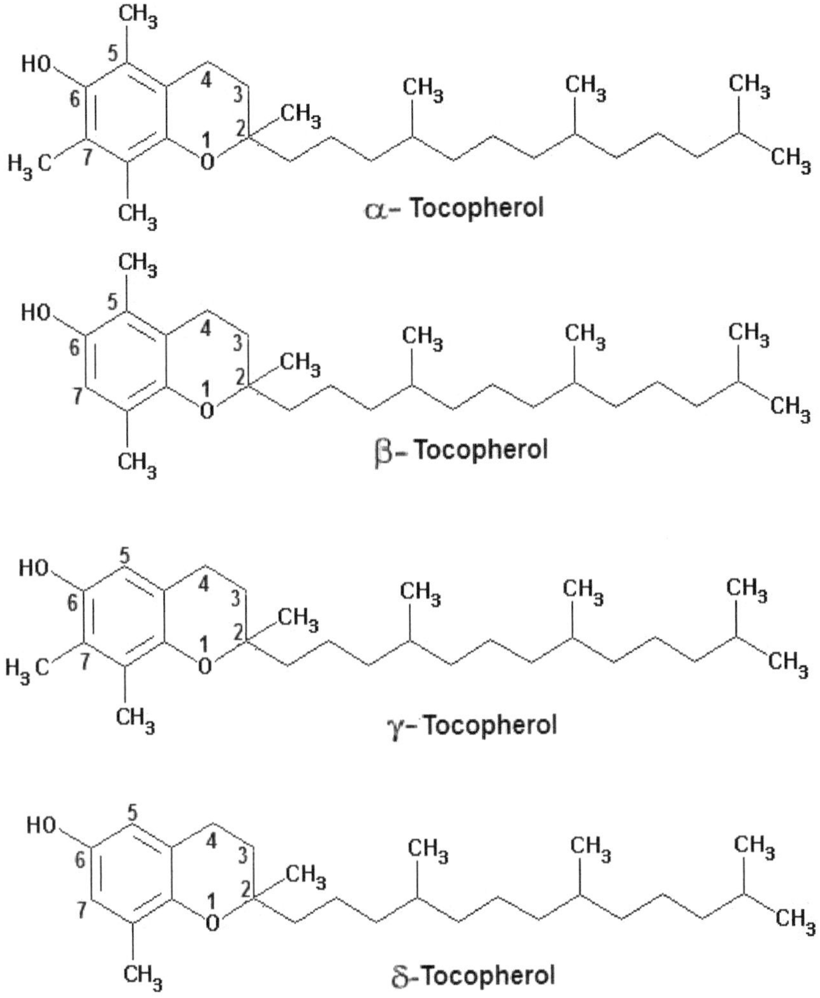

Figure 5. Chemical structure of the tocopherols.

Deficit: The deficiency of vitamin E can be due to two causes: not consuming a certain food that contains it, or poor fat absorption, due to that vitamin E is a liposoluble vitamin, that is, it is diluted in fats for its absorption in the intestine in the micelles.

Vitamin E is essential for humans. Its deficiency is not frequent even with persons who consume diets that are relatively poor in this vitamin, and could develop in cases of intense malabsorption of fats, cystic fibrosis, some forms of chronic liver disease, and congenital abetalipoproteinemia. The newborn, fundamentally the premature infant, is particularly vulnerable to vitamin E deficiency because of its deficient body reserves. The majority of vitamin E deficiency-associated sequelae are subclinical. Neuropathological alterations have been described in at-risk patients and the most frequent manifestations comprise diverse grades of areflexia, walk proprioception disorders, diminution of vibratory sensations, and ophthalmoplegia [1].

With regard to the relationship of vitamin E deficiency and the development of cardiovascular disease and cancer, there are no conclusive results to date [1,19].

The existence of a lack of vitamin E is rare. If this occurs, it is manifested in specific cases, that is, in the following three situations:

a.   Persons with a difficulty of absorbing or secreting bile or who suffer from fat metabolism-related disease (celiac disease or cystic fibrosis)

b.   Premature infants (with Very low birth weight, VLB) who weigh <1,500 grams at birth

c.   Persons with genetic abnormalities in alpha-tocopherol transporter proteins.

Likewise, vitamin E levels can fall due to a zinc deficiency.

Lipid-absorption disorders can present in adults. From 3 years on, lack of absorption presents neurological conditions. The deficiency appears in less time due to the infants' not possessing so great a vitamin-E reserve.

2.2.1. Symptoms of vitamin E deficiency

Irritability, Fluid retention, Hemolytic anemia (destruction of red globules), Ocular alteration

Damage to the nervous system, Difficulty in maintaining equilibrium, Tiredness, apathy

Inability to concentrate, Alterations in the walk and Diminished immune response.

2.2.2. Vitamin E deficiency-related diseases

Encephalomalacia. This is due to the lack of vitamin E, which does not avoid PFA oxidation of the ration of the vitamin; consequently, hemorrhages and edema are produced in the cerebellum.

Exudative diathesis. This is due to deficient rations of vitamin E and selenium. The disease can be prevented with the administration of selenium, which acts on vitamin E as an agent that favors the storage of selenium in the organism.

Nutritional white muscle or muscular dystrophy. Rations with a scarcity of vitamin E, selenium, and azo-containing amino acids and a high content of polyunsaturated fats cause muscle degeneration in chest and thighs.

Ceroid pigmentation. This corresponds to the yellowish-brown coloration of adipose tissue in the liver due to the oxidation *in vivo* of lipids.

Erythrocytic hemolysis. The FR attack membrane and erythrocyte integrity; thus, these are also hemolysis-sensitive.

This produces sterility in some animals and certain disorders associated with reproduction, death, and fetal reabsorption in females and testicular degeneration in males.

The excess of vitamin E does not appear to produce noxious toxic effects.

Obtaining Vitamin E: Tocopherol-rich dietary sources include the following: alfalfa flour; wheat germ flour (125–100 mg/kg); hen's egg (egg yolk); polished rice (100–75 mg/kg); rice bran; mediator wheat (75–50 mg/kg); dry yeast; dry distillery solubles; barley grains; whole soy flour; corn grains; ground wheat residues (50–25 mg/kg); corn gluten flour; wheat bran; rye grains; sorghum; fish flour; oatmeal; sunflower seed flour; cotton seed flour (25–10 mg/kg); almonds; hazelnuts; sunflower seeds; nuts, and peanuts. Other sources include all vegetable oils and green vegetable harvests, above all those with green leaves, sweet chile peppers, avocado, fresh potatoes, celery, cabbage, fruits, chicken, fish, and butter [19, 20].

1 International unit (IU) of vitamin E = 1 mg αlpha-tocopherol, and 1 IU of vitamin E = 0.67 mg of vitamin E. In adults, the Minimum daily requirement (MDR) for vitamin E is 15 mg/day, and up to 200–600 mg/day would not cause any disorder.

The principal sources are vegetable oils and wheat germ. Hydrogenation of the oils does not produce a very important loss of tocopherols in terms of their content in the original oil; thus, margarine and mayonnaise contain this vitamin, in lesser amounts.

One hundred percent of the MDR of vitamin E can be covered with two tablespoons of sunflower seed or corn oil.

Absorption: The absorption of vitamin E in the intestinal lumen depends on the process necessary for the digestion of fats and uptake by the erythrocytes. In order to liberate the free fatty acids from the triglycerides the diet requires pancreatic esterases. Bile acids, monoglycerides, and free fatty acids are important components of mixed micelles. Esterases are required for the hydrolytic unfolding of tocopherol esters, a common form of vitamin E in dietary supplements. Bile acids, necessary for the formation of mixed micelles, are indispensable for the absorption of vitamin E, and its secretion in the lymphatic system is deficient. In patients with biliary obstruction, cholestasic disease of the liver, pancreatitis, or cystic fibrosis, a vitamin E deficiency presents as the result of malabsorption. Vitamin E is transported by means of plasma lipoproteins in an unspecific manner. The greater part of vitamin E present in the body is localized in adipose tissue [19, 20].

The four forms of tocopherol are similarly absorbed in the diet and are transported to the peripheral cells by the kilomicrons. After hydrolysis by the lipoprotein lipases, part of the tocopherols is liberated by the kilomicrons of the peripheral tissues [19].

Vitamin E accumulates in the liver as the other liposoluble vitamins (A and D) do, but different from these, it also accumulates in muscle and adipose tissue.

Toxicity: High doses of vitamin E can interfere with the action of vitamin K and also interfere with the effect of anticoagulants: hemorrhages.

Since 2001, it was calculated that 70% of the U.S. population occasionally consumes dietary supplements and that 40% do so on a regular basis. In 2002, Montuiler and collaborators informed, in a population of physicians, that 64% consumed doses of >400 IU/day of vitamin E and that the average obtained from food sources is 9.3 mg of alpha-tocopherol per day (approximately 14 IU/day). In 2005, Ford and coworkers found that 11.3% of the U.S. population consumes at least 400 IU/day of vitamin E and that median daily ingestion is 8.8 IU/day.

Part of the potential danger of consuming high doses of vitamin E could be attributed to its effect on displacing other soluble antioxidants in fats and breaking up the natural balance of the antioxidant system. This can also inhibit the Glutathione S-transferase (GST) cytosolic enzymes, which contribute to the detoxification of drugs and endogenous toxins. In fact, one study on alpha-tocopherol and β-carotene demonstrated a significant increase in the risk of hemorrhagic shock among study participants treated with vitamin E. Other data suggest that vitamin E could also affect the conversion of β-carotene into vitamin E and the distribution of the latter in animal tissues. Vitamin E possesses anticoagulant properties, possibly on interfering with the mechanisms mediated by vitamin K. In recent studies conducted *in vitro*, it was demonstrated that vitamin E potentiates the antiplatelet effects of acetylsalicylic acid; therefore, one should be alert to this effect when both substances are consumed [19].

2.3. Vitamin A

This is a term that is employed to describe a family of liposoluble compounds that are essential in the diet and that have a structural relationship and share their biological activity. It is an antioxidant vitamin that eliminates Free radicals (FR) and protects the DNA from their mutagenic action, thus continuing to halt cellular aging. Their oxygen sensitivity is due to the large amount of double loops present in their structure. Their biological activity is attributed to all-*trans* retinol, but from the nutritional viewpoint, they should be included in under the denomination of A provitamins, certain carotenoids, and similar compounds, the carotenals, which have the capacity to give rise to retinol from the organism.

Vitamin A is a hydrosoluble alcohol that is soluble in fats and organic solvents. It is stable when exposed to heat and light, but is destroyed by oxidation; thus, cooking in contact with the air can diminish the vitamin A content in foods. Its bioavailability increases with the presence of vitamin E and other antioxidants [21].

Function: In its different forms, vitamin A, also known as an antixerophthalmic, is necessary in vision, normal growth (its deficiency causes bone growth delay), reproduction, cellular proliferation, differentiation (which confers upon it a role in processes such as spermatogenesis, fetal development, immunological response, etc.), fetal development, and the integrity of the immune system. Others of these include its being an antioxidant, amino acid metabolism, the structure and function of other cells, reproduction, and epithelial tissues.

Vitamin A participates in the synthesis of glycoproteins, which contributes to maintaining the integrity of epithelial tissue in all of the body's cavities. Epithelial dissection especially affects the conjunctivae of the eye (xerophthalmia), which renders the cornea opaque and causes crevices, producing blindness and facilitating eye infections.

Sources: Retinol is only found in the lipidic part of foods of animal origin as follows; whole milk; lard; cream; cheese; egg yolk; eels, and fatty fish, due to their self-storage in the liver and in the oils extracted from the liver. The latter, as well as the oils extracted from the liver (veal and pork), comprise an important source of vitamin A. Cod liver oil constitutes source richest in vitamin A, although this cannot be considered a food in the strictest sense. In the case of skim/low-fat milk, this vitamin would be eliminated, but by law it is restituted to its original content; examples of these include manchego cheese, margarine, and butter.

Vegetables contain only provitamins or carotenes (all of these coloring pigments, such as alpha, beta, and gamma carotene). Garden vegetables (spinach and similar vegetables), carrots, sweet chile peppers, potatoes, tomatoes, and red and yellow fruits are the main suppliers. We must bear on mind that there are numerous carotenes that do not possess any provitamnic A activity, such as lycopene from the tomato, although it does act as a neutralizer of FR [21, 22].

Structure of Vitamin A: This vitamin is a diterpene ($C_{20}H_{32}$) that can present in the following various molecular forms:

Retinol (See Figure 6), when the side chain terminal is an alcohol group ($-CH_2OH$)

Retinal, the carbon terminal, is an aldehyde ($-CHO$)

Retinoic acid, when the terminal group is acidic ($-COOH$)

Retinyl-palmitate, in the case of lengthening of a side chain from esterification with palmitic acid ($-CH_2O-CO-(CH_2)_{14}-CH_3$).

Absorption: The metabolism of the vitamin responds to the same general mechanisms of digestion and absorption as those of other lipidic substances. Absorption is carried out in the form of carotenes or similar substances at the intestinal level within the interior of the micelles and quilomicrons, together with other fats.

Retinol esters are absorbed from 80–90%, while the beta-carotenes are absorbed at only 40–50%. Factors in the diet that affect carotene absorption include the origin and the concentration of the fat in the diet, the amount of carotenoids, and the digestibility of the foods. Vitamin A is first processed in the intestine, and afterward it arrives at the liver via portal, the liver being the main storage organ. In addition, the liver is responsible for regulating the secretion of the retinol bound to the retinoid-binding protein. Carotene absorption in particular is very inefficient in raw foods, and its content in lipids in the diet is low. The efficiency of conversion into retinol, which is quite variable and, in general, low, depends not only on the structure of the carotenoids, but also on their proteinic ingestion. Thus, when carotene ingestion is very high, those which have not been transformed into retinol in the retinal mucosa are absorbed unaltered, bind with the lipoproteins, and are deposited in the skin and the mucosa, on which they confer a typical yellowish color, constituting hypercarotenosis [21].

Toxicity: Both the deficiency as well as the excess of vitamin A causes fetal malformations. Ingestion of large amounts of this vitamin can give rise to skin alterations (scaling), hair fall, weakness, choking, vomiting, etc. In extreme cases, great amounts accumulate in the liver, producing hepatic disorders that end up as fatty liver.

It is noteworthy that the administration of vitamin A in chronic form and at doses higher than the recommended doses those can produce a clinical condition of toxicity characterized by fatigue, irritability, cephalea, febricula, hemorrhages in different tissues, and cutaneous alterations.

In children, this can trigger the early closing of the long bones, which causes the height to descend. Megadoses of vitamin A can produce acute intoxication that will be characterized by clinical features of sedation, dizziness, nausea, vomiting, erythema, pruritis, and generalized desquamation of the skin. We should also point out that in the elderly, the safety margin when we administer this vitamin is small; thus, we must be especially cautious and adjust the dose well [21].

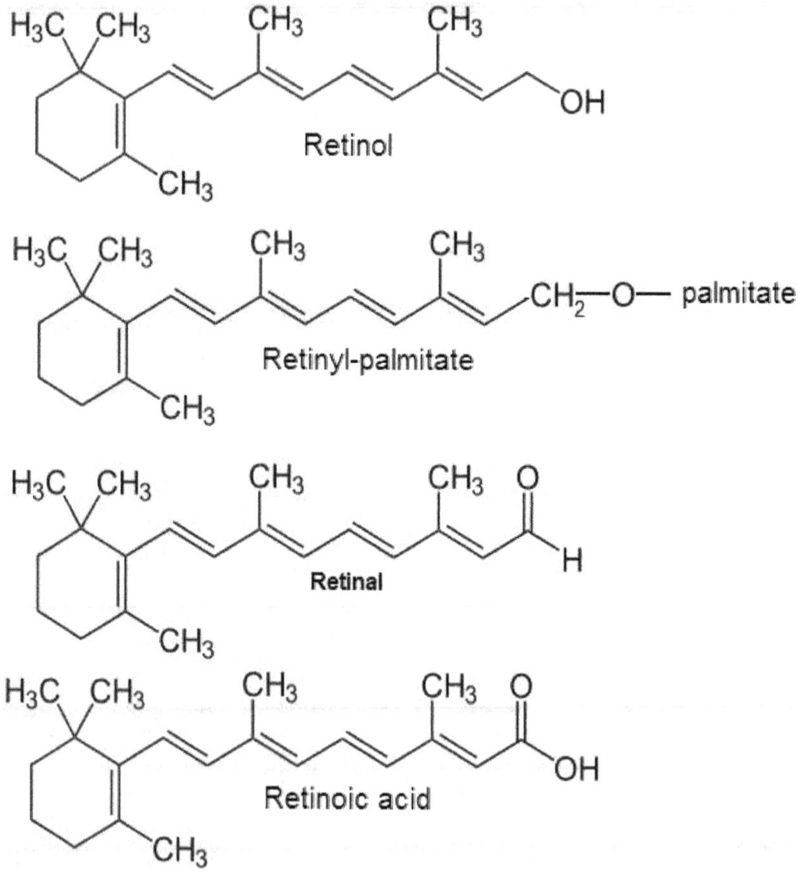

Figure 6. Molecular forms of vitamin A.

2.4. Flavonoids and their antioxidant actions

Flavo comes from the Latin *flavus* and means the color found between yellow and red, such as that of honey or of gold, and flavonoid refers to an aromatic group, with heterocyclic pigments that contain oxygen, which are widely distributed among plants, constituting the majority of yellow, red, and blue fruits. Consequently, flavonoids are found in abundance in grapes, apples, onions, cherries, and cabbage, in addition to forming part of the ginkgo biloba tree and *Camellia sinensis* (green tea). On consuming these, we obtain the anti-inflammatory, antimicrobial, antithrombotic, antiallergic, antitumor, anticancerigenous, and antioxidant properties. With regard to the latter properties, these lie within its function in the nervous system, because a protector relationship has been observed with regard to neurodegenerative diseases [22].

Flavonoids are Low-molecular-weight (LMW) compounds that share a common skeleton with diphenylpyrenes (C6-C3-C6); a flavonoid is a 2-phenyl-ring (A and B) compound linked through the pyrene C ring (heterocyclic). The carbon atoms in the C and A rings are numbered from 2–8, while those of the B ring are numbered from 2'–6'12 (Figure 7). The activity of flavonoids as antioxidants depends on the redox properties of their hydroxy phenolic groups and on the structural relationship among the different parts of their chemical structure[22].

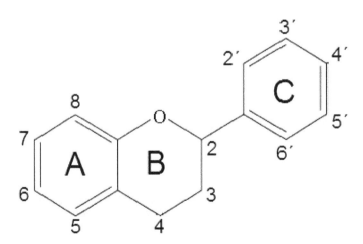

Figure 7. Base structure of the flavonoids

Thanks to the variations of pyrene, the flavonoids achieved classification, as shown in Table 1 (Antiatherogenic properties of flavonoids: Implications for cardiovascular health, 2010) [24].

Name	Structure	Description	Family members	Dietary sources
Flavanones		Carbonyl group at position 4 and an -OH group in position 3 of ring C	Quercetin, myricetin, isorhamnetin, kaempferol, pachypodol, rhamnazin	Onions, apples, broccoli, cranberries, berries, grapes, parsley, spinach
Flavan-3-ols		With an -OH group in position 3 of ring C	Catechins, epigallocatechin gallate, epicatechins, epicatechin gallate	Tea, red wine, cocoa, grapes, plums, fruits, legumes
Flavones		Have a carbonyl group in position 4 of ring C and lacking the hydroxyl group at position C3	Apigenin, nobiletin, tangeritin, luteolin	Celery, lettuce, parsley, citrus fruits, beets, bell peppers, spinach, Brussels sprouts, thyme
Anthocyanidins		Carbonyl group at position 4 and an -OH group in position 3 of ring C	Cyanidin, delphinidin, peonidin, malvidin, pelargonidin	Red wine, blueberries cranberries, black currants, plums, red onions, red potatoes

Table 1. Classification of flavonoids

Distribution: The flavonoids are widely distributed among the higher plants, with the ruta-ceous, polygonaceous, compound, and umbelliferous plant families the principal ones con-taining flavonoids. Flavonoids abound, above all, in young, aerial plant parts and in those most exposed to the sun, such as the leaves, fruits, and flowers, because solar light favors their synthesis, controlling the levels of the auxins (vegetables hormones), which are growth regulators.

These compounds are important for the plant, similar to what occurs with the greater part of secondary metabolites, in addition to being responsible for the coloration of many flowers, fruits, leaves, and seeds, achieving >5,000 distinct flavonoids, because these can be found in the following groups:

a.   Elegiac acid: found in fruits such as grapes and in vegetables

b.   Anthocyanidines: the pigment responsible for the reddish-blue and red color of cherries

c.   Catechins: found in black and green tea

d.   Citroflavonoids: such as quercetin, lemonene, pyridine, rutin, and orangenine. The bit-ter flavor of the orange, lemon, and grapefruit confers orangenine on these fruits, while lemonene has been isolated from the lime and the lemon.

e. Isoflavonoids: such as genestein and daidzein, present in soy foods such as tofu, soy milk, soybeans, soy vegetable protein, tempeh/fermented soybeans, miso/soybean paste, and soy flour

f. Kaempherol: found in broccoli, leeks, endives, red beets, and radishes

g. Proanthocyanidines: these appear in grape seeds, sea pine bark extract, and in red wine.

These merit incorporation into the group of essential nutrients. The mean value of the ingestion of flavonoids is 23 mg/day. The main flavonoid consumed is quercetin, tea being its main source [22].

Properties: The flavonoids are white or yellowish solid, crystallized substances. Their heretosides are soluble in hot water, alcohol, and polar organic dissolvents, being insoluble in apolar organic dissolvents. However, when they are in their free state, they are scarcely hydrosoluble, but are soluble in more or less oxygenated organic substances, depending on their polarity.

On the other hand, these are easily oxidizable substances; thus, they exert an antioxidant effect because they are oxidized more rapidly than other types of substances [23].

Pharmacological activity: Pharmacologically, flavonoids are prominent due to their low toxicity, presenting in general activity on the vascular system with P vitaminic action (protector effect of the vascular wall due to the diminution of permeability and to the increase of capillary resistance). Likewise, they possess an antioxidant effect, can inhibit lipid peroxidation, have antimutagenic effects, and possess the capacity to inhibit diverse enzymes [23, 24].

Antioxidant functions: The flavonoids' antioxidant action depends mainly on their sequestering capacity of FR and on their chelant properties of metals such as iron, impeding the catalytic actions of FR, and they also act by inhibiting the enzyme systems related with vascular functionality, such as the following: Catechol-O-methyl transferase (COMT), with which it increases the duration of the action of the catecholamines, thus inciding in vascular resistance; histidine decarboxylase, thus affecting the histamine's action, and the phosphodiesterases, thus inhibiting platelet aggregation and adhesiveness, in addition to the following oxidases: lipo-oxygenase; cyclo-oxygenase; myeloperoxidase, and xanthinic oxide, therefore avoiding the formation of Reactive oxygen species (ROS) and organic hydroperoxides.

In addition to this, it has been observed that they also indirectly inhibit oxidative processes, such as phospholipase A_2, at the same time stimulating others with recognized antioxidant properties, such as catalase and SOD.

With respect to their structure, flavonoids are their hydroxylic constituents in positions 3′ and 4′; in the B ring, they demonstrate more action as antioxidants and this effect is potentiated by the presence of a double loop between carbons 2 and 3 and a free OH group in position 4. Additionally, the glycols show to be the most potent in their antilipoperoxidative actions than in their corresponding glycosidic actions.

As previously mentioned, quercetin is the flavonoid that unites the requisites for exercising an effective antioxidant function, because it is five times higher than vitamins A and C and additionally possesses a hydrosolubility similar to that of the latter. Therefore, rutin (quercetin-3-b-D-rutinoside) is, to date, the sole flavonoid with a pharmacological presence in Mexico.

There is a synergic effect with all of the vitamins to which we have alluded. This is due to that ascorbic acid reduces the oxidation of quercetin in such a way that combines with it and allows the flavonoids to maintain their functions for a longer time. For its part, quercetin protects vitamin E from oxidation.

The flavonoids remove reactive oxygen, especially in the form of SOD, hydroxyl radicals, hydroperoxides, and lipid peroxides, blocking the harmful effects of these substances on the cell, in which antioxidant protection of flavonoids has been corroborated in the following: queratinocytes; dermal fibroblasts; sensory lymph nodes; the endothelium; nervous tissue, and LDL.

On the other hand, the flavonoids exercise other actions as follows: diuretic; antispasmodic; anti-gastriculcerous, and anti-inflammatory.

In phytotherapy, the flavonoids are mainly employed in cases of capillary fragility as venotonics, although they are also utilized in proctology, metrorrhages, and retinopathies [22].

2.5. Pro-oxidant mechanisms

Due to the structural characteristics of some flavonoids, such as the anthocyanidines, these cause low oxidation potentials (EP/2), which permits them to reduce Fe^{3+} and Cu^{2+} in order for them to undergo auto-oxidation or even to become involved in the redox recycling process, acting in this manner as pro-oxidant agents, which explains the mutagenic and genotoxic effects of some flavonoids.

Some of these mechanisms include the temporary reduction of Cu (II) to CU (I), auto-oxidation of the aroxyl radical and generating the superoxide anion (O2–) that, on following its general sequence, becomes the harmful hydroxyl radical (HO.), as well as the affectation of the functions of the components of the nuclear antioxidant defense system: glutathione, and glutathione-S-transferase.

What determines the antioxidant or pro-oxidant character is the redox stability/lability of the radical compound forming part of the original flavonoid. The pro-oxidant actions only appear to be produced when the flavonoid doses are excessively high [25].

Under this heading, we will present a brief review of the remaining antioxidants present in our diet, their activity, and the foods that supply them.

2.6. Lycopene

Lycopene is the carotenoid that imparts the red color to the tomato and watermelon and that it not converted into vitamin A in the human organism, which does not impede it from possessing very high antioxidant properties.

The highest concentrations of lycopene are found in prostatic tissue. High consumption of lycopene has been related with the prevention of some cancer types, precisely that of the prostate.

Although the tomato is the greatest source of lycopene, there are also other vegetables and fruits that present intense colors, such as watermelons, papayas, apricots, and pink grapefruit. The tomato is the food that concentrates the greatest amount of lycopene, and it should be considered that there are factors that affect its assimilation into the organism, such as its maturity, the distinct varieties, or the manner of cooking, all of which exert an influence on the amount and degree of exploitation of lycopene.

Of all of these, the fried tomato is that which best assimilates this substance, frying being the best way of cooking because, in addition to the heat, there is a certain amount of fat involved, which renders better assimilation of lycopene (fat-soluble). In concrete fashion, its presence in the fried tomato is some 25 µg per 100 g, while in the fresh tomato, this is around 2 µg per 100 g [6, 26, 27].

3. Minerals

Other potent antioxidants include minerals such as copper, manganese, selenium, zinc, and iron. These minerals exercise their antioxidant function in diverse processes and metabolic steps in the organism [6, 26, 27].

3.1. Zinc

Zinc intervenes in >200 enzymatic reactions and its deficit increases the production of oxidant species and Oxidative stress (OS) [6, 26, 27].

3.2. Copper

Copper participates in functions with antioxidant features of the enzyme family denominated Superoxide dismutase (SOD), which is responsible for eliminating the superoxide anion.

It empowers the immune system, participates in the formation of enzymes, proteins, and brain neurotransmitters (cell renovation and stimulation of the nervous system) and is an anti-inflammatory and anti-infectious agent.

Similarly, it facilitates the synthesis of collagen and elastin (necessary constituents of the good state of the blood vessels, lungs, and the skin).

In addition, it acts as an antioxidant, protecting the cells from the toxic effects of FR, and it facilitates calcium and phosphorous fixation [6, 26, 27].

3.3. Manganese

Manganese also intervenes in this family of enzymes, concretely, in enzymes localized within the mitochondria [6, 26, 27].

3.4. Selenium

Selenium intervenes in the synthesis of enzymes related with the oxidative function, such as glutathione peroxidase, which, as its name indicates, eliminates peroxide groups, including oxygen peroxide.

This mineral is incorporated into proteins in the form of selenoproteins and, in this manner, aids in the prevention of cell damage. Epidemiological studies related the lack of selenium in the diet with the incidence of lung, colorectal, and prostate cancer.

The selenium content in the diet is directly related with the selenium content of the soil in which the food was grown. Thus, selenium-deficient soils give rise to a deficit of this element in the population, as in the case of China.

In this specific latter case, the method-of-choice comprises supplementing the diet with contributions of selenium, preferably in the form of selenomethionide, which is the analog, organic form of selenium and which easily increases selenium levels in the blood [6, 26, 27].

3.5. Iron

Iron forms part of the organism's antioxidant system because it contributes to eliminating the peroxide groups. However, its capacity to change valence with ease (2+/3+) renders that it can also intervene, depending on the environment, in the formation of Free radicals (FR) [6, 26, 27].

3.6. Co-enzyme Q

Co-enzyme Q10 or ubiquinone is a liposoluble compound that can be carried in many foods, although it can also be synthesized in the human organism. Co-enzyme Q10 diminishes with age; thus, the metabolic processes in which it has been found implicated are also co-enzyme Q10-sensitive.

Given its liposolubility, its absorption is very los, especially when the diet is poor in fats.

Its principal antioxidant activity resides in that, in its reduced form, it is a liposoluble antioxidant that inhibits lipid peroxidation in LDL. It is also found in the mitochondria, where it could protect protein membranes and the DNA from the oxidative damage that accompanies lipid peroxidation in these membranes.

It additionally acts as an immune system stimulant and through this stimulation also functions as an anticancerigen. In addition, it is capable of directly regenerating alpha-tocopherol [6, 26, 27].

4. Lipoic acid

Lipoic acid or thioctic acid is also a compound that forms part of the antioxidant capital of the organism.

Numerous studies have shown the protector effect of red globules and of the fatty acids of oxidative damage typical of intense exercise and excessive exposure to the sun's UV rays.

It is synthesized by plants and animals, as well as by the human organism, although in the latter case, in very small amounts. Lipoic acid is considered a very good regenerator of potent antioxidants such as vitamin C, vitamin E, glycation, and co-enzyme Q10. It is liposoluble and hydrosoluble, which means that it can act on any part of the organism.

It is found in spinach and similar green-leafed vegetables, broccoli, meat, yeast, and in certain organs (such as kidney and heart) [6, 26, 27].

4.1. Naringenine

The hypolipidemic and anti-inflammatory activities *in vivo* as well as *in vitro* of the flavonoids of citric fruits have been widely demonstrated. Among the flavonoids, naringenine, one of the compounds that causes the bitter taste of grapefruit, has been studied extensively in recent years. In a recently conducted clinical assay, it was found that naringenine reduced Low-density-lipoprotein (LDL) levels in the circulation of 17% of patients with hypercholesterolemia. Additionally, the reducer effects of cholesterol in rabbits and rats were demonstrated, in addition to the reducer effects of Very-low-density-lipoprotein (VLDL) levels through the inhibition of key proteins for their assembly. Other studies reported that naringenine activates enzymes that are important for the oxidation of fatty acids, such as CYP4A1 [28].

5. Conclusion

A good diet influences the development and treatment of diseases, it is increasingly evident. After that epidemiological studies have shown the association between moderate consumption of certain foods and reduced incidence of various diseases at the rate of these observations has attracted considerable interest in studying the properties of substances inherent in the chemical composition of food. Among the characteristics of these substances is the antioxidant activity, associated with the elimination of free radicals and therefore to the prevention of early stages which can trigger degenerative diseases. In this regard it is important to continue the study of dietary antioxidants on the activity may have on human diseases, paying attention to the substances primarily natural antioxidants of food and synthetic way to assess its protective effect on the body.

Author details

Alejandro Chehue Romero*, Elena G. Olvera Hernández, Telma Flores Cerón and Angelina Álvarez Chávez

*Address all correspondence to: chehue_alex@yahoo.com, chehuea@uaeh.edu.mx

Autonomous University of Hidalgo State, Mexico

References

[1]   Criado D.C. y Moya M.M.S.(2009). Vitaminas y antioxidantes. Actualizaciones El Medico. Comisión Nacional de Formación Continuada del Sistema Nacional de Salud, Madrid.

[2]   Harman D. Aging (1956): a theory based on free radical and radiation chemistry. J Gerontol 1956; 11(3):298-300.

[3]   *Rosa Mayor-Oxilia R.* Estrés Oxidativo y Sistema de Defensa Antioxidante. R. *Rev. Inst. Med. Trop. 2010;5(2):23-29.*

[4]   Olivares-Corichi IM, Guzmán-Grenfell AM, Sierra MP, Mendoza RS y Hicks Jj.Perspectivas del uso de antioxidantes como coadyuvantes en el tratamiento del asma. Rev Inst Nal Enf Resp Mex 2005; Volumen 18 - Número 2 , Páginas: 154-161.

[5]   Carlsen MH. The total antioxidant content of more than 3100 foods, beverages, spices, herbs and supplements used worldwide. Nutrition Journal 2010; 9:3. http://www.nutritionj.com/content/9/1/3 (accessed 5 May 2012).

[6]   Torrades S. Aportes extras de vitaminas. ¿Son realmente Necesarios? O F F A R M. Ámbito Farmacéutico Bioquimica 2005; Vol 24 Núm 6, Junio.

[7]   Goodman, L.S; Gilman, A. Las Bases Farmacológicas da Terapéutica. 10ª ed., Editora McGraw Hill Interamericana, 1996. p. 1647.

[8]   Gamboa C. Vitamina A, Guias alimentarias para la educacion nutricional en costa rica, http://www.ministeriodesalud.go.cr/index.php/inicio-menu-principal-comisiones-ms/140 (accessed 20 March 2012).

[9]   Nutricion K, ABC de la Nutricion, Kellogg's, http://www.kelloggs.es/nutricion/index.php?donde=abc (accessed 5 April 2012).

[10] World Health Organization. Vitamin and mineral requirements in human nutrition. 2nd Edition.2004. http://www.who.int/nutrition/publications/micronutrients/9241546123/en/index.html (accessed 15 June 2012).

[11] Sandoval H, Sharon D. Cuantificación de Ácido Ascórbico (Vitamina C) en Néctares de Melocotón y Manzana Comercializados en Supermercados de la Ciudad Capital.Report of Thesis Presented, Universidad de San Carlos de Guatemala. Facultad de Ciencias Químicas y Farmacia; 2010.

[12] García GA, Cobos C, Rey CA, et al. Biología, patobiología y bioclínica de la actividad de oxidorreducción de la vitamina C en la especie humana. Universitas Médica 2006; VOL. 47 Nº 4.

[13] Latham M C. Carencia de vitamina C y escorbuto. In: Food and Agriculture Organization of the United Nations (ed.) Nutrición Humana en el Mundo en Desarrollo, *Colección FAO: Alimentación y nutrición N° 29*, Roma, 2002. Capitulo 19. http://www.fao.org/docrep/006/W0073S/w0073s0n.htm#TopOfPage (accessed 20 March 2012).

[14] Valdés F. Vitamina C. Actas Dermosifiliogr 2006 ; 97(9):557-68.

[15] *Taylor S.* Dietary Reference Intakes for vitamin C, vitamin E, Selenium and Carotenoids. Food and Nutrition Board, Institute of Medicine, National Academy Press 2000; Advance Copy, 3;6-7;97.

[16] World Health Organization and Food and Agriculture Organization of the United Nations. Vitamin and mineral requirements in human nutrition, Second edition. 2004.

[17] Anitra C. and Balz F. Does vitamin C act as A Pro-Oxidant?, The FASEB Journal. Vol. 13 June 1999.

[18] Proteggente vAR, Rehman A, Halliwell B, Rice-Evans CA. (2000) Potential problems of ascorbate and iron supplementation: pro-oxidant effect in vivo?. Biochem Biophys Res Commun 2000; 2;277(3):535-40.

[19] Blé-Castillo JL, Díaz-Zagoyab JC and Méndez JD. Suplementación con vitamina E, ¿benéfica o dañina?. Gac Méd Méx 2008; Vol. 144 No. 2.

[20] Sayago A, Marín MI, Aparicio R y Morales MT. Vitamina E y aceites vegetales, GRASAS Y ACEITES 2007; 58 (1), ENERO-MARZO, 74-86.

[21] Pérez RM y Ruano A. Vitaminas y salud. Aportación vitamínica. O F F A R M. Ámbito Farmacéutico Nutricion 2004; Vol 23 Núm 8, Septiembre.

[22] Christopher Isaac Escamilla Jiménez, E. Y. Flavonoides y sus acciones antioxidantes. *Rev. Fac. Med 2009; UNAM*, 73-75.

[23] Retta E D. . Marcela,A promising medicinal and aromatic plant from Latin America: A review. *Indusrial Crops and Products 2012; 38*, 27-38.

[24] Mulvihill, M. W. Antiatherogenic properties of flavonoids: Implications for cardiovascular health. *Can J Cardiol 2010; Vol 26 Suppl A March*, 17A-21A.

[25] Trueba GP. Los Flavonoides: Antioxidantes o Prooxidantes. *Rev. Cubana Biomed 2003; 22(1)*, 48-57.

[26] Vilaplana M. Antioxidantes presentes en los alimentos O F F A R M. Ámbito Farmacéutico Bioquimica 2007; Vol 26 Núm 10, Noviembre.

[27] García-Álvarez JL, Sánchez-Tovar T. y García-Vigil JL. Uso de antioxidantes para prevenir enfermedad cardiovascular. Metaanálisis de ensayos clínicos. Rev Med Inst Mex Seguro Soc 2009; 47 (1): 7-16.

[28] Goldwasser J, Cohen PY, Yang E, Balaguer P, Yarmush ML, et al. Transcriptional Regulation of Human and Rat Hepatic Lipid Metabolism by the Grapefruit Flavonoid Naringenin: Role of PPARα, PPARγ and LXRα. PLoS ONE 2010; 5(8). http://www.plosone.org/article/info:doi/10.1371/journal.pone.0012399 (accessed 15 January 2012)

Chapter 2

Cell Nanobiology

María de Lourdes Segura-Valdez,
Lourdes T. Agredano-Moreno,
Tomás Nepomuceno-Mejía,
Rogelio Fragoso-Soriano,
Georgina Álvarez-Fernández, Alma Zamora-Cura,
Reyna Lara-Martínez and Luis F. Jiménez-García

Additional information is available at the end of the chapter

1. Introduction

1.1. Cell nanobiology

We define cell nanobiology as an emergent scientific area trying to approach the study of the *in situ* cell processes ocurring at the nanoscale. Therefore, it is part of cell biology but mainly deals with an interphase between analytical methods such as X-ray crystallography producing models at atomic or molecular resolution, and direct nanoscale imaging with high resolution microscopes such as scanning probe microscopes, electron microscopes and super-resolution microscopes. Several cell structures are involved in nanoscale processes (Figure 1).

1.1.1. An overview of cell structure under a genomic approach of gene expression

The main flow of genetic information represented as the so-called central dogma of molecular biology *in situ*, illustrates the major secretory pathway in the cell (Figure 2).

During this pathway, nanoscale particles represent substrates of different moments. During transcription, nuclear particles are involved in transcription and processing of RNA, both, pre-mRNA and pre-rRNA. pre-mRNA is transcribed and processed in the nucleoplasm while pre-rRNA is transcribed and processed within the nucleolus, the major known ribonucleoproteins structure where ribosome biogenesis and other functions of eukaryotic cell take place. Once in the cytoplasm, translation takes place in the ribosome, also a major ribonucleoprotein particle of 10-15 nm in diameter. When the synthesized protein contains a signal peptide, it is translocated into the rough endoplasmic reticulum, helped by the signal recognition particle or SRP, another major and conserved ribonucleoprotein. The transport to Golgi apparatus by the intermediated zone and the TGN producing the three derivatives from the Golgi apparatus are mediated by vesicles [see 1].

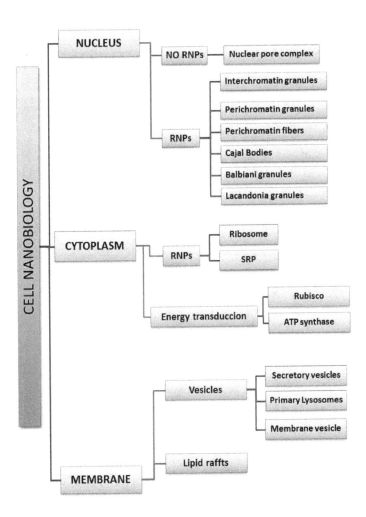

Figure 1. Cell nanobiology proposes to study cell structures using *in situ* high resolution microscopical approaches as electron and atomic force microscopy that could complement molecular and biochemical data to better understand a physiological role at the nanoscale.

a. Semenogelin

Semenogelin is the most abundant protein in the semen of mammals. It is a glycosylated protein that is responsible for properties such as density. As an example, the semenogelin of the tamarin *Saguinus oedipus* is used to show how the signals in the nucleic acids and proteins determine the intracellular pathways associated to that expression. Its expression includes intranuclear events as transcription by RNA polymerase II from a split gene consisting of 3 exones and 2 introns, processing of the transcript as 5′ end methylation, 3′ polyadenylation and splicing. All of them are associated to nuclear particles. Once in the cytoplasm, the mature transcript or mRNA associates to a ribosome that in turns translates the transcript. If the resulting protein contains a signal peptide, the signal recognition particle or SRP -a very well conserved RNA+protein complex- binds to it and associates to the rough endoplasmic reticulum, giving rise to the translocation process that introduces that protein to the lumen. Once there, N-glycosylation takes place at several asparagine residues following the basic rule of adjacent aminoacids showing a basic rule as Asn-X(except proline)-Ser or Asn. In *S. oedipus* semenogelin, there are 14 N-glycosylation sites. The protein then continues flowing through the Golgi apparatus or complex and at the TGN a secretory vesicle forms containing the protein that finally is secreted by the epithelial cell of seminal vesicle. The analysis of the gene sequences, as well as the transcription, processing, translation and post-translational products can predict the cell structures involved in the process [see 1].

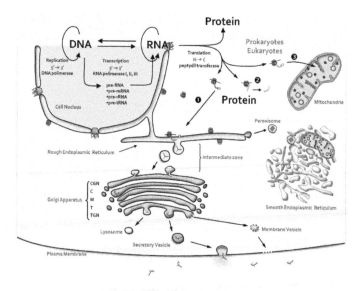

Figure 2. A general overview of the cell structure and function. The diagram illustrates the *in situ* flow of genetic information of a secretory protein encoded in the genome within the cell nucleus. A gene is copied into a pre-mRNA that is processed to mRNA within the nucleus. mRNA in the cytoplasm may contain a signal sequence that allows entrance to rough endoplasmic reticulum and further to Golgi complex. The protein inside a vesicle is secreted out of the cell.

1.1.2. Some nanoscale cell structures

There are many cell structures or products made by cells that could be analyzed under the present approach. Some of them are indicted in Figure 1, but there are others as extracellular matriz components, cytoskeleton elements, etc.; virus are also nanometric structures associated always to cell organelles. Here we will give an overview of some of the cell components, as examples.

a.   Nuclear particles

In eukaryote cells, transcription and processing mainly takes place within the cell nucleus, associated to nuclear particles that are well known since a method for ribonucleoprotein (RNP) structures was described in 1969 [2]. These particles are few nanometers in diameter or lengh. To date, several nuclear RNPs have been described including involved in mRNA metabolism: perichromatin fibers, perichromatin granules, interchromatin granules in mammals. In insects, Balbiani ring granules are well known structures [3]. In 1992 Lacandonia granules were described for some plants [4]. In addition, other nuclear bodies around 300-400 nm in diameter have been described involved in gene expression. As for rRNA transcription and processing, the nucleolus is a nuclear organelle containing pre-ribosomes in the granular component that are about 10-20 nm in diameter.

b.   Rough endoplasmic reticulum particles

i.   The ribosome

Ribosomes are the universal ribonucleoprotein particles that translate the genetic code into proteins. The shape and dimensions of the ribosome were first visualized by electron microscopy [6-8]. Ribosomes have diameters of about 25 nanometers in size and are roughly two-thirds RNA and one-third protein. All ribosomes have two subunits, one about twice the mass of the other. The ribosome basic structure and functions are well-known. There are 70S ribosomes common to prokaryotes and 80S ribosomes common to eukaryotes. The bacterial ribosome is composed of 3 RNA molecules and more than 50 proteins. In humans, the small ribosome unit has 1 large RNA molecule and about 32 proteins; the large subunit has 3 RNA molecules, and about 46 proteins. Each subunit has thousands of nucleotides and amino acids, with hundreds of thousands of atoms. The small subunit (0.85 MDa) initiates mRNA engagement, decodes the message, governs mRNA and tRNA translocation, and controls fidelity of codon–anticodon interactions and the large subunit catalyzes peptide bond formation.

In 1980, the first three-dimensional crystals of the ribosomal 50S subunit from the thermophile bacterium *Geobacillus stearothermophilus* were reported [9]. Since then, ribosome crystallography advanced rapidly. To date crystal structures have been determined for the large ribosomal subunit from the archaeon *Haloarcula marismortui* at 2.4 Å [10] and the 30S ribosomal subunit from *Thermus thermophilus* [11]. The structure of the entire 70S ribosome in complex with tRNA ligands (at 5.5Å resolution) emerged shortly after the structures of the initial subunits [12].

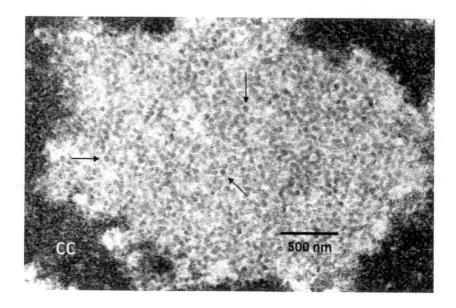

Figure 3. Lacandonia granules are nanoscale (32 nm in diameter) and abundant particles (arrows) in the nucleoplasm of the plant *Lacandonia schismatica*. Transmission electron microscopy image of a cell in the teguments of the flower. Cc, compact chromatin.

Snapshots of ribosome intermediates provided by cryo-EM and x-ray crystallography, associated translation factors, and transfer RNA (tRNA) have allowed dynamic aspects of protein translation to be reconstructed. For example, recent cryo-EM reconstructions of translating ribosomes allowed direct visualization of the nascent polypeptide chain inside the ribosomal tunnel at subnanometer resolution [13-15]. The dimension of the ribosomal tunnel in bacterial, archaeal, and eukaryotic cytoplasmic ribosomes is conserved in evolution [16 -18]. The ribosomal tunnel in the large ribosomal subunit is ~80 Å long, 10–20 Å wide, and predominantly composed of core rRNA [19]. The tunnel is clearly not just a passive conduit for the nascent chain, but rather a compartment in a dynamic molecular dialogue with the nascent chain. This interplay might not only affect the structure and function of the ribosome and associated factors, but also the conformation and folding of the nascent chain [20]. As the nascent polypeptide chain is being synthesized, it passes through a tunnel within the large subunit and emerges at the solvent side, where protein folding occurs.

Peptide bond formation on the bacterial ribosome and perhaps on the ribosomes from all organisms is catalyzed by ribosomal RNA as well as ribosomal protein and also by the 2'-OH group of the peptidyl-tRNA substrate in the P site. The high resolution crystal structures of two ribosomal complexes from T. thermophilus [21] revealed that ribosomal proteins L27 and L16 of the 50S subunit stabilize the CCA-ends of both tRNAs in the peptidyl-transfer reaction, suggesting that peptide chains from both these proteins take part in the catalytic mechanism of peptide bond formation.

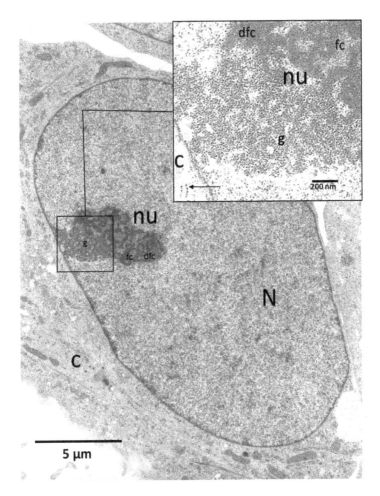

Figure 4. Nucleolus of a PtK2 cell. Within the cell nucleus (N), the nucleolus displays three different components named fibrillar center (fc), dense fibrillar component (dfc) and granular components (g). In the inset, a high magnification of the nucleolus shows granular particles or pre-ribosomes in the granular component (g). In the cytoplasm (c), ribosomes are also visible (arrow).

Ribosomes mediate protein synthesis by decoding the information carried by messenger RNAs (mRNAs) and catalyzing peptide bond formation between amino acids. When bacterial ribosomes stall on incomplete messages, the trans-translation quality control mechanism is activated by the transfer-messenger RNA bound to small protein B (tmRNA–SmpB ribonucleoprotein complex). Trans-translation liberates the stalled ribosomes and triggers degradation of the incomplete proteins. The cryo-electron microscopy structures of tmRNA–SmpB accommodated or translocated into stalled ribosomes demonstrate how tmRNA–SmpB crosses the ribosome and how as the problematic mRNA is ejected, the tmRNA resume codon is placed onto the ribosomal decoding site by new contacts between SmpB and the nucleotides upstream of the tag-encoding sequence [22]. Recently, the crystal structure of a tmRNA fragment, SmpB and elongation factor Tu bound to the ribosome shows how SmpB plays the role of both the anticodon loop of tRNA and portions of mRNA to facilitate decoding in the absence of an mRNA codon in the A site of the ribosome [23].

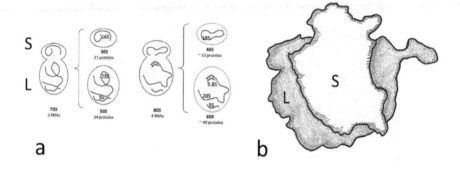

Figure 5. Representation of a prokaryotic (a) and a eukaryotic ribosome (a). Each one is an RNP is constituted by two subunits, each containing rRNA and proteins. b) a model to show the nanoscale morphology of a mammalian cytoplasmic ribosome (small [S] and large [L] subunits).

The structure of the ribosome at high resolution reveals the molecular details of the antibiotic-binding sites, explain how drugs exercise their inhibitory effects. Also, the crystal structures help us to speculate about how existing drugs might be improved, or novel drugs created, to circumvent resistance [24]. Recently, ribosome engineering has emerged as a new tool to promote new crystal forms and improve our knowledge of protein synthesis. To explore the crystallization of functional complexes of ribosomes with GTPase, a mutant 70S ribosomes were used to crystallize and solve the structure of the ribosome with EF-G, GDP and fusidic acid in a previously unobserved crystal form [25].

In contrast to their bacterial counterparts, eukaryotic ribosomes are much larger and more complex, containing additional rRNA in the form of so-called expansion segments (ES) as well as many additional r-proteins and r-protein extensions [26]. The first structural models for the eukaryotic (yeast) ribosome were built using 15-A° cryo–electron microscopy (cryo-EM) maps fitted with structures of the bacterial SSU [11] and archaeal LSU [10], thus identifying the location of a total of 46 eukaryotic r-proteins with bacterial and/or archaeal homologs as well as many ES [27].

Ribosome biogenesis is regulated by the conserved protein kinase TOR (target of rapamycin), a member of the ATM-family protein. TOR up-regulates transcription of rRNA and mRNA for ribosomal proteins in both yeast and mammals [28-30]. Recent results indicate that in yeast, conserved kinases of the LAMMER/Cdc-like and GSK-3 families function downstream of TOR complex 1 to repress ribosome and tRNA synthesis in response to nutrient limitation and other types of cellular stress [31].

ii.   The signal recognition particle (SRP)

The Signal Recognition Particle (SRP) is an evolutionarily conserved rod-shaped 11S ribonu-cleoprotein particle, 5–6 nm wide and 23–24 nm long [32]. It comprises an essential component of the cellular machinery responsible for the co-translational targeting of proteins to their proper membrane destinations [33].

Although SRP is essential and present in all kingdoms of life maintaining its general function, structurally it shows high diversity. Vertebrates SRP consists of a single ~ 300-bp RNA (SRP RNA or 7S RNA) and six polypeptides designated SRP9, SRP14, SRP19, SRP54, SRP68 and SRP72. It can be divided into two major functional domains: the Alu domain (comprising the proteins SRP9 and -14) and the S domain (SRP19, -54, -68, and -72). The S domain functions in signal sequence recognition and SR interaction, whereas the Alu domain is required for translational arrest on signal sequence recognition [34]. In Archaea and Eucarya, the conserved ribonucleoproteic core is composed of two proteins, the accessory protein SRP19, the essential GTPase SRP54, and an evolutionarily conserved and essential SRP RNA [35]. SRP54, comprises an N-terminal domain (N, a four-helix bundle), a central GTPase domain [G, a ras-like GTPase fold, with an additional unique α-β- α insertion box domain (IBD)], and a methionine-rich C-terminal domain [36-37]. The N and G domains are structurally and functionally coupled; together, they build the NG domain that is connected to the M domain through a flexible linker [38]. The M domain anchors SRP54 to SRP RNA and carries out the principal function of signal sequence recognition [39-41]. The NG domain interacts with the SR in a GTP-dependent manner [43].

SRP is partially assembled in the nucleus and partially in the nucleolus. In agreement with that, nuclear localization for SRP proteins SRP9/14, SRP68, SRP72 and SRP19 has been determined [44]. After the transport into the nucleus the subunits bind SRP RNA and form a pre-SRP which is exported to the cytoplasm where the final protein, Srp54p, is incorporated [45-47]. Although this outline of the SRP assembly pathway has been determined, factors that facilitate this and/or function in quality control of the RNA are poorly understood [48]. SRP assembly starts during 7S RNA transcription by RNA polymerase III in the nucleolus, by binding of the SRP 9/14 heterodimer and formation of Alu-domain. Prior to transportation to the nucleus SRP9 and SRP14 form the heterodimer in the cytoplasm, a prerequisite for the binding to 7S RNA [49].

The signal recognition particle displays three main activities in the process of cotranslational targeting: (I) binding to signal sequences emerging from the translating ribosome, (II) pausing of peptide elongation, and (III) promotion of protein translocation through docking to the membrane-bound SRP receptor (FtsY in prokaryotes) and transfer of the ribosome nascent chain complex (RNC) to the protein-conducting channel [50]. Despite the diversity of signal sequences, SRP productively recognizes and selectively binds them, and this binding event serves as the critical sorting step in protein localization within the cell. The structural details that confer on SRP this distinctive ability are poorly understood. SRP signal sequences are characterized by a core of 8–12 hydrophobic amino acids that preferentially form an α-helix, but are otherwise highly divergent in length, shape, and amino acid composition [51-52]. This and the unusual abundance of methionine in the SRP54 M-domain led to the 'methionine bristle' hypothesis, in which the flexible side chains of methionine provide a hydrophobic environment with sufficient plasticity to accommodate diverse signal sequences [53].

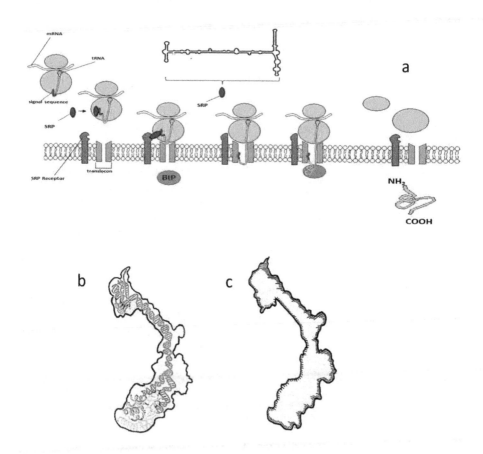

Figure 6. The signal recognition particle. During protein synthesis of the secretory pathway, the signal peptide binds to SRP, an RNP particle containing a small RNA and 6 different proteins (b, [modified from 60]). A model for SRP at nanoscale is shown in (c).

In the SRP pathway, SRP binds to the ribosome synthesizing the polypeptide, and subsequently also binds an SRP receptor, located next to the machinery that transfers proteins across the membrane and out of the cell. This process begins when a nascent polypeptide carrying a signal sequence emerges from the translating ribosome and is recognized by the SRP. The ribosome-nascent chain complex is delivered to the target membrane via the interaction of SRP with the SRP receptor. There, the cargo is transferred to the Sec61p (or secYEG in archaea and bacteria) translocon, which translocates the growing polypeptide across the membrane or integrates it into the membrane bilayer. SRP and SR then dissociate from one another to enter subsequent rounds of targeting.

During the last years, several structures have been solved by crystallography and cryo-electron microscopy that represent distinct functional states of the SRP cycle. On this basis, the first structure-based models can be suggested that explain important aspects of protein targeting, such as the SRP–ribosome [54], SRP-SRP receptor [55] and SRP–SR interactions. The snapshots obtained by single-particle EM reconstructions enable us to follow the path of a nascent protein from the peptidyl-transferase center, through the ribosomal tunnel, to and across the translocon in the membrane. With new developments in image processing techniques it is possible to sort a biological homogenous sample into different conformational states and to reach subnanometer resolution such that folding of the nascent chain into secondary structure elements can be directly visualized [56].

Molecular biology, biochemistry, and cryo-electron microscopy, have been combined to study the ribosome-protein complexes involved in protein assembly, folding and targeting. These approaches led to obtain structural snapshots of entire pathways by which proteins are synthesized and targeted to their final positions. The link between SRP and its receptor is usually transient and chemically unstable, for this reason, engineered SRP receptor bind more stably to SRP, then introduced to ribosomes and observed the resulting complexes using cryo-electron microscopy (cryo-EM). Cryo-EM can be performed in roughly physiological conditions, providing a picture that closely resembles what happens in living cells. This picture can then be combined with higher-resolution crystallography data and biochemical studies [57-58].

c. Peroxisomes

The oxidative stress (EO) is a disorder where reactive oxygen species (ROS) are produced. These compounds, that include free radicals and peroxides, play important roles in cell redox signaling. However, disturbances in the balance between the ROS production and the biological system can be particularly destructive. For example, the P450 oxide reductase activity produces H_2O_2 as a metabolite. This enormous family of enzymes is present in the mitochondrial and smooth endoplasmic reticulum (SER) membranes and catalyzes several reactions in the pathway of the biogenesis of steroid hormones [59] and in the detoxification process or in the first stage of drugs or xenobiotics hydrolysis, converting them in the SER, in water-soluble compounds for its excretion in the urine [60-62].

Peroxisomes are single membrane organelles present in practically every eukaryotic cell. Matrix proteins of peroxisomes synthesized in free polyribosomes in the cytoplasm and imported by a specific signal, are encoded in genes present in the cell nucleus genome.

Peroxisomal membrane-bound PEX proteins, also encoded in the nuclear genome, are synthetized by ribosomes associated to rough endoplasmic reticulum since they display signal peptide. Therefore the peroxisome as an organelle derives from the rough endoplasmic reticulum. These organelles participate in ROS generation, as H_2O_2, but also in cell rescue from oxidative stress by catalase activity. In several biological models for pathological processes involving oxygen metabolites, the role of peroxisomes in prevention of oxidative stress is strongly suggested by de co-localization of catalase and H_2O_2, and the induction of peroxisomes proliferation [63].

d.   Mitochondrion and chloroplast particles

i.   ATP synthase: A rotary molecular motor

To support life, cells must be continuously supplied with external energy in form of light or nutrients and must be equipped with chemical devices to convert these external energy sources into adenosine triphosphate (ATP). ATP is the universal energy currency of living cells and as such is used to drive numerous energy-consuming reactions, e.g., syntheses of biomolecules, muscle contraction, mechanical motility and transport through membranes, regulatory networks, and nerve conduction. When performing work, ATP is usually converted to ADP and phosphate. It must therefore continuously be regenerated from these compounds to continue the cell energy cycle. The importance of this cycle can be best illustrated by the demand of 50 Kg of ATP in a human body on average [64].

Prokaryotes use their plasma membrane to produce ATP. Eukaryotes use instead the specialized membrane inside energy-converting organelles, mitochondria and chloroplasts, to produce most of their ATP. The mitochondria are present in the cells of practically all eukaryotic organisms (including fungi, animals, plants, algae and protozoa), and chloroplasts occur only in plants and algae. The most striking morphological feature of both organelles, revealed by electron microscopy, is the large amount of internal membrane they contain. This internal membrane provides the framework for an elaborate set of electron-transport processes, mediated by the enzymes of the Respiratory Chain that are essential to the process of Oxidative Phosphorylation which generate most of the cell's ATP.

In eukaryotes, oxidative phosphorylation occurs in mitochondria and photophosphorylation in chloroplasts. In the mitochondria, the energy to drive the synthesis of ATP derive from the oxidative steps in the degradation of carbohydrates, fats and amino acids; whereas the chloroplasts capture the energy of sunlight and harness it to make ATP [60].

ii.   The Chemiosmotic Model of Peter Mitchell

Our current understanding of ATP synthesis in mitochondria and chloroplasts is based on the chemiosmotic model proposed by Peter Mitchell in 1961 [60], which has been accepted as one of the great unifying principles of twentieth century. According with this model, the electrochemical energy inherent in the difference in proton concentration and the separation of charge across the inner mitochondrial membrane (the proton motive force) drives the synthesis of ATP as protons flow passively back into the matrix through a proton pore associated with the ATP synthase (Fig. 7).

Under aerobic conditions, the major ATP synthesis pathway is oxidative phosphorylation of which the terminal reaction is catalyzed by F_oF_1-ATP synthase. This enzyme is found widely in the biological world, including in thylakoid membranes, the mitochondrial inner membrane and the plasma membrane of bacteria, and is the central enzyme of energy metabolism in most organisms [65].

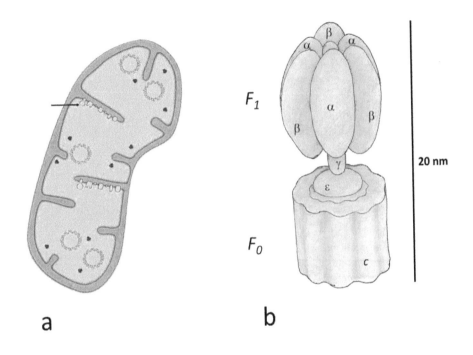

Figure 7. In the mitochondrion (a), ATP synthase (arrow in a; b) is part of the respiratory chain.

Like many transporters the F_oF_1-ATP synthase (or F-type ATPase) has been fascinating subject for study of a complex membrane-associated process. This enzyme catalyzes ATP synthesis from adenosine diphosphate (ADP) and inorganic phosphate (Pi), by using the electrochemical potential of protons (or sodium ions in some bacteria) across the membrane, i.e. it converts the electrochemical potential into its chemical form. ATP synthase also functions in the reverse direction (ATPase) when the electrochemical potential becomes insufficient: it catalyzes proton pumping to form an electrochemical potential to hydrolyze ATP into ADP and Pi. Proton translocation and ATP synthesis (or hydrolysis) are coupled by a unique mechanism, subunit rotation. Electrochemically energy contained in the proton gradient is converted into mechanical energy in form of subunit rotation, and back into chemical energy as ATP (Nakamoto RK, et al, 2008).

Mitochondrial ATP synthase is an F-type ATPase similar in structure and mechanism to the ATP synthases of chloroplasts and bacteria. This large complex of the inner mitochondrial membrane, also called Complex V, catalyzes the formation of ATP from ADP and Pi, accompanied by the flow of protons from P (positive) side to N (negative) side of the membrane [66].

iii. F_oF_1-ATP Synthase Structure and Function

ATP synthase is a supercomplex enzyme with a molecular weight of 500 kDal and consists of two rotary motors. One is F_1 subcomplex (~380 kDal), which is the water-soluble part of ATP synthase. F_1 was identified and purified by Efraim Racker and his colleagues in the early 1960s. When isolated from the membrane portion, it acts as an ATP-driven motor: it rotates its inner subunit to hydrolyze ATP and is therefore term F_1 ATPase. The other rotary motor of ATP synthase is the membrane-embedded Fo subcomplex (~120 kDal) through which the protons flow.

In the simplest form of the enzyme, in bacteria like *Escherichia coli*, F_1 is composed of five different subunits, in a stoichiometry of $\alpha_3\beta_3\gamma\delta\varepsilon$, and F_o consists of three distinct subunits in a stoichiometry of ab_2c_{10-15}. A newer more mechanically-based division differentiates between the "rotor" (in *E. coli*,$\gamma\varepsilon c_n$) and the "stator" ($\alpha_3\beta_3\delta ab_2$). The $\alpha_3\alpha_3$ ring of the stator contains the three catalytic nucleotide sites, on the β subunits at the interphase to the adjacent α subunit. The a subunit contains the static portion of the proton traslocator machinery. $\alpha_3\beta_3$ and a are held together by the "stator stalk" (or "peripheral stalk"), consisting of $b_2\gamma$ [65].

The crystallographic determination of the F_1 structure by John Walker and colleagues [67] revealed structural details very helpful in explaining the catalytic mechanism of the enzyme. The three α- and β- subunits that constitute the hexameric stator ring are alternately arranged like the sections of an orange. The rotor shaft is the γ-subunit, which is accommodated in the central cavity of the $\alpha_3\beta_3$-ring. The ε-subunit binds onto the protruding part of the γ-subunit and provides a connection between the rotor parts of F_1 and F_o. The δ-subunit acts as a connector between F_1 and F_o that connects the stator parts.

Catalytic reaction centers for ATP hydrolysis/synthesis reside at the three of the α-β interfaces, whereas the non-catalytic ATP-binding sites reside on the other α/β interfaces. While the catalytic site is formed mainly with amino acid residues from γ-subunit, the non- catalytic sites are primarily within the α-subunit. Upon ATP hydrolysis on the catalytic sites, F_1 rotates the γ-subunit in the anticlockwise direction viewed from the F_o side [68].

As mentioned before, F_o subcomplex (*o* denoting oligomycin sensitive) consists of ab_2c_{10-15} subunits. The number of c subunits varies among the species and form a ring complex by aligning in a circle. It is widely thought that the c-ring and the a subunit form a proton pathway. With the downhill proton flow through the proton channel, the c-ring rotates against the ab_2 subunits in the opposite direction of the γ-subunit of the F_1 motor [69]. Thus, in the F_oF_1 complex, F_o and F_1 push each other in the opposite direction. Under physiological condition where the electrochemical potential of the protons is large enough to surpass the free energy of ATP hydrolysis, F_o forcibly rotates the γ-subunit in the clockwise direction and then F_1 catalyzes the reverse reaction, *i.e.* ATP synthesis which is the principle function of ATP synthase. In contrast, when the electrochemical potential is small or decreases, F_1 forces F_O to rotate the c-ring in the reverse direction to pump protons against the electrochemical potential.

The c subunit of the F_O complex is a small (Mr 8,000), very hydrophobic polypeptide, consisting almost entirely of two membrane-spanning α-helices, that are connected by a small loop extending from the matrix side of the membrane. The crystal structure of the yeast F_OF_1, solved in 1999, shows the arrangement of the subunits. The yeast complex has 10 c subunits, each with two transmembrane helices roughly perpendicular to the plane of the membrane and arranged in two concentric circles. The inner circle is made up of the amino-terminal helices of each c subunit; the outer circle, about 55 Å in diameter, is made up of the carboxyl-terminal helices. The ε and γ subunits of F_1 form a leg-and-foot that projects from the bottom (membrane) side of F_1 and stands firmly on the ring of c subunits. The a subunit is a very hydrophobic protein that in most models is composed of five transmembrane helices. Ion translocation takes place through subunit a and its interface with subunit c. The b subunits are anchored within the membrane by an N-terminal α-helix and extend as a peripheral stalk all the way to the head of the F_1 domain. According to cross-linking studies, the b subunits contact de C-terminal part of the c subunit and the loop between helices 4 and 5 of the a subunit at the periplasmic surface. The δ-subunit forms a strong complex with the α-subunit. In mitochondria, the peripheral stalk consists of more subunits named OSCP (Oligomycin Sensitive Conferring Protein), b, δ and F_6 [64].

iv.   Structure of F_1 and binding-change mechanism for ATP Synthesis.

The classic working model for F1 is the "binding-change mechanism" proposed by Paul Boyer [70]. The early stage of this model postulated an alternating transition between two chemical states, assuming two catalytic sites residing on F_1. It was later revised to propose the cyclic transition of the catalytic sites based on the biochemical and electron microscopic experiments that revealed that F_1 has the three catalytic sites [71-73]. One important feature of this model is that the affinity for nucleotide in each catalytic site is different from each other at any given time, and the status of the three β-subunits cooperatively change in one direction accompanying γ rotation. This hypothesis is strongly supported by X-ray crystallographic studies performed by Walker's group [67] that first resolved crystal structure of F_1, which revealed many essential structural features of F_1 at atomic resolution. Importantly, the catalytic β-subunits differ from each other in conformation and catalytic state: one binds to an ATP analogue, adenosine 5'-(β,γ-imino)-triphosphate (AMP-PNP), the second binds to ADP and the third site is empty. Therefore, these sites are termed βTP, βDP and βEmpty, respectively. While βTP and βDP have a close conformation wrapping bound nucleotides on the catalytic sites, βEmpty has an open conformation swinging the C-terminal domain away from the binding site to open the cleft of the catalytic site. These features are consistent with the binding-change mechanism. Another important feature found in the crystal is that while the N-terminal domains of the α- and β-subunits form a symmetrical smooth cavity as the bearing for γ rotation at the bottom of the $\alpha_3\beta_3$-ring, the C-terminal domains of the β-subunit show distinct asymmetric interactions with the γ-subunit. Therefore, the most feasible inference is that the open-to-closed transition of the β-subunits upon ATP binding pushes γ, and the sequential conformational change among β- subunits leads the unidirectional γ rotation.

One strong prediction of the binding-change model of Boyer is that the γ subunit should rotate in one direction when $F_O F_1$ is synthesizing ATP and in the opposite direction when the enzyme is hydrolyzing ATP. This prediction was confirmed in elegant experiments in the laboratories of Masasuke Yoshida and Kazuhiko Kinosita Jr. [74]. The rotation of γ in a single F_1 molecule was observed microscopically by attaching a long, thin, fluorescent actin polymer to γ and watching it move relative to $\alpha_3 \beta_3$ immobilized on a microscope slide, as ATP was hydrolyzed [see 75]. Lately the unidirectional γ rotation was visualized in simultaneous imaging of the conformational change of the β-subunit and the γ rotation.

v.   New approaches for studying biological macromolecules.

The Atomic Force Microscope (AFM) is a powerful tool for imaging individual biological molecules attached to a substrate and place in aqueous solution. This technology allows visualization of biomolecules under physiological conditions. However, it is limited by the speed at which it can successively record highly resolved images. Recent advances have improved the time resolution of the technique from minutes to tens of milliseconds, allowing single biomolecules to be watch in action in real time. Toshio Ando and his coworkers at Kanazawa University have been leading innovators in this so-called High-Speed Atomic Force Microscope (HS-AFM) technology [76]. This technology allows direct visualization of dynamic structural changes and dynamic processes of functioning biological molecules in physiological solutions, at high spatial-temporal resolution. Dynamic molecular events appear in detail in an AFM movie, facilitating our understanding of how biological molecules operate to function.

In this regard, the Ando group showed a striking example of molecular motor action in their AFM movies of the isolated subcomplex of the rotary motor protein F_1-ATPase. Previous single-molecule experiments on parts of this enzyme had measured rotation, but they could only be done if at least one subunit of the rotor was attached. The AFM, however, could visualize the conformational change that the β subunits of the stator undergo when they bind ATP. By imaging at 12.5 frames/s, the authors followed the time dependence of these conformational changes, leading to the surprising conclusion that, contrary to what was widely assumed before, the catalysis on the enzyme maintains its sequential rotary order even in absence of the rotor subunits.

"To directly observe biological molecules at work was a holy grail in biology. Efforts over the last two decades at last materialized this long-quested dream. In high-resolution AFM movies, we can see how molecules are dynamically behaving, changing their structure and interacting with other molecules, and hence we can quickly understand in stunning detail how molecules operate to function. This new approach will spread over the world and widely applied to a vast array of biological issues, leading to a number of new discoveries. The extension of high-speed AFM to a tool for imaging live cells, which allows direct in situ observation of dynamic processes of molecules and organelles, remains an exciting challenge but will be made in the near future because it is a right and fruitful goal" [77].

e.   Lipid rafts

Cell membranes are dynamic assemblies of a variety of lipids and proteins. They form a protective layer around the cell and mediate the communication with the outside world. The original fluid mosaic model [78] of membranes suggested a homogenous distribution of proteins and lipids across the two-dimensional surface, but more recent evidence suggests that membranes themselves are not uniform and that microdomains of lipids in a more ordered state exist within the generally disorder lipid milieu of the membrane. These clusters of ordered lipids are now referred to as lipid rafts [79] (Pike LJ 2009).

Lipid rafts (LRs), consist of cell membrane domains rich in cholesterol, sphingolipids and lipid-anchored proteins in the exoplasmic leaflet of the lipid bilayer. Because of their ability to sequester specific lipids and proteins and exclude others, rafts have been postulated to perform critical roles in a number of normal cellular processes, such as signal transduction [80], membrane fusion, organization of the cytoskeleton [81-83], lipid sorting, and protein trafficking/recycling, as well as pathological events [84].

LRs are too small to be resolved by standard light microscopy - they range from 10 to about 200 nm - with a variable life span in the order of milliseconds (msec). Detergent resistant membranes, containing clusters of many rafts, can be isolated by extraction with Triton X-100 or other detergents on ice. However, this method involves breaking up the membrane and has limitations in terms of defining the size, properties, and dynamics of intact microdomains [85-88].Thus, a variety of sophisticated techniques have recently been used to analyze in detail open questions concerning rafts in cell and model membranes including biochemical, biophysical, quantitative fluorescence microscopy, atomic force microscopy and computational methodologies [89-90].

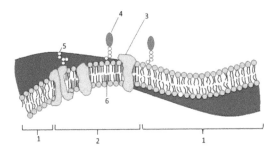

Figure 8. Components of a lipid raft. (1) non raft membrane, (2) lipid raft, (3) lipid raft associated transmembrane protein, (4) GPI-anchored protein, (5) glycosylation modifications (glycoproteins and glycolipids).

The raft affinity of a given protein can be modulated by intra- or extracellular stimuli. Saturated fatty acids are preferentially enriched in the side chains of the membrane phospholipids, which allows closer packing and thus increased rigidity, more order and less fluidity of the LRs compared to the surrounding membrane [91-92]. Proteins with raft affinity include glycosyl-phosphatidylinositol (GPI)-anchored proteins [93-94], doubly acylated proteins, such as Src-family kinases or the α-subunits of heterotrimeric G proteins8, cholesterol-linked and palmitoylated proteins such as Hedgehog9, and transmembrane proteins, particularly palmitoylated ones [92-95].

Different subtypes of lipid rafts can be distinguished according to their protein and lipid composition. Caveolae are types of rafts that are rich in proteins of the caveolin family (caveolin-1, -2 and -3) which present a distinct signaling platform [96]. The caveolae are enriched in cholesterol, glycosphingolipids, and sphingomyelin. They are the site of several important protein–protein interactions, for example, the neurotrophin receptors, TrkA and p75(NTR), whose respective interactions with caveolin regulates neurotrophin signaling in the brain. Caveolins also regulate G-proteins, MAPK, PI3K, and Src tyrosine kinases.

The most important role of rafts at the cell surface may be their function in signal transduction. Lipid rafts have been implicated as the sites for a great number of signaling pathways.They form concentrating platforms for individual receptors, activated by ligand binding [86]. If receptor activation takes place in a lipid raft, the signaling complex is protected from non-raft enzymes such as membrane phosphatases that otherwise could affect the signaling process. In general, raft binding recruits proteins to a new micro-environment, where the phosphory-lation state can be modified by local kinases and phosphatases, resulting in downstream signalling. Individual signaling molecules within the raft are activated only for a short period of time.

Immobilization of signaling molecules by cytoskeletal actin filaments and scaffold proteins may facilitate more efficient signal transmission from rafts [97]. Current evidence supports a role for lipid rafts in the initiation and regulation of The B-cell receptor signaling and antigen trafficking [98-100]. The importance of lipid raft signalling in the pathogenesis of a variety of conditions, such as Alzheimer's, Parkinson's, cardiovascular and prion diseases, systemic lupus erythematosus and HIV, has been elucidated over recent years[101] and makes these specific membrane domains an interesting target for pharmacological approaches in the cure and prevention of these diseases [102]. Rafts serve as a portal of entry for various pathogens and toxins, such as human immunodeficiency virus 1 (HIV-1). In the case of HIV-1, raft microdomains mediate the lateral assemblies and the conformational changes required for fusion of HIV-1 with the host cell [103]. Lipid rafts are also preferential sites of formation for pathological forms of the prion protein (PrPSc) and of the β-amyloid peptide associated with Alzheimer's disease {104].

Plasma membranes typically contain higher concentrations of cholesterol and sphingomyelin than do internal membranous organelles [105-106]. Thus, along the secretion pathway, there are very low concentrations of cholesterol and sphingolipids in the endoplasmic reticulum, but the concentrations of these lipids increase from the cis-Golgi to the trans-Golgi and then to the plasma membrane [107-108]. On the contrary, recent evidence suggests that mitochon-dria do not contain lipid rafts, and lipid rafts do not contain mitochondrial proteins [109].

Lipid raft domains play a key role in the regulation of exocytosis [110]. The association of SNAREs protein complexes with lipid rafts acts to concentrate these proteins at defined sites of the plasma membrane that are of functional importance for exocytosis [111-114].

f.   The nucleolus

The cell nucleus contains different compartments that are characterized by the absence of delineating membranes that isolate it from the rest of the nucleoplasm [5]. Due to the high concentration of RNA and proteins that form it, the nucleolus is the most conspicuous nuclear body in cycling cells observed by light and electron microscopy. Nucleoli are formed around nucleolar organizer regions (NORs), which are composed of cluster of ribosomal genes (rDNA) repeat units [115-121]. The number of NOR-bearing chromosomes varies depending on the species, can be found 1 in haploid yeast cells to 10 in human somatic cells (short arms of chromosomes 13, 14, 15, 21 and 22). Nucleolus is the organelle of rDNA transcription by RNA polymerase I, whose activity generates a long ribosomal precursor (pre-rRNA), this molecule is the target of an extensive process that includes removing or cutting the spacers and 2'-O-methylation of riboses and coversions of uridine residues into pseudouridines. The net result of these reactions is the release of mature species of ribosomal RNA (rRNA) 18S, 5.8S and 28S. These particles are assembled with approximately 82 ribosomal proteins and rRNA 5S (synthesized by RNA polymerase III) to form the 40S and 60S subunits; both of these subunits are then exported separately to the cytoplasm and are further modified to form mature ribosomal subunits. Currently, it is widely accepted that nucleolar transcription and early pre-rRNA processing take place in the fibrillar portion of nucleolus while the later steps of processing and ribosome subnits assembly occurs mainly in the granular zone. The architecture of the nucleolus reflects the vectorial maturation of the pre-ribosomes. The nucleolar structure is organized by three canonical subdomains that are morphologically and biochemically different. The fibrillar centers (FC), dense fibrillar component (DFC) and granular component (GC). The FCs are structures with a low electron density, often circular shape of ~0.1 to 1μm in diameter. The FCs are enriched with rDNA, RNA polymerase I, topoisomerase I and upstream binding factor (UBF). DFCs are a compact fibrillar region containing a high concentration of ribonucleoprotein molecules that confer a high electrodensity. This component entirely or partially surrounds the FCs. DFCs contains important proteins such as fibrillarin and nucleolin as well as small nucleolar RNAs, pre-rRNA and some transcription factors. FCs and DFCs are embedded in the GC, composed mainly of granules of 15 to 20nm in diameter with a loosely organized distribution. In the GC are located B23/nucleophosmin, Nop 52, r-proteins, auxiliary assembly factors, and the 40S and 60S subunits that the GC is itself composed of at least two distintc molecular domains. Considering the species, cell type and physiological state of the cell, there is considerable diversity in the prevalence and arrangement of the three nucleolar components.

On the other hand, the current eukaryotic nucleolus is involved in the ribosomal biogenesis but has been described as a multifunctional entity. Extra ribosomal functions include biogenesis and/or maturation of other ribonucleoprotein machines, including the signal recognition particle, the spliceosomal small nuclear RNPs and telomerase, processing or export of some mRNAs and tRNA, cell cycle and cell proliferation control, stress response and apoptosis [116]. The plurifunctional nucleolus hypothesis is reinforced by the description of nucleolar proteome of several eukaryotes. A proteomic analysis has identified more than 200 nucleolar proteins in Arabidopsis and almost 700 proteins in the nucleolus of HeLa cells. A comparison of nucleolar proteome from humans and budding yeast showed that ~90% of human nucleolar proteins have yeast homologues. Interestingly, only 30% of the human nucleolar proteome is intended for ribosomal biogenesis [120, 122].

1.1.3. Microscopy

Fundamental to approach the cell at the nanoscale in cell nanobiology are the classical and also remarkably new types of microscopy. Three different epochs characterize microscopy: 1) Light microscopy, developed since ca. 1500, where glass lenses and light as source of illumination are used to get resolution of up to 0.2 μm. Different types such as bright field, phase contrast, differential interference contrast (Nomarsky), dark field, polarization, fluorescence, confocal, and super-resolution, are variants of this type of microscopy. 2) Electron microscopy, developed since early 1930s, where electromagnetic lenses and electrons as source of illumination are used to get resolution of up to nm or A°. Transmission and scanning electron microscopy –including the environmental and high resolution modes- are the two forms of this microscopy. 3) Scanning probe microscopy, developed in the early 1980s, where no lenses or illuminations are used, but instead the microscope consists of a fine tip interacting with the samples to potentially obtaining atomic resolution. Scanning tunneling microscopy and atomic force microscopy are the major variants of this type of modern microscopy. Because atomic force microscopy may produce images at high resolution even under liquid, we have been using such microscopy for imaging the cell components. To test this approach, we used several cell types and generated images at low magnification (Figure 9a). Nuclear particles *i.e.* Lacandonia granules were already visualized using this approach (Figure 9b).

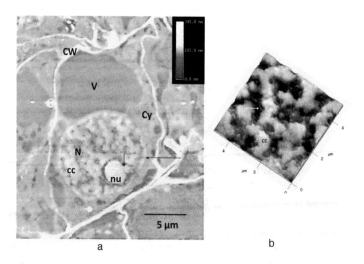

Figure 9. a) Atomic force microscopy image of a cell from the tegument of the plant *Lacandonia schismatica*. Cell wall (CW), vacuole (V), cytoplasm (Cy), Within the nucleus (N), compact chromatin (cc), nucleolus (nu), nucleolar organizer (small arrow) and nuclear pore (large arrow). b) Atomic force microscopy of Lacandonia granules within the nucleus of a tegument cell of the plant *Lacandonia schismatica*. Three dimensional displaying shows compact chromatin (cc) and associated particles (arrow).

1.1.4. Further research

Further research in our laboratory will focusing in visualizing the nanoscale cell structures involved in fundamental processes as ribosome biogenesis, at a high resolution *in situ* under liquid conditions to perform quantitative analysis.

2. Conclusion

A view of the cell emphasizing vertical resolution obtained by atomic force microscopy may represent a way to understand cell structure and function at the nanoscale, an interphase between molecular biology and cell biology.

Acknowledgements

DGAPA-UNAM PAPIIT IN-227810, PAPIME PE211412, CONACyT 180835.

Rogelio Fragoso-Soriano and Tomás Nepomuceno-Mejía are postdoctoral fellows from ICyTDF and DGAPA-UNAM at Faculty of Sciences-UNAM, respectively. Georgina Alvarez-Fernandez is on-a leave-of-absence from the Department of Biochemistry, Faculty of Medicine, UNAM. Luis and Teresa Jiménez Segura for SRP and ATP synthase figures.

Author details

María de Lourdes Segura-Valdez[1], Lourdes T. Agredano-Moreno[1],
Tomás Nepomuceno-Mejía[2], Rogelio Fragoso-Soriano[2], Georgina Álvarez-Fernández[2],
Alma Zamora-Cura[1], Reyna Lara-Martínez[1] and Luis F. Jiménez-García[1]

1 Laboratory of Cell Nanobiology and Electron Microscopy Laboratory (Tlahuizcalpan), Department of Cell Biology, Faculty of Sciences, National Autonomous University of Mexico (UNAM), Circuito Exterior, Ciudad Universitaria, Coyoacán 04510, México D.F, México

2 Visiting from the Department of Biochemistry, Faculty of Medicine, UNAM, México

References

[1] Jiménez García LFSegura Valdez ML. Biología celular del genoma. México: Las Prensas de Ciencias-UNAM: (2010).

[2] Monneron, A, & Bernhard, W. Fine structural organization in the interphase nucleus in some mammalian cells. Journal of Ultrastructural Research (1969). , 27-266.

[3] Vázquez-nin, G. H, & Echeverría, O. M. The polytene nucleus in morphological, cytochemical, and functional studies of Messenger RNA transcription, processing, and transportation. European Journal of Histochemistry (1996). , 40-7.

[4] Jiménez-garcía, J, Agredano-moreno, L. F, Segura-valdez, L. T, Echeverría, M. L, Martínez, O. M, Ramos, E, & Vázquez-nin, C. H. GH. The ultrastructural study of the interphase cell nucleus of Lacandonia schismatica (Lacandoniaceae:Triuridales) reveals a non typical extranucleolar particle. Biol. Cell (1992). , 75-101.

[5] Spector, D. L. Macromolecular domains within the cell nucleus. Annual Reviews in Cell Biology (1993). , 9-265.

[6] Tischendorf, G. W, Zeichhardt, H, & Stoffler, G. Architecture of the Escherichia coli ribosome as determined by immune electron microscopy. Proceedings of the National Academy of Sciences USA (1975). , 72-4820.

[7] Lake, J. A. (1976). Ribosome structure determined by electron microscopy of Escherichia coli small subunits, large subunits and monomeric ribosomes. Journal of Molecular Biology 1976; , 105-131.

[8] Boublik, M, Hellmann, W, & Kleinschmidt, A. K. Size and structure of Escherichia coli ribosomes by electron microscopy. Cytobiologie (1977). , 14-293.

[9] Yonath, A, Mussig, J, Tesche, B, Lorenz, S, Erdmann, V. A, & Wittmann, H. G. (1980). Crystallization of the large ribosomal subunits from Bacillus stearothermophilus. Biochemistry International 1980; , 1-428.

[10] Ban, P, Nissen, J, Hansen, P, Moore, B, & Steitz, T. A. The complete atomic structure of the large ribosomal subunit at 2.4 A resolution. Science (2000). , 289-905.

[11] Wimberly, B. T, Brodersen, D. E, Clemons, W. M, Morgan-warren, R. J, Carter, A. P, Vonrhein, C, Hartsch, T, & Ramakrishnan, V. Structure of the 30S ribosomal subunit. Nature (2000). , 407-327.

[12] Yusupov, M. M, Yusupova, G. Z, Baucom, A, Lieberman, K, Earnest, T. N, Cate, J. H, & Noller, H. F. Crystal structure of the ribosome at 5.5 A resolution. Nature (2001). , 292-883.

[13] Seidelt, B, Innis, C. A, Wilson, D. N, Gartmann, M, Armache, J. P, Villa, E, Trabuco, L. G, Becker, T, Mielke, T, Schulten, K, et al. Structural insight into nascent polypeptide chain-mediated translational stalling. Science (2009). , 326-1412.

[14] Bhushan, S, Gartmann, M, Halic, M, Armache, J. P, Jarasch, A, Mielke, T, Berninghausen, O, & Wilson, D. N. Beckmann R. α-helical nascent polypeptide chains visualized within distinct regions of the ribosomal exit tunnel. Nature Structural Molecular Biology (2010). , 17-313.

[15] Bhushan, S, Hoffmann, T, Seidelt, B, Frauenfeld, J, Mielke, T, Berninghausen, O, Wilson, D. N, & Beckmann, R. SecM-stalled ribosomes adopt an altered geometry at the peptidyl transferase center. PLoS Biol (2011). e1000581.

[16] Frank, J, Zhu, J, Penczek, P, Li, Y, Srivastava, S, Verschoor, A, Radermacher, M, Grassucci, R, Lata, R. K, & Agrawal, R. K. A model of protein synthesis based on cryo-electron microscopy of the E. coli ribosome. Nature (1995). , 376-441.

[17] Ben-shem, A. Garreau de Loubresse N, Melnikov S, Jenner L, Yusupova G, Yusupov M.The structure of the eukaryotic ribosome at 3.0A ° resolution. Science (2011). , 334-1524.

[18] Klinge, S, Voigts-hoffmann, F, Leibundgut, M, Arpagaus, S, & Ban, N. Crystal structure of the eukaryotic 60S ribosomal subunit in complex with initiation factor 6. Science (2011). , 334-941.

[19] Nissen, P, Hansen, J, Ban, N, Moore, P. B, & Steitz, T. A. The structural basis of ribosome activity in peptide bond synthesis. Science (2000). , 289-920.

[20] Kampmann, M. R, & Blobel, G. Biochemistry. Nascent proteins caught in the act. Perspectives in Biochemistry (2009). , 326, 1352-1353.

[21] Voorhees, R. M, Schmeing, T. M, Kelley, A. C, & Ramakrishnan, V. The mechanism for activation of GTP hydrolysis on the ribosome. Science (2010). , 330-835.

[22] Weis, F, Bron, P, Giudice, E, Rolland, J. P, Thomas, D, Felden, B, & Gillet, R. tmRNA-SmpB: a journey to the centre of the bacterial ribosome. The EMBO Journal (2010). , 29-3810.

[23] Neubauer, C, Gillet, R, Kelley, A. C, & Ramakrishnan, V. Decoding in the Absence of a Codon by tmRNA and SmpB in the Ribosome. Science (2012). , 335-1366.

[24] Poehlsgaard, J, & Douthwaite, S. The bacterial ribosome as target for antibiotics. Nature Reviews in Microbiology (2005). , 3-871.

[25] Selmer, M, Dunham, C. M, Murphy, F. V, Weixlbaumer, A, Petry, S, Kelley, A. C, Weir, J. R, & Ramakrishnan, V. Structure of the 70S ribosome complexed with mRNA and tRNA. Science (2006). , 313-1935.

[26] Wilson, D. N. Doudna Cate JH. The and function of the eukaryotic ribosome. Cold Spring Harbor Perspectives in Biology (2012).

[27] Spahn, C. M, Beckmann, R, Eswar, N, Penczek, P. A, Sali, A, Blobel, G, & Frank, J. Structure of the 80S ribosome from Saccharomyces cerevisiae-tRNA-ribosome and subunit- subunit interactions. Cell (2001). , 107-373.

[28] Powers, T, & Walter, P. Regulation of ribosome biogenesis by the rapamycin-sensitive TOR-signaling pathway in Saccharomyces cerevisiae. Mol Biol Cell (1999). , 10-987.

[29] Hannan, K. M, Brandenburger, Y, Jenkins, A, Sharkey, K, Cavanaugh, A, Rothblum, L, Moss, T, Poortinga, G, Mcarthur, G. A, Pearson, R. B, & Hannan, R. D. mTOR-dependent regulation of ribosomal gene transcription requires S6K1 and is mediated by phosphorylation of the carboxy-terminal activation domain of the nucleolar transcription factor UBF. Molecular Cellular Biology (2003). , 23-8862.

[30] Mayer, C, Zhao, J, Yuan, X, & Grummt, I. mTOR-dependent activation of the transcription factor TIF-IA links rRNA synthesis to nutrient availability. Genes & Development (2004). , 18-423.

[31] Lee, J, Moir, R. D, Mcintosh, K. B, & Willis, I. M. TOR signaling regulates ribosome and tRNA synthesis via LAMMER/Clk and GSK-3 family kinases. Molecular Cell (2012). , 45-836.

[32] Andrews, D. W, Walter, P, & Ottensmeyer, F. P. Structure of the signal recognition particle by electron microscopy. Proceedings of the National Academy of Sciences USA (1985). , 82-785.

[33] Walter, P, & Johnson, A. E. Signal sequence recognition and protein targeting to the endoplasmic reticulum membrane.Annual Reviews in Cell Biology (1994). , 10-87.

[34] Walter, P, & Blobel, G. Translocation of proteins across the endoplasmic reticulum III. Signal recognition protein (SRP) causes signal sequence-dependent and site-specific arrest of chain elongation that is released by microsomal membranes. Journal of Cell Biology (1981). , 91, 557-561.

[35] Andersen, E. S, Rosenblad, M. A, Larsen, N, Westergaard, J. C, Burks, J, Wower, I. K, Wewer, J, Gorodkin, J, Samuelsson, T, & Zwieb, C. The TmRDB and SRPDB resourdes. Nucleic Acid Research (1999). D , 163-188.

[36] Freymann, D. M, Keenan, R. J, Stroud, R. M, & Walter, P. Structure of the conserved GTPase domain of the signal recognition particle. Nature (1997). , 385-361.

[37] Montoya, G, Kaat, K, Moll, R, Schafer, G, & Sinning, I. The crystal structure of the conserved GTPase of SRP54 from the archaeon Acidianus ambivalens and its comparison with related structures suggests a model for the SRP-SRP receptor complex. Structure (2000). , 8-515.

[38] Keenan, R. J, Freymann, D. M, Walter, P, & Stroud, R. M. Crystal structure of the signal sequence binding subunit of the signal recognition particle. Cell (1998). , 94-181.

[39] Zopf, D, Bernstein, H. D, Johnson, A. E, & Walter, P. The methionine-rich domain of the 54 kd protein subunit of the signal recognition particle contains an RNA binding site and can be crosslinked to a signal sequence. The EMBO Journal (1990). , 9-4511.

[40] High, S, & Dobberstein, B. The signal sequence interacts with the methionine-rich domain of the kD protein of signal recognition particle. Journal of Cell Biology (1991). , 54.

[41] Tcke, L, High, H, Romisch, S, Ashford, K, & Dobberstein, A. J. B. The methionine-rich domain of the 54 kDa subunit of signal recognition particle is sufficient for the interaction with signal sequences. The EMBO Journal (1992). , 11-543.

[42] Egea, P. F, Shan, S. O, Napetschnig, J, Savage, D. F, Walter, P, & Stroud, R. M. Substrate twinning activates the signal recognition particle and its receptor. Nature (2004). , 427-215.

[43] Focia, P. J, Shepotinovskaya, I. V, Seidler, J. A, & Freymann, D. M. Heterodimeric GTPase core of the SRP targeting complex. Science (2004). , 303-373.

[44] Politz JC Yarovoi S Kilroy SM Gowda K Zwieb C and Pederson TSignal [44] recognition particle components in the nucleolus. Proceedings of the National Academy of Sciences USA (2000). , 97-55.

[45] Walter, P, & Blobel, G. Disassembly and reconstitution of signal recognition particle. Cell. (1983). , 34-525.

[46] Politz, J. C, Lewandowski, L. B, & Pederson, T. Signal recognition particle RNA localization in the nucleolus differs from the classical site of ribosome synthesis. Journal of Cell Biology (2002). , 159-411.

[47] Grosshans H Deinert K Hurt E and Simos GBiogenesis of the signal recognition particle (SRP) involves import of SRP proteins into the nucleolus, assembly with the SRP-RNA, and Xpo1p-mediated export. Journal of Cell Biology (2001). , 153-745.

[48] Leung, E, & Brown, J. D. Biogenesis of the signal recognition particle. Biochemistry. Society Transactions (2010). , 38-1093.

[49] Strub, K, & Walter, P. Assembly of the Alu domain of the signal recognition particle (SRP): dimerization of the two protein components is required for efficient binding to SRP RNA, Mol Cell Biol. (1990). , 10-777.

[50] Koch HG Moser M and Muller MSignal recognition particle-dependent protein targeting, universal to all kingdoms of life. Reviews of Physiology and Biochemical Pharmacology (2003). , 146, 55-94.

[51] Gierasch, L. M. Signal sequences. Biochemistry (1989). , 28-923.

[52] Zheng, N, & Gierasch, L. M. Signal sequences: the same yet different. Cell , 1996-86.

[53] Bernstein, H. D, Poritz, M. A, Strub, K, Hoben, P. J, Brenner, S, & Walter, P. Model for signal sequence recognition from amino-acid sequence of 54K subunit of signal recognition particle. Nature (1989). , 340-482.

[54] Halic, M, Becker, T, & Pool, M. R. Spahn CMT, Grassucci RA, Frank J and Beckmann R. Structure of the signal recognition particle ineracting with the elongation-arrested ribosome. Nature. (2004). , 427-808.

[55] Ataide, S. F, Schmitz, N, Shen, K, Ke, A, Shan, S, Doudna, J. A, & Ban, N. The crystal structure of the signal recognition particle in complex with its receptor. Science (2011). , 331-881.

[56] Knoops, K, Schoehn, G, & Schaffitzel, C. Cryo-electron microscopy of ribosomal complexes in cotranslational folding, targeting and translocation. Wiley Interdisciplinary Reviews in RNA. (2012). , 3-429.

[57] Schmeing, T. M, & Ramakrishnan, V. What recent ribosome structures have revealed about the mechanism of translation. Nature (2009). , 461-1234.

[58] Estrozi, L. F, & Boehringer, D. Shan Shu-ou,, Ban N, Schaffitzel C. Cryo-EM structure of the E. coli translating ribosome in complex with SRP and its receptor. Nat. Structural & Molecular Biology (2011). , 18-88.

[59] Repetto, M. Marcadores periféricos de estrés oxidativo en pacientes hipogonádicos. Revista Argentina de Endocrinología y Metabolismo (2005). , 42-26.

[60] Alberts, B, Johnson, A, Lewis, J, Raff, M, Roberts, K, & Walter, P. Molecular Biology of the Cell. New York: Garland; (2008).

[61] Chilo, N. H, & El Citocromo, p. y su rol en la hepatotoxicidad inducida por las drogas. Enfermedades del Aparato Digestivo (1999). , 2-34.

[62] Lodish, H, Berk, A, Kaiser, C. A, Kreiger, M, Scott, M, Bretscher, A, Ploegh, H, & Matsudaira, P. New York: Freeman: (2007).

[63] Schrader, M, & Fahimi, H. D. Peroxisomes and oxidative stress. Biochimica et Biophysica Acta (BBA)- Molecular Cell Research (2006). , 1763-1755.

[64] Von Ballmoos, C, Wiedenmann, A, & Dimroth, P. Essentials for ATP synthesis by F1F0 ATP synthases. Annual Review in Biochemistry (2009). , 78-649.

[65] Weber, J. ATP Synthase: Subunit-Subunit Interactions in the Stator Stalk. Biochimical Biophysical Acta (2006). , 1757-1162.

[66] Nelson, D. L, & Cox, M. Lehninger Principles of Biochemistry. New York: Freeman and Company; (2008).

[67] Abrahams, J. P, Leslie, A. G, Lutter, R, & Walker, J. E. Structure at 2.8 A resolution of F1ATPase from bovine heart mitochondria. Nature (1994). , 370-621.

[68] Okuno, D, Riota, I, & Noji, H. Rotation and structure of FOF1-ATP synthase. Journal of Biochemistry (2011). , 149-655.

[69] Diez, M, Zimmermann, B, Borsch, M, Konig, M, Schweinberger, E, Steigmiller, S, Reuter, R, Felekyan, S, Kudryavtsev, V, Seidel, , & Graber, P. . Proton-powered subunit rotation in single membrane-bound F0F1-ATP synthase. Nature Structural Molecular Biology 2004; 11-135.

[70] Boyer, P. D. The ATP synthase-a splendid molecular machine. Annual Reviews in Biochemistry (1997). , 66-717.

[71] Yoshida, M, Sone, N, Hirata, H, & Kagawa, Y. A highly stable adenosine triphosphatase from a thermophillie bacterium. Purification, properties, and reconstitution. Journal of Biological Chemistry (1975). , 250-7910.

[72] Kagawa, Y, Sone, N, Yoshida, M, Hirata, H, & Okamoto, H. Proton translocating ATPase of a thermophilic bacterium. Morphology, subunits, and chemical composition. Journal of. Biochemistry (1976). , 80-141.

[73] Wakabayashi, T, Kubota, M, Yoshida, M, & Kagawa, Y. Structure of ATPase (coupling factor TF1) from a thermophilic bacterium. Journal of Molecular Biology (1977). , 117-515.

[74] Noji, H, Yasuda, R, Yoshida, M, & Kinosita, K. Direct observation of the rotation of F1ATPase. Nature (1977). , 386-299.

[75] http://www.k2.phys.waseda.ac.jp/Movies.html for related experimental set up and video.

[76] Katan, A. J, & Dekker, C. High-Speed AFM Reveals the Dynamics of Single Biomolecules at the Nanometer Scale. Cell (2011). , 147-979.

[77] Ando, T. High-speed atomic force microscopy coming of age. Nanotechnology (2012)., 23-062001

[78] Singer, S. J, & Nicolson, G. L. The fluid mosaic model of the structure of cell membranes. Science. (1972). , 175-720.

[79] Pike, L. J. The challenge of lipid rafts. Journal of Lipid Research (2009). S , 323-328.

[80] Asano, S, Kitatani, K, Taniguchi, M, Hashimoto, M, Zama, K, Mitsutake, S, Igarashi, Y, Takeya, H, Kigawa, J, Hayashi, A, Umehara, H, & Okazaki, T. Regulation of cell migration by sphingomyelin synthases: sphingomyelin in lipid rafts decreases responsiveness to signaling by the CXCL12/CXCR4 pathway. Molecular Cell Biology (2012). , 27-35.

[81] Baumgart, T, Hammond, A. T, Sengupta, P, Hess, S. T, Holowka, D. A, Baird, B. A, & Webb, W. W. Large-scale fluid/fluid phase separation of proteins and lipids in giant plasma membrane vesicles. Proceedings of the National Academy of Sciences USA (2007). , 104-3165.

[82] Su, B, Gao, L, Meng, F, Guo, L. W, Rothschild, J, & Gelman, I. H. Adhesion-mediated cytoskeletal remodeling is controlled by the direct scaffolding of Src from FAK complexes to lipid rafts by SSeCKS/AKAP12. Oncogene (2012).

[83] Viola, A, & Gupta, N. Tether and trap: regulation of membrane-raft dynamics by actin-binding proteins. Nature Reviews in Immunology (2007). , 7-889.

[84] Mañes, S. del Real G, Martínez AC. Pathogens: raft hijackers. Nature Reviews in Immunology (2003). , 3-557.

[85] Douglass, A. D, & Vale, R. D. Single-molecule microscopy reveals plasma membrane microdomains created by protein-protein networks that exclude or trap signaling molecules in T cells. Cell (2005). , 121-937.

[86] Pike, L. J. Rafts defined: a report on the keystone symposium on lipid rafts and cell function. Journal of Lipid Research (2006). , 47-1597.

[87] Munro, S. Lipid rafts: elusive or illusive? Cell (2003). , 115-377.

[88] Lichtenberg, D, Goñi, F. M, & Heerklotz, H. Detergent-resistant membranes should not be identified with membrane rafts. Trends in Biochemical Sciences (2005). , 30-430.

[89] Owen, D. M (b, Williamson, D, Magenau, A, & Gaus, K. Optical techniques for imaging membrane domains in live cells (live-cell palm of protein clustering). Methods in Enzymology (2012). , 504-221.

[90] Anderton, C. R, Lou, K, Weber, P. K, Hutcheon, I. D, & Kraft, M. L. Correlated AFM and NanoSIMS imaging to probe cholesterol-induced changes in phase behavior and non-ideal mixing in ternary lipid membranes. Biochimical and Biophysical Acta (2011). , 1808-307.

[91] Simons, K, & Ikonen, E. Functional rafts in cell membranes. Nature (1997). , 387-569.

[92] Brown, D. A, & London, E. Structure and origin of ordered lipid domains in biological membranes. Journal of Membrane Biology (1998). , 164-103.

[93] Brown, D. A, & London, E. Structure and function of sphingolipid- and cholesterol-rich membrane rafts. Journal of Biological Chemistry (2000). , 275-17221.

[94] Hooper, N. M. Detergent-insoluble glycosphingolipid/cholesterol-rich membrane domains, lipid rafts and caveolae. Molecular Membrane Biology (1999). , 160-145.

[95] Simons, K, & Toomre, D. Lipid rafts and signal transduction. Nature Reviews in Molecular and Cell Biology (2000). , 1-31.

[96] Parton, R. G, & Simons, K. Digging into caveolae. Science (1995). , 269-1398.

[97] Kenichi, G, & Suzuki, N. Lipid rafts generate digital-like signal transduction in cell plasma membranes. Biotechnology Journal (2012). , 7-753.

[98] Cheng, P. C, Dykstra, M. L, Mitchell, R. N, & Pierce, S. K. A role for lipid rafts in BCR signaling and antigen targeting. Journal of Experimental Medicine (1999). , 190-1549.

[99] Chung, J. B, Baumeister, M A, & Monroe, J. G. Differential sequestration of plasma membrane-associated B cell antigen receptor in mature and immature B cells into glycosphingolipid-enriched domains. Journal of Immunology (2001). , 166-736.

[100] Pierce, S. K. Lipid rafts and B-cell activation.Nature Reviews in Immunology (2002). , 2-96.

[101] Hicks, D. A, Nalivaeva, N. N, & Turner, A. J. Lipids rafts and Alzheimer's disease: protein-lipid interactions and perturbation of signaling. Nature (2006). , 233-126.

[102] Michel, V, & Bakovic, M. Lipid rafts in health and disease. Biology of the Cell (2007). , 99-129.

[103] Brugger, B, Glass, B, Haberkant, P, Leibrecht, I, Wieland, F. T, & Krausslich, H. G. The HIV lipidome: A raft with an unusual composition. Proceedings of the National Academy of Sciences USA (2006). , 103-2641.

[104] Abad-rodriguez, J, Ledesma, M. D, Craessaerts, K, Perga, S, Medina, M, Delacourte, A, Dingwall, C, De Strooper, B, & Dotti, C. G. Neuronal membrane cholesterol loss enhances amyloid peptide generation. Journal of Cell Biology (2004). , 167-953.

[105] Keenan, T. W, & Morré, D. J. Phospholipid class and fatty acid composition of golgi apparatus isolated from rat liver and comparison with other cell fractions. Biochemistry (1997). , 9-19.

[106] Fridriksson, E. K, Shipkova, P. A, Sheets, E. D, Holowka, D, Baird, B, & Mclafferty, F. W. Quantitative analysis of phospholipids in functionally important membrane domains from RBL-2H3 mast cells using tandem high-resolution mass spectrometry. Biochemistry. (1999). , 38-8056.

[107] Van Helvoort, A, & Van Meer, G. Intracellular lipid heterogeneity caused by topology of synthesis and specificity in transport. Example: sphingolipids.FEBS Letters (1995). , 369-18.

[108] Gkantiragas, I, Brügger, B, Stüven, E, Kaloyanova, D, Li, X. Y, Löhr, K, Lottspeich, F, Wieland, F. T, & Helms, J. B. Sphingomyelin-enriched microdomains at the Golgi complex. Molecular Biology of the Cell (2001). , 12-1819.

[109] Zheng, Y. Z, Berg, K. B, & Foster, L. J. Mitochondria do not contain lipid rafts, and lipid rafts do not contain mitochondrial proteins Journal of Lipid Research (2009). , 50-988.

[110] Salaün, C, James, D. J, & Chamberlain, L. H. Lipid rafts and the regulation of exocytosis. Traffic (2004). , 5-255.

[111] Lang, T, Bruns, D, Wenzel, D, Riedel, D, & Holroyd, P. SNAREs are concentrated in cholesterol-dependent clusters that define docking and fusion sites for exocytosis. The EMBO Journal (2001). , 20-2202.

[112] Gil, C, Soler-jover, A, Blasi, J, & Aguilera, J. Synaptic proteins and SNARE complexes are localized in lipid rafts from rat brain synaptosomes. Biochemical and Biophysical Research Communications (2005). , 329-117.

[113] Predescu, S. A, Predescu, D. N, Shimizu, K, Klein, I. K, & Malik, A. B. Cholesterol dependent syntaxin-4 and SNAP-23 clustering regulates caveolar fusion with the endothelial plasma membrane. Journal of Biological Chemistry (2005). , 280-37130.

[114] Lang, T. SNARE proteins and ´membrane refts´. Journal of Physiology (2007). , 26-693.

[115] HernandezVerdun D. The nucleolus: a model for the organization of nuclear functions. Histochemistry and Cell Biology (2006). , 126-135.

[116] Pederson, T. The plurifunctional nucleolus. Nucleic Acid Research (1998). , 26-3871.

[117] Scheer, U, & Hock, R. Structure and function of the nucleolus. Curr Opin Cell Biol (1999). , 11-385.

[118] Thiry, M, & Lafontaine, D. Birth of a nucleolus: the evolution of nucleolar compartments. Trends Cell Biology (2005). , 15-194.

[119] Raska, I, Shaw, P. J, & Cmarko, D. Structure and function of the nucleolus in the spotlight. Current Opinion in Cell Biology (2006). , 18-325.

[120] Pederson, T. The nucleolus. Cold Spring Harb Perspect Biol (2010). pii: a000638. doi:cshperspect.a000638.

[121] Sirri, V, Urcuqui-inchima, S, Roussel, P, & Hernandez-verdun, D. Nucleolus: the fascinating nuclear body. Histochememistry and Cell Biology (2008). , 129-13.

[122] Andersen, J. S, Lam, Y. W, Leung, A. K, Ong, S. E, Lyon, C. E, Lamond, A. I, & Mann, M. Nucleolar proteome dynamics. Nature (2005). , 433-77.

Chapter 3

Chemistry of Natural Antioxidants and Studies Performed with Different Plants Collected in Mexico

Jorge Alberto Mendoza Pérez and
Tomás Alejandro Fregoso Aguilar

Additional information is available at the end of the chapter

1. Introduction

1.1. Antioxidants

Oxidation is the transfer of electrons from one atom to another and represents an essential part of both aerobic life and our metabolism, since oxygen is the ultimate electron acceptor in the electron flow system that produces energy in the form of ATP. However, problems may arise when the electron flow becomes uncoupled (transfer of unpaired single electrons), generating free radicals [1]. Antioxidants are important in living organisms as well as in food because they may delay or stop formation of free radical by giving hydrogen atoms or scavenging them. Oxidative stress is involved in the pathology of cancer, atherosclerosis, malaria and rheumatoid arthritis. An antioxidant can be defined in the broadest sense of the word, as any molecule capable of preventing or delaying oxidation (loss of one or more electrons) from other molecules, usually biological substrates such as lipids, proteins or nucleic acids. The oxidation of such substrates may be initiated by two types of reactive species: free radicals and those species without free radicals are reactive enough to induce the oxidation of substrates such as those mentioned. There are three main types of antioxidants:

1. Primary: Prevent the formation of new free radicals, converting them into less harmful molecules before they can react or preventing the formation of free radicals from other molecules. For example:

   - Enzyme superoxide dismutase (SOD) which converts $O_2 \bullet -$ to hydrogen peroxide (H_2O_2)

- Enzyme glutathione peroxidase (GPx), which converts H_2O_2 and lipid peroxides to harmless molecules before they form free radicals.

- Catalases

- Glutathione reductase.

- Glutathione S transferase.

- Proteins that bind to metals (ferritin, transferrin and ceruloplasmin) limit the availability of iron necessary to form the radical OH

2. Secondary: Capture free radicals, preventing the chain reaction (eg vitamin E or alpha-tocopherol, vitamin C or ascorbic acid, beta-carotene, uric acid, bilirubin, albumin, ubiquinol-10, methionine)

3. Tertiary: They repair damaged biomolecules by free radicals (eg DNA repair enzymes and methionine sulfoxide reductase) [2].

It also handles the classification based according to where they perform their activities, their background and their biochemical characteristics. So, antioxidants are also classified into two broad groups, depending on whether they are water soluble (hydrophilic) or lipid (hydrophobic). In general, water soluble antioxidants react with oxidants in the cell cytoplasm and blood plasma, whereas the liposoluble antioxidants protecting cell membranes against lipid peroxidation. In the metabolism it is a contradiction that while the vast majority of life requires oxygen for its existence, oxygen is a highly reactive molecule that damages living organisms by producing reactive oxygen species. Therefore, organisms possess a complex network of antioxidant metabolites and enzymes that work together to prevent oxidative damage to cellular components such as DNA, proteins and lipids. Usually antioxidant systems prevent these reactive species are formed or removed before they can damage vital components of the cell. Reactive oxygen species produced in cells include hydrogen peroxide (H_2O_2), hypochlorous acid (HClO), and free radicals such as hydroxyl radical ($\bullet OH$) and superoxide radical ($O_2 \bullet -$). The hydroxyl radical is particularly unstable and reacts rapidly and non-specifically with most biological molecules. This species produces hydrogen peroxide redox reactions catalyzed by metals such as the Fenton reaction. These oxidants can damage cells starting chemical chain reactions such as lipid peroxidation or by oxidizing DNA or DNA damage proteins. These effects can cause mutations and possibly cancer if not reversed by DNA repair mechanisms, while damage proteins will cause enzyme inhibition, denaturation and degradation of proteins. The use of oxygen as part of the process for generating metabolic energy produces reactive oxygen species. In this process, the superoxide anion is produced as a byproduct of several steps in the electron transport chain. Particularly important is the reduction of coenzyme Q in the compound III as a highly reactive free radical is formed as intermediate ($Q \bullet -$). This unstable intermediate can lead to loss of electrons when these jump directly to molecular oxygen to form superoxide anion instead of moving with well controlled series of reactions of electron transport chain. In a similar set of reactions in plants reactive oxygen species are also produced during photosynthesis under high light intensity. This effect is partly offset by the involvement of carotenoids in photoinhibition, which involves these antioxidants reacting with over-reduced forms of the photo-synthetic reaction centers and thereby prevent the production of superoxide. Another process which produces reactive oxygen species is lipid oxidation that takes place following the production of eicosanoids. However, the cells are provided with mechanisms that prevent unnecessary oxidation. Oxidative enzymes of these biosynthetic pathways are coordinated and highly regulated [3].

1.2. Free radicals

A free radical from the chemical viewpoint, is any species (atom, molecule or ion) containing at least one unpaired electron and its outermost orbital, and which is in turn able to exist independently (Figure 1).

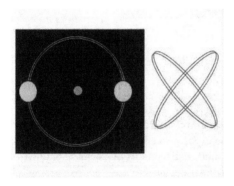

Figure 1. Atomic orbitals

The atoms arrange their electrons in regions called "atomic orbitals" in the form of pairs of electrons. The latter confers stability atom, or low chemical reactivity towards its environment. However, under certain circumstances, it may lose its parity orbital, either giving or capturing an electron. When this occurs, the resulting orbit exhibits an unpaired electron, making the atom in a free radical. The presence of an unpaired electron in an orbital outermost atom latter confers an increased ability to react with other atoms and / or molecules present in the environment, usually, lipids, proteins and nucleic acids (Figure 2). The interaction between free radicals and such substrates results in eventually structural and functional alterations [4].

Free radicals cause damage to different levels in the cell: Attack lipids and proteins in the cell membrane so the cell cannot perform its vital functions (transport of nutrients, waste disposal, cell division, etc.).

The superoxide radical, O_2, which is normally in the metabolism cause a chain reaction of lipid peroxidation of the fatty acids of phospholipids of the cell membrane. Free radicals attack DNA avoiding cell replication and contributing to cellular aging.

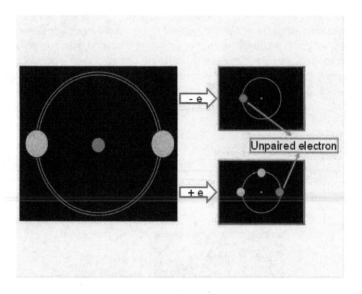

Figure 2. Unpaired electron in a free radical

The normal body processes produce free radicals that involve food metabolism, breathing and exercise. We are also exposed to environmental elements that create free radicals such as industrial pollution, snuff, radiation, drugs, chemical additives in processed foods and pesticides. Not all free radicals are dangerous because, for example, immune cells create free radicals to kill bacteria and viruses, but if there is sufficient control by antioxidants, healthy cell can be damaged.

The reactive oxygen species (ROS) is a collective term, widely used, comprising all the reactive species, whether or not free radicals, focus their reactivity in an oxygen atom. However, often under the designation ROS include other chemical species whose reactivity is focused on other than oxygen atoms [5].

1.3. Metabolites

In general, water soluble antioxidants react with oxidants in the cell cytoplasm and blood plasma, whereas the liposoluble antioxidants protecting cell membranes against lipid peroxidation. These compounds can be synthesized in the body or obtained from the diet (Table 1) [6].

1.4. Antioxidant metabolites in plants

Plants produce many different secondary metabolites some of them are potent antioxidants, some examples of these compounds is shown in Figure 3 [5,7].

Antioxidant metabolite	Solubility	Concentration in human serum (M)	Liver tissue concentration (mol / kg)
Ascorbic acid (vitamin C)	Water	50 to 60	260 (male)
Glutathione	Water	from 325 to 650	6,400 (male)
Lipoic acid	Water	from 0.1 to 0.7	4- 5 (rat)
Uric acid	Water	from 200 to 400	1,600 (male)
Carotene	Lipid	β-carotene: 0.5 to 1	retinol (vitamin A): 1 - 3 5 (male, total carotenoids)
α-tocopherol (vitamin E)	Lipid	10 to 40	50 (male)
Ubiquinol (coenzyme Q)	Lipid	5	200 (male)

Table 1. Biochemical properties of antioxidant metabolites

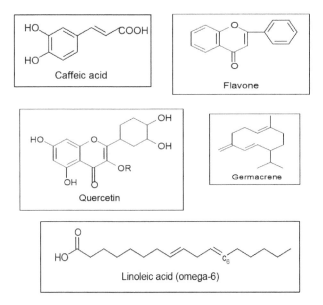

Figure 3. Different kinds of antioxidant metaboilites produced by plants

2. Enzyme systems

As with chemical antioxidants, cells are protected against oxidative stress by a network of antioxidant enzymes. Superoxide released by processes such as oxidative phosphorylation, is first converted into hydrogen peroxide and immediately reduced to give water. This route of detoxification is the result of multiple enzymes with superoxide dismutase catalyzing the first step and then catalases and peroxidases that eliminate several hydrogen peroxide [8].

2.1. Oxidative stress

Free radicals oxidize many biological structures, damaging them. This is known as oxidative damage, a major cause of aging, cancer, atherosclerosis, chronic inflammatory processes and cataracts, which are the most characteristic.

In certain circumstances, production of free radicals can increase uncontrollably, a situation known as oxidative stress. This means an imbalance between the speeds of production and destruction of toxic molecules, leading to an increase in cellular concentration of free radicals. Cells have mechanisms to protect against the harmful effects of free radicals based on a complex defense mechanism consisting of the antioxidants. Oxidative stress has been implicated in over one hundred human disease conditions, such as cancer, cardiovascular disease, aging and neurodegenerative diseases [9]. However, the innate defense in the human body may not be enough for severe oxidative stress. Hence, certain amounts of exogenous antioxidants are constantly required to maintain an adequate level of antioxidants in order to balance the ROS. As an example, epidemiological evidence indicates that the consumption of grapes reduces the incidence of coronary heart disease (CHD), atherosclerosis and platelet aggregation [10]. This greater protection may be due to the phenolic components of grapes, which are particularly abundant since they behave as reactive oxygen species-scavengers and metal-chelators. Polyphenolic substances in grapes and other red fruits are usually subdivided into two groups: flavonoids and nonflavonoids. The most common flavonoids are flavonols (quercetin, kaempferol, and myricetin), flavan-3-ols (catechin, epicatechin, and tannins), and anthocyanins (cyanin). Nonflavonoids comprise stilbenes, hydroxycinnamic acids and benzoic acids. Numerous papers have been published red fruits and their antioxidant properties have been correlated with their polyphenol contents [11,12,13,14,15].

2.2. Oxidative stress and disease

It is thought that oxidative stress contributes to the development of a wide range of diseases including Alzheimer's disease, Parkinson's disease, the pathologies caused by diabetes, rheumatoid arthritis, and neurodegeneration in motor neuron diseases. In many cases, it is unclear if oxidants trigger the disease, or occur as a result of this and cause the symptoms of the disease as a plausible alternative, a neurodegenerative disease may result from defective axonal transport of mitochondria that perform oxidation reactions. A case in which it fits is particularly well understood in the role of oxidative stress in cardiovascular disease. Here, the oxidation of low density lipoprotein (LDL) seems to trigger the process of atherogenesis, which leads to atherosclerosis, and ultimately to cardiovascular disease.

In diseases that have a high impact on the health sector Diabetes Mellitus is one of the most known. The World Health Organization (WHO) estimates that there are just over 180 million diabetics worldwide and likely to double this number for 2030 is quite high. Countries like China, India, United States of America and Mexico are at the top of this pathology [16]. In Mexico, this condition is a major cause of mortality and morbidity are estimated to be approximately 10 million individuals with diabetes, of whom 22.7% did not know they are sick, while 55% do not have good control netheir condition. This pathology is multifactorial, presenting various metabolic problems (polyuria, polyphagia, polydipsia, weight changes). The disorder is characterized by the inadequate use of glucose, due to insufficient production, insulin resistance and some without production of the hormone, resulting in unfavorable a high index of this monosaccharide in the blood. This causes abnormal function of some organs, tissues and systems that can cause kidney failure, vision loss, and amputation of a limb, diabetic coma and even death.

Different factors increase the likelihood of the individual to develop diabetes as are smoking, sedentary lifestyle, lack of exercise coupled with unbalanced diet causes both overweight and obesity. Naturally the body causes the formation of free radicals (highly unstable molecules), these chemical species are responsible for cellular aging, but when there is a greater concentration of these molecules may contribute to the development of various diseases and chronic degenerative neuro Parkinson's, Alzheimer's and diabetes. Obesity increases oxygen consumption and thus the production of free radicals, thus creating the phenomenon known as oxidative stress. Excess fat naturally stored in fat cells, causes the more than normal synthesis of substances called adiposines IL6 or leptines. These substances in higher concentrations also cause insulin resistance [17].

3. Alternative medicine

Due to the current problem in the health issue we propose the use of herbs as an option to improve the style of living of the people, not only for the adjuvant treatment, but because the use of plants offers great nutritional benefits somehow reducing the incidence of such chronic degenerative diseases. This is not intended to impair the option of preventive diagnosis by the health sector does not provide such benefits, but rather the use of plants known to have medicinal activity coupled with the clinical - pharmacology, could present better results, for the treatment of the various degenerative chronic diseases. Given the increasing scientific evidence that the etiology of several chronic degenerative diseases such as diabetes is influenced by factors such as metabolic redox imbalance. Is currently booming studying the formation of metabolites against free radicals that diverse plant species presents. An example of this has been widely documented, is the cranberry, a plant used for treating various diseases and, as has been discovered, is due to its potential antioxidant that has these properties beneficial to health [18, 19].

Similarly, Mexico has focused attention on other plants with potential antioxidant properties and for some years and was used in the treatment of diabetes. In this regard, since 2006, our research work focused on the task of describing the effects of plants such as Noni (*Morinda citrifolia*), Moringa (*Moringa oleifera*), the Guarumbo (*Cecropia obtusifolia Bertolt*), the Musaro

(*Lophocereus sp.*) and Neem (*Azadirachta indica*) in murine models of chemically induced diabetes with streptozotocine. More recently, we began to evaluate the antioxidant properties of some of these plants through *in vitro* techniques [20].

3.1. Antioxidant effects in Mexican plants

The use of traditional medicine is widespread in Mexico and plants are indeed the first source for preparing remedies in this form of alternative medicine. Among the various compounds found in plants, antioxidants are of particular importance because they might serve as leads for the development of novel drugs. Several plants used as anti-inflammatory, digestive, antinecrotic, neuroprotective, and hepatoprotective properties have recently been shown to have and antioxidant and/or antiradical scavenging mechanism as part of their activity [21,22]. The search for natural sources of medicinal products that also have antioxidant and radical scavenging activity is on the rise [23,24]. Among the medicinal properties associated with them are the following: the fruitand bark of *Licania arborea* is used as a soap for hair infections, the latex from *Ficus obtusifolia* is employed as an anti parasitic and also for reducing fever, *Bunchosia cannesens* is prescribed as an antidiarrhoeic, *Sideroxylon capiri* is used for hiccups, as an antiseptic for cleaning wounds, and women use its leaves in a water bath after giving birth. The latex of *Sapium macrocarpum* is used against scorpion stings, fever and some skin problems such as warts; its use as an anti-coagulant is also widespread. The latex of *Ficus cotinifolia* is used in the treatments of urinary infections, vomiting, malaria and against inflammatory pathologies of the spleen. The leaves of *Annona squamosa* are used in cicatrisation of wounds, diarrhoea, ulcers, menstrual disorders, and also to help weight loss. The seeds of this plant are also employed as an insecticide. The leaves of *Vitex molli* are used to treat stomach ache, digestion disorders, nervous alterations, and also scorpion stings. *Piper leucophyllum* is employed for reducing fever and its dried leaves are used for cleaning eyes and as spice in cooking. The leaves and bark of *Gliricidia sepium* are used against high fever, skin infections, urine disorders, malaria, and headache. However, its seeds are reported to be toxic. *Hamelia paten is* used to accelerate wound cicatrisation. The Mexican and Central America native species of *Astianthus viminalis* is used for the curing of diabetes and malaria and to reduce hair loss. *Swietenia humilis* is used as anti parasitic, and it is also utilized for hair care as a shampoo. It is also used with other plants in mixed herbal teas, and used as home remedies. *Stemmandenia bella* is employed for curing wounds; *Rupechtia fusca* is used in some stomach disorders; *Bursera grandifolia* is used as a tooth paste and against digestive disorders; *Ziziphus amole* is prepared as infusion and it is applied for washing wounds and to treat gastric ulcers. The fruit and the latex of *Jacaratia mexicana* are used against ulcers in the mouth and digestive disorders. *Gyrocarpus jathrophifolius* leaves and bark are used as an analgesic. *Pseudobombax ellipticum* is used in respiratory disorders such as cough, and also against fever and as an anti microbial. The stems and flowers of *Comocladia engleriana* are toxic because they produce dermatitis. The flowers and the latex of *Plumeria rubra* can be used for stopping vaginal blood shed, and toothache, and the latex of the plant is used against earache. Infusions are used as an Eye-cleaning liquid [23, 24, 25 & 26].

Polyphenolic compounds are commonly found in both edible and inedible plants, and they have been reported to have multiple biological effects, including antioxidant activity [25]. Herbs are used in many domains, including medicine, nutrition, flavouring, beverages, dyeing, repellents, fragrances, cosmetics [26]. Many species have been recognized to have medicinal properties and beneficial impact on health, e.g. antioxidant activity, digestive stimulation action, antiinflammatory, antimicrobial, hypolipidemic, antimutagenic effects and anticarcinogenic potential [27,28]. Crude extracts of herbs and spices, and other plant materials rich in phenolics are of increasing interest in the food industry because they retard oxidative degradation of lipids and thereby improve the quality and nutritional value of food. The basic flavonoids structure is the flavan nucleus, which consists of 15 carbon atoms arranged in three rings (C6–C3–C6), labelled A, B, and C (Figure 3). Various clases of flavonoid differ in the level of oxidation and saturation of ring C, while individual compounds within a class differ in the substitution pattern of rings A and B. The differences in the structure and substitution will influence the phenoxyl radical stability and thereby the antioxidant properties of the flavonoids. Plant species belong to several botanical families, such as Labiatae, Compositae, Umbelliferae, Asteracae, Polygonacae and Myrtacae. Many spices have been investigated for their antioxidant properties for at least 50 years [29,30].

4. DPPH experiments with different plants collected in México

Herein it is presented a brief description of two experiments to evaluate the antioxidant properties of some plants collected in México, one conducted in the Moringa tree (*Moringa oleifera*) and the other in the Neem tree (*Azadirachta indica*).

1. Antioxidant properties of *Moringa oleifera*: In this experiment were collected fresh leaves of *M. oleifera* in the Municipality of Apatzingan in the state of Michoacan, Mexico, the leaves are brought to the facilities of the National School of Biological Sciences, IPN where allowed to air dry and then were macerated and placed in containers containing methanol. After one week the solvent was decanted and concentrated under reduced pressure using a rotary evaporator. The crude methanol extract was stable in distilled water obtained the following experimental concentrations: 50, 25, 12.5 and 6.25 mg / mL to which they evaluated the antioxidant capacity.

It was used as standard test the unstable radical 2,2-diphenyl-picrylhydrazyl (DPPH), which originally has a purple and when it is placed against a substance having antioxidant properties in a UV spectrophotometer at 517 nm, changes to yellow colour yellow (Figure 4). It was prepared a calibration curve of DPPH type in methanol at concentrations of 40, 120, 160 and 200 ug. Subsequently, aliquots of the above concentrations of methanol extract of *M. oleifera* and combined with DPPH in methanol for measuring its absorbance in the UV spectrophotometer at 0, 10, 30, 60 45 min., and then self-assess their antioxidant capacity [31].

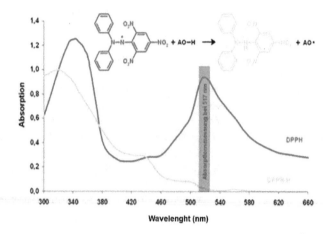

Figure 4. Graph showing the colour change of DPPH from purple to yellow when it is exposed to an antioxidant substance

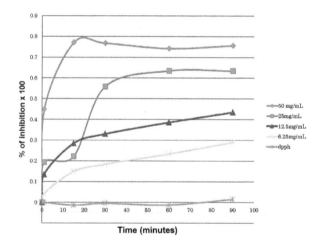

Figure 5. Antioxidant activity of different concentrations of methanol extract of *M. oleifera*

It was found that higher concentrations *of M. oleifera* antioxidant activity showed a concentration dependance, ie the higher the concentration of the extract metabolic higher antioxidant capacity (Figure 5). It was noted that the highest concentrations (50 mg / mL and 25 mg / mL) reached 50% of its antioxidant activity (assessed by the percentage of inhibition of purple fading to yellow against DPPH respective concentration of the methanol extract of *M. oleifera*) within 10 min. the reaction is initiated, which is indicative of the high antioxidant capacity of this plant and that could explain its therapeutic efficacy in the treatment of diabetes in mouse models that are currently underway [32].

2. Comparison of the antioxidant properties of *Azadirachta indica* with other species and a commercial product. This experiment followed a similar protocol to that described for *M. oleifera*, with the exception that the samples were evaluated for their antioxidant capacity at times 1, 15, 30, 60 and 90 minutes. The protocol is divided into two parts, one of which was evaluated for antioxidant activity from four different extracts from leaves of *Azadirachta indica*: a) methanol, b) infusion, c) ethyl acetate and d) ethanol. This was done by measuring the percent inhibition of loss of the color purple to yellow the respective front DPPH extract (Figure 6). It was found that infusion of Neem showed the highest antioxidant activity (80% inhibition) than the other extracts even from the first minute after initiating the reaction. This would correspond to the ethnomedical use that people from rural zones done with this tree, and then take it as a tea before the first food of the day. The methanol extract of Neem leaves also exhibited a high antioxidant capacity, as a percentage inhibition of the radical DPPH greater than 50% during the entire reaction time [33].

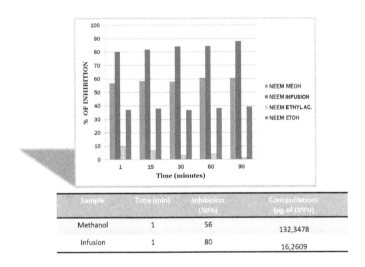

Sample	Time (min)	Inhibition (50%)	Concentrations (µg of DPPH)
Methanol	1	56	132,3478
Infusion	1	80	16,2609

Figure 6. Assessment of% inhibition of DPPH radical from 4 extracts of Neem (*Azadirachta indica*).

In the second part of this protocol, we chose a fraction of the methanol extract of Neem leaves and their antioxidant activity was compared against a commercial preparation, a kind of juice containing: Pomegranate (*Punica granatum*) green tea (*Camel sinensis*), cranberry (*Vaccinum mytillus*), red grape (*Vitis vinifera*) and methanol extract of leaves, seed and fruit peel of passion fruit (*Passiflora edulis*). It was found that the commercial preparation showed the highest antioxidant activity throughout the reaction time (Figure 7), presented as a% inhibition greater than 50% from the first minute (68% inhibition), reaching end of the reaction with values higher than 85% inhibition of the presence of radical DPPH. Compared to the various organs of *Passiflora edulis*, the Neem leaf extract showed higher antioxidant capacity from the reaction started, it displayed a% inhibition of DPPH radical by 66% to reach 15 minutes of reaction, and reaching values greater than 85% inhibition at 60 minutes, falling a little activivty (63% approximately) at 90 min. the reaction is initiated. This would confirm previous data of different authors and from our laboratory (unpublished data), in the sense of the effectiveness of Neem tree leaves for the treatment of chronic degenerative diseases such as diabetes, it has proven effective in significantly reducing blood glucose levels in streptozotocin-treated mice, an effect that may be due to the presence of secondary metabolites of the steroid type saponins, flavonoids and phenols among others that seem to owe much its hypoglycemic action, thanks to its antioxidant properties. In this regard, our laboratory is conducting a more rigorous characterization of secondary metabolites found in these plants by using spectroscopic techniques (e.g, Infrared Spectroscopy with Fourier Transform, Proton Nuclear Magnetic Resonance, etc.) trying to isolate, purify these metabolites and test its therapeutic efficacy in animal models of diabetes. However, we should mention that, although the future is promising with regard to therapy options offered by these plants in Mexico (and other developing countries), yet to be made deeper studies and comparative scale phytochemical, toxicological and drug to confirm everything mentioned in this chapter [34].

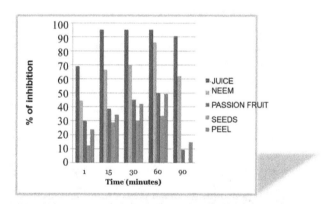

Sample	Time (min)	Inhibition (50%)	Concentration (µg of DPPH)
Comercial juice	1	68%	56
Neem	15	66%	60

Figure 7. Assessment of% inhibition of DPPH radical of Neem (Azadirachta indica) with respect to a commercial preparation and different organs of Passion fruit (Passiflora edulis).

Another experiment with Noni (*Morinda citrifolia*) showed that the leaves of this plant also have a high antioxidant effect at low concentrations reaching a maximum effect at a dose of 5mg/mL (Figure 8). These results disagreed with the effect observed with the Noni fruit, where it was reported that the antioxidant effect of protection tends to increase with respect to higher concentrations in tests performed with lyophilized juice extracts [35].

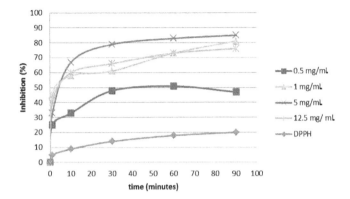

Figure 8. Results from DPPH method for testing the free-radical inhibition effect of different concentrations of methanolic extracts of Noni leaves.

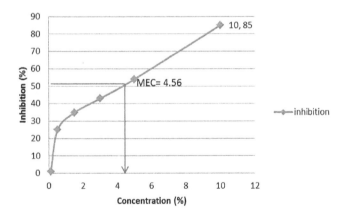

Figure 9. Shows the curve concentration vs. inhibition (%) of the radical 1,1-diphenyl-2-picrylhydrazyl (DPPH ●), in which it is observed that the mean effective concentration (MEC) of noni leaf extract was 4,56%, at this concentration is carried out the 50% inhibition of the unpaired radical.

Furthermore a methanolic solution was prepared with a concentration of 860 mmol / mL of DPPH. From this concentration the experiment was performed with a duplicated calibration curve at different concentrations, which was read at 517nm. An average concentration of those used for standard curve analysis was chosen for the initial concentration of the metha-nol extract from leaves of *Morinda citrifolia*. From this methanol extract different percentage dilutions were prepared. After obtaining the extract dilutions 1000μL of each one was mixed with 860μL methanol and 140μL of DDPH and after that it were allowed to stand for 10 mi-nutes at 30°C and then absorbance was read (517nm) at time intervals of 10, 30, 45 and 60 minutes in a UV-Vis spectrophotometer HACH DR 5000. Through a mathematical analysis of linear regression curve was obtained concentration vs. inhibition (%), and finally deter-mined the mean effective antioxidant concentration (Figure 9) [36].

5. Phytochemical results

5.1. Moringa oleifera

Extraction by maceration from *M. Oleifera* leaves was carried out in methyl alcohol for the phytochemical sieve, the results are shown in the following table (Table 2), and qualitative tests that were performed for each secondary metabolite are named in the same table.

	Reagent	Test
Alkaloids	Dragendorf	++
	Mayer	+
	Wagner	+
	Sonneshain	+
	Silicotungstine	+
Tanines	ferric chloride	-
phenols	Potassium ferricianide	+
reducing sugars	Fehling	-
	Benedic	-
Coumarins	ammonium hydroxide	-
	Erlich	+
Flavonoids	Sodium Hydroxide	+
Sesquiterpenlactones	hidroxylamine chlorehydrate	-
Saponines	Libermann Bouchard	+
cardiotonic Glycosides	Kedde, Baljet, legal	-
cyanogenic Glycosides	Guignar	-

Table 2. Phytochemical sieve of methanolic extract from *M. olerifera* leaves.

Metabolites	Reagent	Test			
		Azadirachta indica	Cecropia obtusifolia Bertolt	Lophocereus sp	Morinda citrifolia
Alkaloids	Dragendorf	++	-	++	-
	Mayer	+	+	+	+
	Wagner	+	-	+	+
	Sonneshain	+	-	-	-
	Silicotungstine	+	-	-	-
Tanines	ferric chloride	-	+	-	-
phenols	Potassium ferricianide	+	+	+	+
reducing sugars	Fehling	-	-	+	+
	Benedic	-	-	+	+
Coumarins	ammonium hydroxide	-	-	+	+
	Erlich	+	+	+	+
Flavonoids	Sodium Hydroxide	+	+	-	+
Sesquiterpenlactones	hidroxylamine chlorehydrate	-	+	-	-
Saponines	Libermann Bouchard	+	+	+	+
cardiotonic Glycosides	Kedde, Baljet, legal	-	+	-	+
cyanogenic Glycosides	Guignar	-	+	+	-

Table 3. Phytochemical sieve of different plants collected in Mexico. All the tests were performed with methanolic extracts of the leaves.

The results of this phytochemical analysis presented in Table 2, shows qualitatively the secondary metabolites found in the methanol extract of leaves of *M. oleifera*. The more abundant compounds in the leaves of this tree and also the must reported in several articles are:

i.   Alkaloids

ii.   Coumarins

iii. Phenolics

iv.   Flavonoids

v.   Saponins

In the case of metabolites such as tannins, cardiac glycosides, cyanogenic glycosides, the tests results were negative for the methanol extract of leaves of *M. oleifera* [37, 38].

Phytochemical results for the extracts obtained from leaves of Guarumbo (Cecropia obtusifolia Bertolt), Musaro (Lophocereus sp.), Neem (Azadirachta indica) and Noni (Morinda citrifolia) are showed in Table 3. Several of those metabolites are highly active antioxidants.

6. Description of those plants collected in México and presenting high content of antioxidant compounds

6.1. *Azadirachta indica*

Active compounds of Neem have been identified while others have not, and analyzed the most common are: nimbin; nimbidin; ninbidol; gedunin; sodium nimbinate; queceretin; salannin and azadirachtin.

Parts used and their uses:

Neem bark is bitter, astringent, is used to treat diseases of the mouth, teeth, loss of appetite, fever, cough and intestinal parasites (Figure 10)

Figure 10. Stem of the Neem tree.

The leaves help to neuromuscular problems, eliminate toxins, purify the blood, are also used to treat snake bites and insect (Figure 11)

Figure 11. Branch and leaves of Neem or Azadirachta indica.

The Neem fruit is bitter, and it is used as a purgative and for hemorrhoids (Figure 12)

Figure 12. Fruits from Neem Tree

The flowers are astringent and expectorant and also Seed oil is extracted. (Figure 13).

Figure 13. Flowers from Neem Tree

Neem kills some infectious organisms, contributes to the immune response at various levels, this increases the possibility that the body fight bacterial infections alone, viral and fungal. Neem increases the production of antibodies, improves the response of immune cells that release mediators white blood cells. For Diabetes: Neem extract orally reduces insulin requirements by between 30% and 50% to people who are insulin dependent. With Cancer: polysaccharides and limonoids found in the bark, leaves and Neem oil reduces tumors and cancer [39].

6.2. *Passiflora edulis Sims*

Another plant which has been described with therapeutic applications is the genus Passiflora, whics comprises about 500 species and is the largest in the family Passifloraceae. *Passiflora edulis Sims* is native from the Brazilian Amazon, known by the common name for passion fruit [40, 41]. The word passion comes from the Portuguese- Brazilian passion fruit, which means food prepared in Totuma [41, 42].

Passiflora edulis (Figure 14) is a widely cultivated species in tropical and subtropical countries, there are two varieties: *Passiflora edulis Sims* var. *flavicarpa*, whose fruits are yellow and *Passiflora edulis Sims* var. purple, with purple fruits and adapts to higher ground [42].

Passion fruit (Figure 15) is a woody perennial, climbing habit and rapid development, which can reach up to 10 m long, the leaves are simple, alternate, and a tendril conestipules in the armpit, with serrated margins, the flowers (Figure 16) are solitary and axillary, fragrant and showy, the fruit is a spherical berry, globose or ellipsoid, measuring 10 cm in diameter and weighs up to 190 g, yellow or purple, with a highly aromatic pulp [43].

Figure 14. Passion fruit

Figure 15. Leaves and flower of passion fruit

The ethnopharmacological information reveals that *Passiflora edulis Sims* has been used in traditional medicine around the world. In India, the fresh leaves of this plant are boiled in small amount of water and the extract is drunk to treat dysentery and hypertension, and the fruits are eaten to relieve constipation. In South America, native people drink the tea leaves and flowers as a sedative, infusion of the aerial parts is used in the treatment of tetanus, epilepsy, insomnia and hypertension is also indicated as a muscle relaxant, diuretic, to treat stomach aches, fever and intestinal tumors[44].

The phytochemical study of *Passiflora edulis Sims* (Passifloraceae) shows the presence of glycosides, including passiflorine, flavonoid glycosides: luteolin-6-Cchinovóside, cyanogenic glycosides, alkaloids harman, triterpenes and saponins, phenols, carotenoids, anthocyanins, L-ascorbic acid, γ-lactones, esters, volatile oils, eugenol, amino acids, carbohydrates and minerals [45].

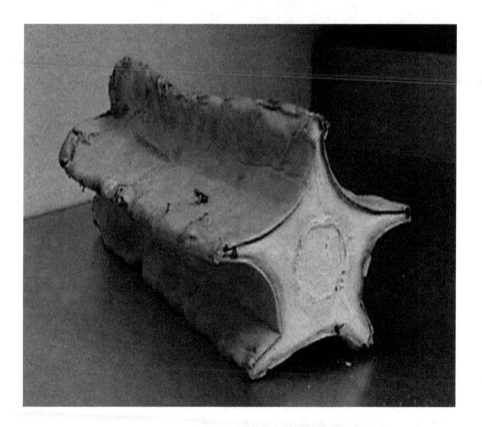

Figure 16. Five ribs cladode from Musaro (*Lophocereus* sp)

6.3. *Lophocereus* sp

Gender *Lophocereus* sp. (Figure 17) develops as a succulent plant which presents a plus size and can reach 7m tall. These cacti have a columnar development right, has ribbed stems that branch with age. The flowers are nocturnal, appear only on copies of more than two meters high and are colored green, but in the four seasons can take a pink color, the stem proliferation and leads to masses of slender spines long they can get to cover it completely. Multiply by seeds in spring or summer. These plants are drought resistant so as to avoid daily watering and exposure to damp, besides being very sensitive to frost. But they need full sun and well drained soils. It is endemic to Baja California Sur, Baja California and Sonora in Mexico and Arizona in the United States It is a common species that has spread efficiently over the worldwide. The fruits are edible, but are hard to come by competition with birds and insects. This plant has curative properties. Southeastern Mexican indians prepare a tea from the pulp and skin of the cactus to relieve arthritis. Some members of this ethnic group say the plants with five ribs are very good for treating cancer. For diabetes: using the Musaro with a seven-pointed cladode, boiled in 1 liter of water, strain and drink three glasses of cooking, one before each meal. It is also widely used to treat ulcers, wounds, and stomach diseases.

Musaro or gooseberry is the tea made by slicing sections fifteen or twenty two inches long from the stems of cactus. These cuts are then placed in a container large enough to contain five gallons of water and boiled for eight to ten hours until the liquid is reduced to approximately one gallon. People take this treatment for serious stomach diseases but they should drink tea in large amount. From extracts of *Lophocereus sp* two compounds have been isolated: pylocereine and lophocine a dimeric alkaloid with cytotoxic activity [46, 47].

6.4. *Cecropia obtusifolia Bertolt*

Family: Cecropiaceae

Common name: The names that are known are "guarumbo", "chancarro", "hormiguillo", "chiflón" and "koochlle" among others. This plant is widely used by people suffering from type 2 diabetes. It is known mainly in rural areas.

Botanical description: A tree 20 m tall, commonly grows in tropical rain forests. Its trunk is straight and presents a cavity where you can find some ants inside it, with branches that grow along this horizontally. The leaves are placed in a spiral on the branches. They have a deep green on the upper surface of the leaf and gray on the back (Figure 18).

Distribution in Mexico: located along the coasts of Tamaulipas and San Luis Potosi to Tabasco on the side of the Gulf of Mexico and Sinaloa to Chiapas Pacific side. Traditionally the dried leaves (15 g) are heated in water (500 mL) and the result is an infusion which is then filtered and taken as "daily water", cold infusion is often consumed in hot weather. States where is often used as traditional medicine this plant are: Hidalgo, Guerrero, Veracruz, Yucatan, Campeche, Tabasco, Mexico State, Oaxaca and Chiapas. It is reported that this plant contains β-sitosterol, stigmasterol, 4-ethyl-5-(n-3 valeroil)-6-hexahydrocoumarins. From butanol extracts were isolated chlorogenic acid and isorientin [48, 49].

Figure 17. Guarumbo (*Cecropia obtusifolia Bertolt*) leaves.

7. Conclusion

The plant kingdom has been the best source of remedies for curing a variety of disease and pain. This is why medicinal plants have played a key role in the worldwide maintenance of health. Traditional herbal medicine is intimately related to the Mexican popular culture; its use has origins based on ancestral knowledge. Natural products of higher plants are an important source of therapeutic agents; therefore, many research groups are currently screening the different biological activities of plants. Mexico has an extensive variety of plants; it is the fourth richest country worldwide in this aspect. Some 25 000 species are registered, and it is thought that there are almost 30 000 not yet described. Natural antioxidants that are present in herbs and spices are responsible for inhibiting or preventing the deleterious consequences of oxidative stress. Spices and herbs contain free radical scavengers like polyphenols, flavonoids and phenolic compounds, having antioxidant activities, Indeed all these plant studied have several biological effects and they could also be used as a source of natural antioxidants. Further pharmacological studies are underway to identify the active constituents of the plant extracts responsible for the showed activities. As a final comment, compounds in plants are of great importance for the treatment of several chronic and degenerative diseases like diabetes and cancer, among others. For that reason the use in traditional medicine is of great interest in order to know the activity and the mechanism of action of these compounds which could be used for the treatment and prevention of that mentioned diseases, which are associated nowadays with stress, fast food diets and lack of daily exercise, just to name a few factors.

Acknowledgments

The authors wants to thank the ENCB-IPN for the support received for this work through the SIP projects 20120789 and 20120899.

Author details

Jorge Alberto Mendoza Pérez[1] and Tomás Alejandro Fregoso Aguilar[2*]

1 Department of Environmental Systems Engineering at National School of Biological Sciences-National Polytechnic Institute. Mexico, D.F., Mexico

2 Department of Physiology at National School of Biological Sciences-National Polytechnic Institute. Mexico, D.F., Mexico

References

[1] PIETTA, P. G. (2000) Flavonoids as Antioxidants, *Journal Natural Product*, 63: 1035–1042.

[2] Rivera Arce E, Morales González J A, Fernández Sánchez A M, Bautista Ávila M, Vargas Mendoza N, Madrigal Santillán E O. (eds.) Chemistry of Natural Antioxidants and Studies Performed with Different Plants Collected in Mexico, (2009) 227-238

[3] Devasagaya T P, Tilak J C, Boloor K K. Free Radicals and Antioxidants in Human Health: Currens Status and Future Prospects. J. Assoc. Physicians India. 2004; 52:794-804.

[4] ACS. Chemistry. España:Reverte; 2010. p.460-65

[5] Camacho Luis A, Mendoza Pérez J A, The Ephemeral Nature of Free Radicals: Chemistry and Biochemistry of Free Radicals. In: Morales González J A, Fernández Sánchez A M, Bautista Ávila M, Vargas Mendoza N, Madrigal Santillán E O. (eds.) Antioxidants and Chronic Degenerative Diseases. México: Ciencia al Día; 2009. p. 27-76

[6] VanderJagt T J, Ghattas R, VanderJagt D J, Crossey M, Glew R H. Comparison of the Total Antioxidant Content of 30 Widely Used Medicinal Plants of New Mexico. Life Sci. 2002; 70: 1035-1040.

[7] Bors W, Heller W, Michael C, Saran, M. Radical Chemistry of Flavonoids Antioxidants. Advances in Experimental Medicine and Biology. 1990; 264: 165–170.

[8] Ballester M. Antioxidants, Free Radicals and Health. A Physic-Organic Chemistry View. *Med Clinc (Barc)*.1996;107:509-515.

[9] Bagchi D, Bagchi M, Stohs S J, Das D K, Ray S D, Kuszynski C A. Free Radicals and Grape Seed Proanthocyanidin Extract: Importance in Human Health and Disease Prevention. Toxicology. 2000; 148:187–197.

[10] Tedesco I, Russo M, Russo P, Iacomino G, Russo G L, Carraturo A. Antioxidant Effect of Red Wine Polyphenols on Red Blood Cells. The Journal of Nutritional Biochemistry. 2000; 11(2):114–119.

[11] Fernandez-Pachon, M S, Villano D, Garcia-Parrilla M C, Troncoso A M. Antioxidant Activity of Wines and Relation with their Polyphenolic Composition. Analytica Chimica Acta.2004; 513(1):113–118.

[12] Fernandez-Pachon, M. S., Villano, D., Troncoso, A. M., Garcia-Parrilla, M. C. Determination of the Phenolic Composition of Sherry and Table White Wines by Liquid Chromatography and their Relation with Antioxidant Activity. Analytica Chimica Acta. 2006; 563(1–2):101–108.

[13] Cimino, F., Sulfaro, V., Trombetta, D., Saija, A., Tomaino, A. Radicalscavenging Capacity of Several Italian Red Wines. Food Chemistry. 2007;103(1):75–81.

[14] Arnous, A., Makris, D. P., Kefalas, P. Correlation of Pigment and Flavanol Content with Antioxidant Properties in Selected Aged Regional Wines from Greece. Journal Of Food Composition And Analysis. 2002; 15(6):655–665.

[15] Minussi, R. C., Rossi, M., Bologna, L., Cordi, L., Rotilio, D., Pastore, G. M. Phenolic Compounds and Total Antioxidant Potential of Commercial Wines. Food Chemistry. 2003; 82(3): 409–416.

[16] Nava Ch. G, Veras G. M. E. Epidemiology of diabetes In: Morales G.J.A., Madrigal S.E.O., Nava Ch.G., Durante M.I., Jonguitud F.A., Esquivel S. J. (eds.) Diabetes 2nd. ed. Universidad Autónoma del Estado de Hidalgo: México; 2010. p.57 – 63.

[17] Fernández S. A.M. (2009). Antioxidants and diseases: Obesity. In: Morales González J A, Fernández Sánchez A M, Bautista Ávila M, Vargas Mendoza N, Madrigal Santillán E O. (eds.) Antioxidants and Chronic Degenerative Diseases. México: Ciencia al Día; 2009. p.411-25

[18] Posmontier B. The Medicinal Qualities of Moringa Oleifera. USA:Lippincott Williams & Wilkin; 2011. p. 80-87

[19] Fragoso Antonio S. Antioxidant and Antigenotoxic Effects of Cranberry Juice. MsC Thesis. National School of Biological Sciences-National Polytechnic Institute; 2011.

[20] Castilla, D. Phytochemical and Biological Studies of Leaf Extract and Fruit of *Morinda citrifolia* (Noni) in a Mouse Model of Diabetes. Bachelor Thesis. National School of Biological Sciences-National Polytechnic Institute; 2012.

[21] Lin C C, Huang P C. Antioxidant and Hepatoprotective Effects of *Acanthopanax senticosus*. Phytother. Res. 2002; 14: 489-494.

[22] Perry E K, Pickering A T, Wang W W, Houghton P J, Perru N S Medicinal Plants and Alzheimer's Disease:from Ethnobotany to Phytotherapy . J. Pharm. Pharmacol. 1999; 51: 527-534.

[23] Schinella G R, Tournier H A, Prieto J M, Mordujovich de Buschiazzo P, Ríos J L. Antioxidant Activity of Antiinflammatory Plant Extracts. Life Sci. 2002; 70: 1023-1033.

[24] Velázquez E, Tournier HA, Mordujovich de Buschiazzo P, Saavedra G, Schinella GR. Antioxidant activity of Paraguayan plant extracts. Fitoterapia. 2003; 74: 91-97.

[25] Moure A, Franco D, Sineiro J, Domínguez H, Núñez JM, Lema MJ. Evaluation of Extracts from *Gevuina avellana* Hulls as Antioxidants. J. Agric. Food Chem. 2000; 48: 3890-3897.

[26] Singleton VL, Rossi JA Colorimetry of Total Phenolics with Phosphomolybdic-Phosphotungstic Acid Reagents. Am. J. Enol. Vitinicult. 1965; 16: 55-61.

[27] Takao T, Kitatani F, Watanabe N, Yagi A, Sakata K A Simple Screening Method for Antioxidants and Isolation of Several Antioxidants Produced by Marine Bacteria from Fish and Shellfish. Bioscience. Biotechnol. Biochem. 1994; 58: 1780-1783.

[28] Miller H E. A Simplified Method for the Evaluation of Antioxidants. J. Am. Oil Chem. Soc. 1971; 48: 92-97.

[29] Monroy-Ortíz C, Castillo E P. Medicinal Plants Used in Morelos State. UAEM: México, 2000. p. 20-50

[30] Morales-Cifuentes C, Gómez-Serranillos MP, Iglesias I, Villar del Fresno AM. Neuropharmacological profile of ethnomedicinal plants of Guatemala J. Ethnopharmacol. 2001; 76: 223-228.

[31] Robak J, Gryglewski R. J. Flavonoids are scavengers of superoxide anions.Biochemical Pharmacology.1988; 37(5):837–841.

[32] Brandwilliams W, Cuvelier M E, Berset C. Use of a Free-Radical Method to Evaluate Antioxidant Activity. Food Science and Technology-Lebensmittel- Wissenschaft & Technologie.1995; 28(1): 25–30.

[33] Cimino F, Sulfaro V, Trombetta D, Saija A, Tomaino A. Radicalscavenging Capacity of Several Italian Red Wines. Food Chemistry. 2007; 103(1): 75–81

[34] Sreelatha S, Padma P.R. Antioxidant Activity and Total Phenolic Content of Moringa Oleifera Leaves in Two Stages of Maturity. Plant Foods Hum Nutr. 2009; 64:303–311.

[35] De Las Heras B, Slowing K, Benedí J, Carreto E, Ortega T, Toledo C, Bermejo P, Iglesias I, Abad MJ, Gómez-Serranillos P, Liso PA, Villar A, Chiriboga X. Antiinflamatory and Antioxidant Activity of Plants Used in Traditional Medicine in Ecuador. J. Ethnopharmacol. 1998; 61: 161-166.

[36] Faustino R S, Clark T. A, Sobrattee S, Czubryl M. P, Pierce G. N. Differential Antioxidant Properties of Red Wine in Water Soluble and Lipid Soluble Peroxyl Radical Generating Systems. Molecular And Cellular Biochemistry. 2004; 263(1):211–215.

[37] Dinis T. C. P, Madeira V. M. C, Almeida L. M. Action of Phenolic Derivates (Acetoaminophen, Salicylate, And 5-Aminosalicylate) as Inhibitors of Membrane Lipid Peroxidation and as Peroxyl Radical Scavengers. Archive Of Biochemistry And Biophysics. 1994; 315:161–169.

[38] Huang D. J, Ou B. X, Hampsch-Woodill M, Flanagan J. A, Prior R. L. High-Throughput Assay of Oxygen Radical Absorbance Capacity (ORAC) Using a Multichannel Liquid Handling System Coupled with a Microplate Fluorescence Reader in 96-Well Format. Journal of Agricultural and Food Chemistry. 2002; 50(16):4437–4444.

[39] Abu Syed Md. Mosaddek and Md. Mamun Ur Rashid. A Comparative Study of the Anti-Inflammatory Effect of Aqueous Extract of Neem Leaf and Dexamethasone. Bangladesh J Pharmacol. 2008; 3: 44-47

[40] Arteaga S, Andrade-Cetto A, Cárdenas R. *Larrea tridentata* (Creosote bush), an Abundant Plant of Mexican and US-American Deserts and its Metabolite Nordihydroguaiaretic Acid. J. Ethnopharmacol. 2005; 98: 231 – 239.

[41] Mareck U, Herrmann K, Galensa R, Wray V. The 6-Cchinovoside and 6-C-Fucoside of Luteolin from *Passiflora edulis*. Phytochemistry. 1991; 30: 3486-7.

[42] Christensen J, Jaroszewski J. Natural Glycosides Containing Allopyranose from the Passion Fruit Plant Circular Dichroism of Benzaldehyde Cyanohydrin Glycosides. Org Lett. 2001; 3:2193-5.

[43] Seigler D, Pauli G, Nahrstedt A, Leen R. Cyanogenic Allosides and Glucosides from *Passiflora edulis* and *Carica papaya*. Phytochemistry. 2002; 60: 873-82.

[44] Slaytor M, McFarlane I. The Biosynthesis and Metabolism of Harman in *Passiflora edulis*. Phytochemistry. 1968; 7:605-11.

[45] Shawan K, Dhawan S, Sharma A. *Passiflora*: a Review Update. Journal of Ethnopharmacology. 2004; 94: 1-23.

[46] Andrade, C. A, Heinrich B M. Mexican Plants with Hypoglycemic Effect Used in the Treatment of Diabetes. Journal Of Ethnopharmacology. 2005; 99:325–348.

[47] Andrade, C. Hipoglicemic Effect of Cecropia Obtusifolia on Streptozotocin Diabetic Rats. Journal Of Ethnopharmacology. 2001; 78(2-3):145-149.

[48] Hicks S. Desert Plants and the People. The Naylor Company Book:San Antonio, Texas; 2009. p. 35-37

[49] Escalante J.S. Our Plants from the Sonora State http://www.apnsac.org. (Accesed 15 July 2012)

Chapter 4

Foods or Bioactive Constituents of Foods as Chemopreventives in Cell Lines After Simulated Gastrointestinal Digestion: A Review

Antonio Cilla, Amparo Alegría, Reyes Barberá and María Jesús Lagarda

Additional information is available at the end of the chapter

1. Introduction

Epidemiological studies on the relationship between dietary habits and disease risk have shown that food has a direct impact on health. Indeed, our diet plays a significant role in health and well-being, since unbalanced nutrition or an inadequate diet is known to be a key risk factor for chronic age-related diseases [1]. An example that illustrates this fact is the protective effect of the so-called Mediterranean diet. The lower occurrence of cancer and cardiovascular disease in the population located around the Mediterranean sea has been linked to the dietary habits of the region, in which the components of the diet contain a wide array of molecules with antioxidant and antiinflammatory actions [2].

Many diseases with a strong dietary influence include oxidative damage as an initial event or in an early stage of disease progression [3]. In fact, Western diets (typically dense in fat and energy and low in fiber) are associated with disease risk [4]. Therefore, dietary modification, with a major focus on chronic age-related disease prevention through antioxidant intervention, could be a good and cost-effective strategy [5]. The intake of whole foods and/or new brand developed functional foods rich in antioxidants would be suitable for this purpose. In this sense, dietary antioxidants such as polyphenols, carotenoids and peptides, as well as other bioactive chemopreventive components such as fiber and phytosterols have been regarded to have low potency as bioactive compounds when compared to pharmaceutical drugs, but since they are ingested regularly and in significant amounts as part of the diet, they may have noticeable long-term physiological effects [6].

For decades, the beneficial role of antioxidants was related to the reduction of unwanted and uncontrolled production of reactive oxygen species (ROS), leading to a situation referred to as oxidative stress [7]. Nowadays, the term "antioxidant" has become ambiguous, since it has different connotations for distinct audiences. For instance, for biochemists and nutritionists, the term is related to the scavenging of metabolically generated ROS, while for food scientists the term implies use in retarding food oxidation or for the categorization of foods or substances according to *in vitro* assays of antioxidant capacity, such as the ORAC and TEAC tests [8]. The antioxidant values provided by these assays sometimes have been misinterpreted by both food producers and consumers due to the fact that health claims advertised on the package labeling are directly associated with benefits that include slowing of the aging process and decreasing the risk of chronic disease. Nevertheless, contemporary scientific evidence indicates that total antioxidant capacity measured by currently popular chemical assays may not reflect the actual activity *in vivo*, since none of them take biological processes such as bioavailability, uptake and metabolism into account [9]. Therefore, no *in vitro* assay that determines the antioxidant capacity of a nutritional product describes *in vivo* outcomes, and such testing should not be used to suggest such a connection. In this sense, it is currently recognized that the mechanisms of action of antioxidants *in vivo* might be far more complex than mere radical scavenging - involving interactions with specific proteins central to intracellular signaling cascades [10], and in the specific case of cancer cells there might be a direct antioxidant effect, antiproliferation and anti-survival action, the induction of cell cycle arrest, the induction of apoptosis, antiinflammatory effects and the inhibition of angiogenesis and metastasis [11].

In order to determine and verify the action of these bioactive compounds, it is clear that data from human intervention studies offer the reference standard and the highest scientific evidence considering the bioavailability and bioactivity of a food component, while *in vitro* methods are used as surrogates for prediction [12]. From a physiological perspective, food after consumption undergoes a gastrointestinal digestion process that may affect the native antioxidant potential of the complex mixture of bioactive compounds present in the food matrix before reaching the proximal intestine. *In vitro* methods which apply human simulated digestion models (including or not including colonic fermentation) are considered valuable and useful tools for the estimation of pre-absorptive events (i.e., stability, bioaccessibility) of different food components from distinct food sources, and also for determining the effect which processing may have upon food components bioavailability [13]. In addition, *in vitro* assays combining a simulated gastrointestinal digestion process and cell cultures as pre-clinical models can be useful for unraveling mechanisms of action and for projecting further *in vivo* assays [9]. Nevertheless, in most cases these *in vitro* studies are unrealistic, because they involve single compounds used at high concentrations (pharmacological and not dietary concentrations) far from the low micromolar or nanomolar concentrations detected *in vivo*, or use the bioactive compounds "as they are in food" versus the metabolites or derivatives considered to be the true bioactive compounds, over an extended period of time (up to 120 h). As a result, biological activity may be overestimated, since no account is taken of the possible transformation of these compounds during gastrointestinal digestion with or without colonic fermentation [6]. Likewise, the use of single or

crude compounds instead of whole foods impedes the detection of synergistic and/or antagonistic actions among bioactive chemopreventive compounds [14, 15].

Taking this background together, and in order to obtain a more precise view of the *in vivo*
situation, we propose the use of whole foods or related target bioactive constituents subjected to a human simulated gastrointestinal digestion including or not including colonic fermentation, depending on the nature of the studied compounds, in order to gain better
insight from a nutritional/functional point of view of the chemopreventive action derived
from foods and bioactive compounds in cell models of disease.

This review introduces the main features of the different *in vitro* gastrointestinal digestion
(solubility and dialysis) and colonic fermentation procedures (batch, continuous and continuous with immobilized feces) for studying the bioaccessibility and further bioavailability
and bioactivity of nutrients and bioactive compounds. It also includes a definition of the
terms: bioavailability including bioaccessibility and bioactivity. Likewise, the main advantages and disadvantages of these *in vitro* methods versus *in vivo* approaches, the improvement of these models with the inclusion of cell lines, and a short comment on the main
effects that digestion and/or fermentation have on bioactive compounds are included. On
the other hand, a short description is provided of the studies involving the use of human
simulated gastrointestinal digestion and/or colonic fermentation procedures, and of the subsequent bioactivity-guided assays with cell line models.

2. Simulated gastrointestinal digestion assays

Bioavailability is a key concept for nutritional effectiveness, irrespective of the type of food
considered (functional or otherwise). Only certain amounts of all nutrients or bioactive compounds are available for use in physiological functions or for storage.

The term bioavailability has several working conditions. From the nutritional point of view,
bioavailability is defined as the proportion of a nutrient or bioactive compound can be used for
normal physiological functions [16]. This term in turn includes two additional terms: bioaccessibility and bioactivity. Bioaccessibility has been defined as the fraction of a compound that is
released from its food matrix in the gastrointestinal tract and thus becomes available for intestinal absorption. Bioaccessibility includes the sequence of events that take place during food
digestion for transformation into potentially bioaccessible material, absorption/assimilation
through epithelial tissue and pre-systemic metabolism. Bioactivity in turn includes events
linked to how the bioactive compound is transported and reaches the target tissue, how it interacts with biomolecules, the metabolism or biotransformation it may undergo, and the generation of biomarkers and the physiologic responses it causes [12]. Depending on the *in vitro*
method used, evaluation is made of bioaccessibility and/or bioactivity.

In vitro methods have been developed to simulate the physiological conditions and the sequence of events that occur during digestion in the human gastrointestinal tract. In a first
step, simulated gastrointestinal digestion (gastric and intestinal stages, and in some cases a
salivary stage) is applied to homogenized foods or isolated bioactive compounds in a closed

system, with determination of the soluble component fraction obtained by centrifugation or dialysis of soluble components across a semipermeable membrane (bioaccessible fraction). Simulated gastrointestinal digestion can be performed with static models where the products of digestion remain largely immobile and do not mimic physical processes such as shear, mixing, hydration. Dynamic models can also be used, with gradual modifications in pH and enzymes, and removal of the dialyzed components – thereby better simulating the actual *in vivo* situation. All these systems evaluate the aforementioned term "bioaccessibility", and can be used to establish trends in relative bioaccessibility.

The principal requirement for successfully conducting experimental studies of this kind is to achieve conditions which are similar to the *in vivo* conditions. Temperature, shaking or agitation, and the chemical and enzymatic composition of saliva, gastric juice, duodenal and bile juice are all relevant aspects in these studies. Interactions with other food components must also be taken into account, since they can influence the efficiency of digestion [12, 17]. A recent overview of the different *in vitro* digestion models, sample conditions and enzymes used has been published by Hur et al. [13]. En lipophilic compounds such as carotenoids and phytosterols, it is necessary to form mixed micelles in the duodenal stage through the action of bile salts, phospholipases and colipase. This allows the compounds to form part of the micelles, where they remain until uptake by the enterocytes [18]. In the case of lycopene, during digestion isomerization of trans-lycopene may occur with the disadvantage that trans-isomers are less soluble in bile acid micelles [19]. Salivary and gastric digestion exert no substantial effect on major phenolic compounds. However, polyphenols are highly sensitivity to the mild alkaline conditions in pancreatic digestion, and a good proportion of these compounds can be transformed into other unknown and/or undetected forms [20].

Bioactive compounds such as dietary fiber, carotenoids, polyphenols and phytosterols undergo very limited absorption, and may experience important modifications as a result of actions on the part of the intestinal microbiota. Small intestine *in vitro* models are devoid of intestinal microbes, and are designed to only replicate digestion and absorption processes; as a result, they are unable to provide information on intestinal fermentation processes. The incorporation of colonic/large intestine fermentation offers a better approximation to the *in vivo* situation, and allows us to study the effect/interaction between these compounds and the intestinal microbiota.

In vitro colonic fermentation models are characterized by the inoculation of single or multiple chemostats with fecal microbiota (of rat or human origin) and operated under physiological temperature, pH and anaerobic conditions. There are two types of colonic fermentation models: batch culture and continuous cultures. Batch culture describes the growth of pure or mixed bacterial suspensions in a carefully selected medium without the further addition of nutrients in closed systems using sealed bottles or reactors containing suspensions of fecal material under anaerobic conditions. The advantages of batch fermentation are that the technique is inexpensive, easy to set up, and allows large number of substrates of fecal samples to be tested. However, these models have their weakness in microbiological control and the need to be of short duration in order to avoid the selection of non-representative microbial populations. The technique is useful for fermentation studies, for the investigation of metabolic profiles of short chain fatty acids arising from the active metabolism of dietary compounds by the gut micro-biota, and especially for substrate digestion evaluation studies [21, 22]. Several of the publications in this field are based on a European interlaboratory study for estimation of the fermentability of dietary fiber *in vitro* [23].

Continuous cultures allow us to control the rate and composition of nutrient feed, bacterial metabolism and the environmental conditions. These models simulate proximal (single-state models) or proximal, transverse and distal colonic regions (multistage models). Continuous cultures are used for performing long-term studies, and substrate replenishment and toxic product removal are facilitated - thereby mimicking the conditions found *in vivo*. The most variable factor in these models is the technique used for fecal inoculation. The use of liquid fecal suspension as inoculum, where the bacterial populations are in the free-cell state, produces rapid washout of less competitive bacteria; as a result, the operation time is less than four weeks. The formation of fecal beads from the immobilization of fecal microbiota in a porous polysaccharide matrix allows release of the microbiota into the culture medium, with better reproduction of the *in vivo* flora and longer fermentation times [21, 22].

Artificial continuous models including host functions/human digestive functions have been developed. Models of this kind control peristaltic movement, pH and gastrointestinal secretions. The SHIME model (Simulated Human Intestinal Microbial Ecosystem) comprises a 5-step multi-chamber reactor simulating the duodenum and jejunum, ileum, cecum and the ascending colon, transverse colon and descending colon [24]. In turn, TIM-1 is an intestinal model of the stomach and small intestine, while TIM-2 is a proximal colon simulator model developed by TNO (*Netherlands Organization for Applied Scientific Research*). These models have been validated based on human and animal data [25]. They incorporate some host functions; however, they do not reproduce immune modulating and neuroendocrine responses. A remaining challenge is the difficulty of establishing a representative human gut microbiota *in vitro*. Other difficulties are the availability of the system, its cost, the prolonged time involved, its laboriousness, the use of large working volumes, and long residence times.

Combined systems that include the fractions obtained from simulated human digestion (gastrointestinal and/or colonic fermentation) and the incorporation of cell culture-based models allow us to evaluate bioaccessibility (estimate the amount of bioactive compounds assimilated from the bioaccessible fraction by cell culture) and to conduct bioactivity studies. The Caco-2 cell model is the most widely used and validated intestinal epithelium or human colon carcinoma cell model. Although colonic in origin, Caco-2 cells undergo spontaneous differentiation in cell culture to form a monolayer of well-polarized cells at confluence, showing many of the functional and morphological properties of mature human enterocytes (with the formation of microvilli on the brush border membrane, tight intercellular junctions and the excretion of brush border-associated enzymes) [26]. However it must be mentioned that this cell line differs in some aspects from *in vivo* conditions. For example, it does not reproduce the different populations of cells in the gut, such as goblet, Paneth and crypt cells, which are less organized and therefore leakier. Likewise, the model lacks regulatory control by neuroendocrine cells and through the blood [27].

The advantage of these systems versus those which only evaluate the influence of digestion is their greater similarity to the *in vivo* conditions. The combination of *in vitro* human intestinal cell models with *in vitro* digestion models in turn creates an advanced *in vitro* model system where samples obtained from host responses lacking in *in vitro* digestion models can be directly applied to monolayer cell models for host function studies [21].

3. Bioactivity of digested/fermented foods or related target bioactive compounds in cell lines

The chemopreventive properties of bioactive compounds have been investigated in cultured cells exposed to individual compounds. However, gut epithelial cells are more likely to be exposed to complex food matrixes containing mixtures of bioactive and antioxidant *in vivo* compounds [6]. In addition, food matrixes undergo a digestion process that may affect the structure and properties of the bioactive compounds. Therefore, the *in vitro* protective effects of antioxidant bioactive compounds do not necessarily reflect *in vivo* chemoprotection, which is more likely due to the combined effects of all the bioactive components present in the food [28].

A potential cell culture model for cancer or cardiovascular chemoprevention research involving dietary antioxidants (polyphenols, carotenoids and peptides) and other bioactive chemopreventive components such as phytosterols, should include some of the proposed mechanisms of action: inhibition of cell proliferation, induction of tumor suppressor gene expression, induction of cell cycle arrest, induction of apoptosis, antioxidant enzyme induction, and enhanced detoxification, antiinflammatory activities and the inhibition of cholesterol absorption [9, 15, 29, 30]. In addition, other mechanisms of chemoprevention could involve protection against genotoxic compounds or reactive oxygen species [31].

It recently has been stated that the measurement of cellular bioactivity of food samples coupled to *in vitro* digestion can provide information close to the real-life physiological situation [32]. In this sense, we surveyed more than 30 studies conducted in the past 10 years, involving human simulated gastrointestinal digestion and/or colonic fermentation procedures and subsequent bioactivity-guided assays with cell line models. These studies are presented in Tables 1, 2 and 3, which correspond to the mechanism of action related to chemoprevention of digested, fermented or digested plus fermented foods or bioactive constituents in cell lines, respectively.

The chemopreventive effect of digested foods or bioactive constituents in cell lines is summarized in Table 1. From the 22 studies surveyed, and according to the digestion method used, it can be seen that most of them involve solubility (n = 17) versus dialysis (n = 5). Samples used are preferably of vegetal origin (n = 15), the target compounds responsible for the chemopreventive action being polyphenols, antioxidants (in general), antioxidant peptides, lycopene and phytosterols. Furthermore, these compounds are mainly studied in colon-derived cells (as a cancer model when not differentiated, or as an intestinal epithelial model when differentiated). Concentrations tested are physiologically achievable in colon cells, since the bioaccessible fractions obtained after digestion are considered to be fractions that can pass through the stomach and small intestine reaching the colon, where they can exert antioxidant activity *in situ* [33]. In addition, polyphenols are studied in neuronal cells, liver-derived cells and lymphocytes. In the case of neuronal cells, the concentrations used (0-6 µM polyphenols) are similar to those reported for dietary polyphenolic-derived metabolites found in plasma (0-4 µM) [34], but for lymphocytes and liver, the concentrations are unknown or higher than expected *in vivo*, respectively. Another aspect to bear in mind is the time of cell exposure to the digested food or bioactive constituents. The range found in these studies is from 30 min to 120 h (this latter time-point not being expectable from a physiological standpoint).

Bioactive compounds of digested foods present four different but in some cases complementary modes of action: (1) inhibition of cholesterol absorption (phytosterols), and (2) antiproliferative, (3) cytoprotective and (4) antiinflammatory activities (polyphenols and general antioxidants).

1. The inhibition of cholesterol absorption has been reported to be mainly due to competition between phytosterols and cholesterol for incorporation to the micelles as a previous step before absorption by the intestinal epithelial cells [35].

2. Antiproliferative activity has been linked to cell growth inhibition associated to polyphenols [28, 32, 36-38] and lycopene [39], which is mainly regulated by two mechanisms: cell-cycle arrest and apoptosis induction. The cell cycle can be halted at different phases: G_0/G_1 with down-regulation of cyclin D_1 [39], S with down-regulation of cyclins D_1 and B_1 [28, 37] and G_2/M [36]. Apoptosis induction in turn occurs as a result of caspase-3 induction and down-regulation of the anti-apoptotic proteins Bcl-2 and Bcl-xL [39].

3. The cytoprotective effect of polyphenols, peptides and antioxidants against induced oxidative stress is related to the preservation of cell viability [40-47], an increase in the activity of antioxidant enzymes (such as catalase, glutathione reductase or glutathione peroxidase) [41, 43, 47, 48], the prevention of reduced glutathione (GSH) depletion [46, 47, 49], a decrease in intracellular ROS content [46, 50, 51], the maintenance of correct cell cycle progression [41, 43, 47, 52], the prevention of apoptosis [43], and the prevention of DNA damage [42, 51, 52].

4. The antiinflammatory action of peptides and polyphenols is derived from the decrease in the release of proinflammatory cytokines such as IL-8 when cells are stimulated with stressors such as H_2O_2 or TNFα [53, 54].

Studies on the chemopreventive effect of foods or isolated bioactive constituents following colonic fermentation or gastrointestinal digestion plus colonic fermentation in cell lines are shown in Tables 2 and 3, respectively. The colonic fermentation procedure used in these assays has always been a batch model, except for one study combining batch and dynamic fermentation. In turn, when gastrointestinal digestion is involved, dialysis has been the method used. Foods of plant origin rich in fiber, and short chain fatty acids (mainly butyrate) and polyphenols as the target compounds have been used in such studies. The use of colon-derived cell lines is common in these assays, which have been performed using physiologically relevant concentrations and time periods of exposure of samples to cells ranging between 24 h and 72 h.

The mechanism of action underlying the treatment of cells with colonic fermented foods or isolated bioactive constituents (see Table 2) mainly comprises antiproliferative activity (i) and/or cytoprotective action (ii). In the first case, antiproliferative activity (i) has been attributed to cell growth inhibition [55-59], mainly due to apoptosis induction [58-59] and/or the up-regulation of genes involved in cell cycle arrest (*p21*) and apoptosis (*WNT2B*) [59]. Studies referred to a cytoprotective effect against oxidative damage (ii) in turn have been linked to the prevention of DNA damage [55, 56] and to the induction of antioxidant enzymes such as glutathione-S-transferase (GST) [56].

The bioactivity observed with the incubation of cells lines with foods or isolated bioactive constituents following gastrointestinal digestion plus colonic fermentation (see Table 3) is derived from antiproliferative activity (i) regulated by cell growth inhibition [60-62], cell cycle arrest [60] and/or apoptosis induction [60, 62], or by a cytoprotective effect against induced oxidative stress (ii) as a result of preservation of cell viability [63], protection against DNA damage [31, 61, 63] and/or induction of antioxidant enzymes such as CAT, GST and sulfotransferase (SULT2B1) [31].

4. Conclusions and future perspectives

From the data here reviewed in disease cell models, it can be concluded that gastrointestinal digestion/colonic fermentation applied to whole foods or isolated bioactive constituents may have potential health benefits derived from cell growth inhibition through the induction of cell-cycle arrest and/or apoptosis, cytoprotection against induced oxidative stress, antiinflammatory activity and the reduction of cholesterol absorption.

Studies conducted with single bioactive compounds are unrealistic from a nutritional and physiological point of view, since they do not take into account physicochemical changes during digestion and possible synergistic activities. Thus, a combined model of human simulated digestion including or not including colonic fermentation (depending on the nature of the studied compounds) with cell lines should be carried out if *in vitro* bioactivity assays with whole foods or bioactive chemopreventive compounds for the prevention of oxidative stress-related diseases are planned.

Although digested/fermented bioactive compounds appear as promising chemopreventive agents, our understanding of the molecular and biochemical pathways behind their mechanism of action is still limited, and further studies are warranted. In addition, the need for harmonization of the *in vitro* methods: (i) conditions of the gastrointestinal procedure, (ii) cell line used, (iii) concentrations of bioactive compounds used (usually much higher than those achievable in the human body when the digestion process is not considered), and (iv) time of cell exposure to the bioactive compounds (more than 24 h is unlikely to occur *in vivo*), should be considered for improved study designs more similar to the *in vivo* situation and for allowing comparisons of results among laboratories. This task is currently being carried out at European level within the project "Improving health properties of food by sharing our knowledge on the digestive process (INFOGEST) (2011-2015) (FAO COST Action FA 1005) (http://www.cost-infogest.eu/ABOUT-Infogest)".

Sample (Target compound/s)	Cell type	Cell treatment (Concentrations and time)	Cellular mechanism	References
		Gastrointestinal digestion (dialysis)		
(Polyphenols)				
Chokeberry juice	Caco-2 (human colon carcinoma)	85 to 220 (µM total polyphenols) 2 h a day for a 4-day period	Cell growth inhibition Viability decrease Cell cycle arrest at G_2/M phase Up-regulation of tumor suppression gene *CEACAM1*	Bermúdez-Soto et al. (2007) [36]
Raspberries	HT29, Caco-2 and HT115 (human colon carcinoma)	3.125 to 50 (µg/mL) 24 h	Prevention of H_2O_2 (75µM/5min)-induced DNA damage and decrease in G_1 phase of cell cycle (HT29 cells) No effect on epithelial integrity (Caco-2 cells) Inhibition of colon cancer cell invasion (HT115 cells)	Coates et al. (2007) [52]
Green tea	Differentiated PC12 (model of neuronal cells)	0.3-10 µg/mL (for H_2O_2) and 0.03-0.125 µg/mL (for $A\beta_{(1-42)}$) Pretreatment 24 h and stressed 24 h	Protection against H_2O_2 and $A\beta_{(1-42)}$ induced cytotoxicity (only at low concentrations)	Okello et al. (2011) [44]
Blackberry (*Rubus sp.*)	SK-N-MC (neuroblastoma cells)	1.5-6 µM total polyphenols 24 h	Preservation of cell viability against H_2O_2 (300 µM- 24 h) –induced oxidative stress (not related to modulation of ROS nor GSH levels)	Tavares et al. (2012a) [45]

CECAM1: Carcinoembryonic antigen-related cell adhesion molecule 1. Aβ(1-42): β-amyloid peptide 1-42. ROS: reactive oxygen species. GSH: reduced glutathione.

Table 1. Mechanisms involved in the chemopreventive effect of *in vitro* digested foods or bioactive constituents in cell lines.

The *in vitro* simulation of the conditions of gastrointestinal digestion represents an alternative to *in vivo* studies for evaluating the bioavailability and/or functionality of bioactive components of foods. *In vitro* studies do not replace *in vivo* studies; rather, both complement each other. *In vitro* methods need to be improved and validated with more *in vivo* studies. Thus, caution is mandatory when attempting to extrapolate observations obtained *in vitro* in cell line studies to humans.

Sample (Target compound/s)	Cell type	Cell treatment (Concentrations and time)	Cellular mechanism	References
Wild blackberry species	SK-N-MC (neuroblastoma cells)	0-6 µM total polyphenols 24 h	Preservation of cell viability and mitochondrial membrane potential against H_2O_2 (300 µM -24 h)-induced oxidative stress Decrease of intracellular ROS against H_2O_2 (200 µM -1 h)-induced oxidative stress (only *R. brigantines*) Prevention of GSH depletion against H_2O_2 (300 µM -24 h)-induced oxidative stress Induction of caspase 3/7 activity against H_2O_2 (300 µM -24 h)-induced oxidative stress (preconditioning effect)	Tavares et al. (2012b) [46]
		Gastrointestinal digestion (solubility)		
(Polyphenols)				
Fruit beverages with/without milk and/or iron	Caco-2 (human colon carcinoma)	2%, 5% and 7.5% (v/v) in culture medium (3.4-22.7 mg/mL total polyphenols) 4 hours-4 days or 24 h	Cell growth inhibition (no clear dose-response) Cell cycle arrest at S phase (7.5%) Down-regulation of cyclins D_1 and B1 No apoptosis (cytostatic effect)	Cilla et al. (2009) [28]
Zinc-fortified fruit beverages with/without iron and/or milk	Caco-2 and HT-29 (human colon carcinoma)	7.5% (v/v) in culture medium (~50 µM total polyphenols) 24 h	Cell growth inhibition (without citotoxicity) Cell cycle arrest at S phase No apoptosis and resumption of cell cycle after digest removal (cytostatic effect)	Cilla et al. (2010) [37]
Fruit juices enriched with pine bark extract	Caco-2 (human colon carcinoma)	4% (v/v) in culture medium 24-120 h	Cell growth inhibition	Frontela-Saseta et al. (2011) [38]

ROS: reactive oxygen species. GSH: Reduced glutathione.

Table 1. (continued-I).

Sample (Target compound/s)	Cell type	Cell treatment (Concentrations and time)	Cellular mechanism	References
Feijoada- traditional Brazilian meal	HepG2 (human liver cancer cells)	10-100 mg/mL 72 h (antiproliferatio n) and 1 h (antioxidant)	Antiproliferative activity ("/ 80 mg/mL) Increase in cellular antioxidant activity (0.6 μM quercetin equivalents)	Kremer-Faller et al. (2012) [32]
Culinary herbs: rosemary, sage and thyme	PBL (peripheral blood lymphocytes) and Differentitat ed Caco-2 (model of intestinal epithelium)	1:10 (v/v) in culture medium. Stressors (H_2O_2 2 mM and TNFα 100 μg/mL) Co-incubation 24 h or pre- incubation 3h then stress 24 h	PBL: significant decrease in IL-8 release when co-incubation with H_2O_2 and pre- incubation prior H_2O_2 and TNFα Caco-2: significant decrease in IL-8 release only when co-incubation with TNFα	Chohan et al. (2012) [54]
(Antioxidants)				
Fruit beverages with/without milk and/or iron/zinc	Differentiate d Caco-2 (model of intestinal epithelia)	1:1 (v/v) in culture medium	Preservation of cell viability No alteration of SOD	Cilla et al. (2008) [40]
Fruit beverages with/without milk or CPPs	Differentiate d Caco-2 (model of intestinal epithelia)	1:1 (v/v) in culture medium or CPPs (1.4 mg/mL)	Preservation of cell viability (only fruit beverages)	Laparra et al. (2008) [41]
Beef patties enriched with sage and oregano	Caco-2 (human colon carcinoma)	10-100% (v/v) 24 h	Increase in cell viability at low concentrations (20-40%) but slight decrease at high concentrations (80-100%) Increase in GSH (only sage-enriched samples at 10%) Protection against H_2O_2 (200 μM/1h)- induced GSH depletion (at 10%)	Ryan et al. (2009) [49]

IL-8: Proinflammatory interleukin-8. TNFα: tumor necrosis factor α. SOD: Superoxide dismutase. CPPs: caseinophos- phopeptides. GSH-Rd. glutathione reductase. GSH: reduced glutathione.

Table 1. (continued-II).

Sample (Target compound/s)	Cell type	Cell treatment (Concentrations and time)	Cellular mechanism	References
Ellagic acid-, lutein- or sesamol-enriched meat patties	Caco-2 (human colon carcinoma)	0-20% (v/v) in culture medium 24 h	Viability maintenance against H_2O_2 (500 µM/ 1h)-induced stress Prevention of H_2O_2 (50 µM/30 min)-induced DNA damage	Daly et al. (2010) [42]
Pacific hake fish protein hydrolysates	Caco-2 (human colon carcinoma)	0.625-5 mg/mL 2 h	Inhibition (at non cytotoxic doses) of intracellular oxidation induced by AAPH (50 µM/1-2 h)	Samaranayaka et al. (2010) [50]
Human breast milk	Co-culture of Caco-2 BBE and HT29-MTX (model of human intestinal mucosa)	1:3 (v/v) in culture medium 30 min	Decrease of H_2O_2 (1 mM/30 min)-induced ROS Prevention of H_2O_2 (500 µM/30 min)-induced DNA damage	Yao et al. (2010) [51]
Fruit beverages with/without milk and/or iron/zinc	Differentiated Caco-2 (model of intestinal epithelia)	1:1 (v/v) in culture medium Pre-incubation 24 h then stressed 2h with H_2O_2 5 mM	Preservation of cell viability Increase in GSH-Rd activity (only Fe or Zn with/without milk samples) Prevention of G_1 cell cycle phase decrease induced by H_2O2 Prevention of apoptosis (caspase-3) induced by H2O2	Cilla et al. (2011) [43]
Purified milk hydrolysate peptide fraction from digested human milk	Caco-2 and FHs 74 int (human colon carcinoma and primary fetal enterocytes)	0.31-1.25 g/L (peptide) and 150 µM (tryptophan) 2 h (peptide) and 1-12 h (tryptophan)	Exacerbation of AAPH (50 µM/1-2 h)-induced oxidative stress (peptide) Up-regulation of Nrf-2 and subsequent up-regulation of GSH-Px2 gene as adaptive response to stress (tryptophan)	Elisia et al. (2011) [48]

AAPH: 2,2'-azobis (2-amidinopropane) dihydrochloride. ROS: reactive oxygen species. GSH-Rd: glutathione reductase. Nrf-2: nuclear response factor 2. GSH-Px2: glutathione peroxidase.

Table 1. (continued-III).

Sample (Target compound/s)	Cell type	Cell treatment (Concentrations and time)	Cellular mechanism	References
CPPs from digested cow's skimmed milk	Differentiated Caco-2 (model of intestinal epithelia)	1, 2 and 3 mg/mL Pre-incubation 24 h then stressed 2h with H_2O_2 5 mM	Preservation of cell viability Increase in GSH content and induction of CAT activity Decrease in lipid peroxidation Maintenance of correct cell cycle progression	García-Nebot et al. (2011) [47]
Purified hen egg yolk-derived phosvitin phosphopeptides	Differentiated Caco-2 (model of intestinal epithelia)	0.05-0.5 mg/mL 2 h	Reduced IL-8 secretion in H_2O_2 (1 mM/6 h)-induced oxidative stress	Young et al. (2011) [53]
(Lycopene)				
Tomatoes	HT29 and HCT-116 (human colon carcinoma)	20-100 mL/L 24 h	Cell growth inhibition Cell cycle arrest at G_0-G_1 phase and apoptosis induction (caspase-3) Down-regulation of cyclin D_1 and anti-apoptotic proteins Bcl-2 and Bcl-xL	Palozza et al. (2011) [39]
(Phytosterols)				
Orange juice enriched with fat-free phytosterols	Differentiated Caco-2 (model of intestinal epithelia)	2 mL test medium/well 4 h	Reduced micellarization of cholesterol Decrease in cholesterol accumulation by Caco-2 cells	Bohn et al. (2007) [35]

GSH: reduced glutathione. CAT: catalase. IL-8: proinflammatory interleukin-8.

Table 1. (continued-IV).

Sample (Target compound/s)	Cell type	Cell treatment (Concentrations and time)	Cellular mechanism	References
(SCFA)				
Fibre sources: linseed, watercress, kale, tomato.soya	HT29 (human colon carcinoma)	2.5-25% (v/v) in culture medium 72 h	Cell growth inhibition (all samples except watercress) Prevention of HNE (150µM/30 min)-induced DNA damage (only soya flour)	Beyer-Sehlmeyer et al. (2003) [55]

Sample (Target compound/s)	Cell type	Cell treatment (Concentrations and time)	Cellular mechanism	References
flour, chicory inulin and wheat				
Wheat bran-derived arabinoxylans	HT29 (human colon carcinoma)	0.01-50% (v/v) in culture medium 24-72 h	Cell growth inhibition Prevention of HNE (200µM/30 min)-induced DNA damage (at 25-50%) Induction of GST activity (at 10%)	Glei et al. (2006) [56]
Inulin-type fructans	LT97 and HT29 (human colon adenoma and carcinoma)	1.25-20% (v/v) in culture medium 24-72 h	Cell growth inhibition (at 5-10%) Apoptosis induction (cleavage of PARP) only in LT97 cells (at 5-10%)	Munjal et al. (2009) [58]
Wheat aleurone	LT97 and HT29 (human colon adenoma and carcinoma)	5-10% (v/v) in culture medium 24-72 h	Cell growth inhibition Apoptosis induction (caspase-3) Up-regulation of genes p21 (cell cycle arrest) and WNT2B (apoptosis)	Borowicki et al. (2010a) [59]
(polyphenols)				
Apples	LT97 and HT29 (human colon adenoma and carcinoma)	100-900 µg/mL 24-48 h	Cell growth inhibition (LT97 more sensitive than HT29 cells)	Veeriah et al. (2007) [57]

SCFA: short chain fatty acids. GST: Glutathione-S-Transferase. HNE: 4-Hydroxynonenal. PARP: Poly (ADP-ribose) polymerase. WNT2B: Wingless-type MMTV integration site family member 2.

Table 2. Mechanisms involved in the chemopreventive effect of *in vitro* colonic fermented (in batch) of foods or bioactive constituents in cell lines.

Sample (Target compound/s)	Cell type	Cell treatment (Concentrations and time)	Cellular mechanism	References
(SCFA)				
Resistant starches	Differentiated Caco-2 (model of intestinal epithelia)	10% (v/v) in culture medium 24 h	Preservation of cell viability Prevention of H_2O_2 (75 μM/5 min)-induced DNA damage Maintenance of barrier function integrity (TEER)	Fässler et al. (2007) [63]
Wheat aleurone	HT29 (human colon carcinoma)	10% (v/v) in culture medium 24-72 h	Cell growth inhibition Cell cycle arrest in G_0-G_1 phase Apoptosis induction (caspase-3)	Borowicki et al. (2010b) [60]
Wheat aleurone	HT29 (human colon carcinoma)	5-10% (v/v) in culture medium 24-72 h	Induction of antioxidant enzymes (CAT and GST) Up-regulation of genes CAT, GSTP1 and SULT2B1 Prevention of H_2O_2 (75 μM/5 min)-induced DNA damage	Stein et al. (2010) [31]
(SCFA and polyphenols)				
Bread	HT29 (human colon carcinoma)	2.5-5% (v/v) in culture medium 24-72 h	Cell growth inhibition Prevention of H_2O_2 (75 μM/5 min)-induced DNA damage	Lux et al. (2011) [61]
(butyrate)				
Bread	LT97 (human colon adenoma)	5-20% (v/v) in culture medium 24-72 h	Up-regulation of genes from DNA repair, biotransformation, differentiation and apoptosis Increase in GST activity, GSH content and AP activity (differentiation) Cell growth inhibition Apoptosis induction (caspase-3)	Schörlmann et al. (2011) [62]

SCFA: Shot chain fatty acids. TEER: Trans Epithelial Electrical Resistance. GST: Glutathione-S-Transferase. GSH: Glutathione. CAT: catalase. SULT: Sulfotransferase. AP: Alkaline phosphatase.

Table 3. Mechanisms involved in the chemopreventive effect of *in vitro* digested (dialysis) plus colonic fermented (batch) foods or bioactive constituents in cell lines.

Acknowledgements

This work was partially supported by Consolider Fun-C-Food CSD2007-00063 and the Generalitat Valenciana (ACOMP 2011/195).

Author details

Antonio Cilla*, Amparo Alegría, Reyes Barberá and María Jesús Lagarda

*Address all correspondence to: antonio.cilla@uv.es

Nutrition and Food Science Area. Faculty of Pharmacy, University of Valencia, Avda. Vicente Andrés Estellés s/n, 46100 - Burjassot, Valencia, Spain

References

[1] Millen, B. E., Quatromoni, P. A., Pencina, M., Kimokoti, R., Nam-H, B., Cobain, S., Kozak, W., Appagliese, D. P., Ordovas, J., & D'Agostino, R. B. (2005). Unique dietary patterns and chronic disease risk profiles of adult men: The Framinghan nutrition studies. *J. Am. Diet. Assoc.*, 105, 1723-1734.

[2] Puawels, E. K. J. (2011). The protective effect of the Mediterranean diet: focus on cancer and cardiovascular risk. *Med. Princ. Pract.*, 20, 103-111.

[3] Valko, M., Leibfritz, D., Moncol, J., Cronin, M. T., Mazur, M., & Telser, J. (2007). Free radicals and antioxidants in normal physiological functions and human disease. . Int. J. Biochem. Cell Biol. , 39, 44-84.

[4] Johansson, I., Nilsson, L., Stegmayr, B., Boman, K., Hallmans, G., & Winkvist, A. (2012). Associations among 25-year trends in diet, cholesterol and BMI from 140,000 observations in men and women in Northern Sweden. *Nutr. J.*, 11, 1-40.

[5] Bruce, W. R., Giacca, A., & Medline, A. (2000). Possible mechanisms relating diet and risk of colon cancer. *Cancer Epidemiol. Biomarkers Prev.*, 9, 1271-1279.

[6] Espín, J. C., García-Conesa, M. T., & Tomás-Barberán, F. A. (2007). Nutraceuticals: facts and fiction. *Phytochemistry*, 68, 2986-3008.

[7] Holst, B., & Williamson, G. (2008). Nutrients and phytochemicals: from bioavailability to bioefficacy beyond antioxidants. *Curr. Opin. Biotechnol.*, 19, 73-82.

[8] Finley, J. W., Kong, N.-A., Hintze, K. J., Jeffery, E. H., Ji, L. L., & Lei, X. G. (2011). Antioxidants in foods: state of the science important to the food industry. *J. Agric. Food Chem*, 59, 6837-6846.

[9]   Liu, R. H., & Finley, J. (2005). Potential cell culture models for antioxidant research. *J. Agric. Food Chem.*, 53, 4311-4314.

[10] Stevenson, D. E., & Hurst, R. D. (2007). Polyphenolic phytochemicals- just antioxidants or much more? *Cell. Mol. Life Sci.*, 64, 2900-2916.

[11] Ramos, S. (2008). Cancer chemoprevention and chemotherapy: dietary polyphenols and signaling pathways. *Mol. Nutr. Food Res.*, 52, 507-526.

[12] Fernández-García, E., Carvajal-Lérida, I., & Pérez-Gálvez, A. (2009). In vitro bioaccessibility assessment as a prediction tool of nutritional efficiency. *Nutr. Res.*, 29, 751-760.

[13] Hur, S. J., Lim, B. O., Decker, E. A., & Mc Clements, D. J. (2011). In vitro human digestion models for food applications. *Food Chem.*, 125, 1-12.

[14] Liu, R. H. (2003). Health benefits of fruit and vegetables are from additive and synergistic combinations of phytochemicals. *Am. J. Clin. Nutr.*, 78, 517S-520S.

[15] de Kok, T. M., van Breda, S. G., & Manson, M. M. (2008). Mechanisms of combined action of different chemopreventive dietary compounds: a review. *Eur. J. Nutr.*, 47, 51-59.

[16] Fairweather-Tait, S. J. (1993). Bioavailability of nutrients. . Macrae R, Robinson RK, Sadler MJ, editors. Encyclopaedia of food science, food technology and nutrition. London: Academic Press. , 384-388.

[17] Ekmekcioglu, C. (2002). A physiological approach for preparing and conducting intestinal bioavailbility studies using experimental systems. *Food Chem.*, 76, 225-230.

[18] Yonekura, L., & Nagao, A. (2007). Intestinal absorption of dietary carotenoids. . Mol. Nutr. Food Res. , 51, 107-115.

[19] Parada, J., & Aguilera, J. M. (2007). Food microstructure affects the bioavailability of several nutrients. *J. Food Sci.*, 72, R 21-R32.

[20] Bermúdez-Soto, M. J., Tomás-Barberán, F. A., & García-Conesa, M. T. (2007). Stability of polyphenols in chokeberry (Aronia melanocarpa) subjected to in vitro gastric and pancreatic digestion. *Food Chem.*, 102, 865-874.

[21] Payne, A. N., Zihler, A., Chassard, C., & Lacroix, C. (2012). Advances and perspectives in in vitro human gut fermentation modeling. *Trends Biotech.*, 30, 17-25.

[22] Macfarlane, G. T., & Macfarlane, S. (2007). Models for intestinal fermentation: association between food components, delivery sustems, bioavailability and functional interactions in the gut. *Curr. Opin. Biotechnol.*, 18, 156-162.

[23] Barry, B. J. L., Hoebler, C., Macfarlane, G. T., Macfarlane, S., Mathers, J. C., Reed, K. A., Mortensen, P. B., Norgaard, I., Rowland, I. R., & Rumney, C. J. (1995). Estimation of the fermentability of dietary fibre in vitro: a European interlaboratory study. *Br. J. Nutr.*, 74, 303-322.

[24] Molly, K., Woestyne, M. V., & Verstraete, W. (1993). Development of a 5-step multi-chamber reactor as a simulation of the human intestinal microbial ecosystem. *Appl. Microbiol. Biotechnol.*, 39, 254-258.

[25] Minekus, M., Marteau, P., Havenaar, R., & Huis in't Veld, J. H. J. (1995). A multi compartmental dynamic computer-controlled model simulating the stomach and small intestine. *ATLA.*, 23, 197-209.

[26] Pinto, M., Robine-Leon, S., Appay, M. D., Kedinger, M., Triadou, N., Dussaulx, E., Lacroix, B., Simon-Assmann, P., Haffen, K., Fogh, J., & Zweibaum, A. (1983). Entero-cyte-like differentiation and polarization of the human colon carcinoma cell line Ca-co-2 in culture. *Biol. Cell*, 47, 323-330.

[27] Ekmekcioglu, C., Pomazal, K., Steffan, I., Schweiger, B., & Marktl, W. (1999). Calcium transport from mineral waters across Caco-2 cells. *J. Agric. Food Chem.*, 47, 2594-2599.

[28] Cilla, A., González-Sarrías, A., Tomás-Barberán, F. A., Espín, J. C., & Barberá, R. (2009). Availability of polyphenols in fruit beverages subjected to in vitro gastrointes-tinal digestión and their effects on proliferation, cell-cycle and apoptosis in human colon cancer Caco-2 cells. *Food Chem.*, 114, 813-820.

[29] Bradford, P. G., & Awad, A. B. (2010). Modulation of signal transduction in cancer cells by phytosterols. *Biofactors*, 36, 241-247.

[30] Brüll, F., Mensik, R. P., & Plat, J. (2009). Plant sterols: functional lipids in immune function and inflammation? *Clin Lipidol.*, 4, 355-365.

[31] Stein, K., Borowicki, A., Scharlau, D., & Glei, M. (2010). Fermented wheat aleurone induces enzymes involved in detoxification of carcionogens and in antioxidative de-fence in human colon cells. *Br. J. Nutr.*, 104, 1101-1111.

[32] Kremer-Faller, A. N., Fialho, E., & Liu, R. H. (2012). Cellular antioxidant activity of Feijoada whole meal coupled with an in vitro digestion. *J. Agric. Food Chem.*, 60, 4826-4832.

[33] Halliwell, B., Rafter, J., & Jenner, A. (2005). Health promotion by flavonoids, toco-pherols, tocotrienols, and other phenols: direct or indirect effects? Antioxidant or not? *Am. J. Clin. Nutr.*, 81, 268S-276S.

[34] Manach, C., Williamson, G., Morand, C., Scalbert, A., & Remesy, C. (2005). Bioavaila-bility and bioefficacy of polyphenols in humans. I. Review of 97 bioavailability stud-ies. *Am. J. Clin. Nutr.*, 81, 230S-242S.

[35] Bohn, T., Tian, Q., Chitchumroonchokchai, C., Failla, M. L., Schwartz, S. J., Cotter, R., & Waksman, J. A. (2007). Supplementation of test meals with fat-free phytosterol products can reduce cholesterol micellarization during simulated digestion and cho-lesterol accumulation by Caco-2 cells. . J. Agric. Food Chem. , 55, 267-272.

[36] Bermúdez-Soto, Larrosa. M., García-Cantalejo, J. M., Espín, J. C., Tomás-Barberán, F. A., & García-Conesa, M. T. (2007). Up-regulation of tumor supresor carcinoembryon-

ic antigen-related cell adhesión molecule 1 in human colon cancer Caco-2 cells fol-
lowing repetitive exposure to dietery levels of a polyphenol-rich chokeberry juice. *J. Nutr. Biochem.*, 18, 259-271.

[37]   Cilla, A., Lagarda, Barberá. R., & Romero, F. (2010). Polyphenolic profile and antiproliferative activity of bioaccessible fractions of zinc-fortified fruit beverages in human
colon cancer cell lines. *Nutr. Hosp.*, 25, 561-571.

[38]   Frontela-Saseta, C., López-Nicolás, R., González-Bermúdez, C. A., Peso-Echarri, P.,
Ros-Berruezo, G., Martínez-Graciá, C., Canalli, R., & Virgili, F. (2011). Evaluation of
antioxidant activity and antiproliferative effect of fruit juices enriched with Pycnoge-
nol® in colon carcinoma cells. The effect of in vitro gastrointestinal digestion. *Phytot-
er. Res.*, 25, 1870-1875.

[39]   Palozza, P., Serini, S., Bonisegna, A., Bellovino, D., Lucarini, M., Monastra, G., & Gae-
tani, S. (2007). The growth-inhibitory effects of tomatoes digested in vitro in colon
adenocarcinoma cells occur through down regulation of cyclin D1, Bcl-2 and Bcl-xL.
Br. J. Nutr., 98, 789-795.

[40]   Cilla, A., Laparra, J. M., Alegría, A., Barberá, R., & Farré, R. (2008). Antioxidant effect
derived from bioaccessible fractions of fruit beverages against H2O2-induced oxidative stress in Caco-2 cells. *Food Chem.*, 106, 1180-1187.

[41]   Laparra, J. M., Alegría, A., Barberá, R., & Farré, R. (2008). Antioxidant effect of casein
phosphopeptides compared with fruit beverages supplemented with skimmed milk
against H202-induced oxidative stress in Caco-2 cells. *Food Res. Int.*, 41, 773-779.

[42]   Daly, T., Ryan, E., Aherne, S. A., O'Grady, M. N., Hayes, J., Allen, P., Kerry, J. P., &
O'Brien, N. M. (2010). Bioactivity of ellagic acis-, lutein- or sesamol-enriched meat
patties assessed using an in vitro digestion in Caco-2 cell model system. *Food Res.
Int.*, 43, 753-760.

[43]   Cilla, A., Laparra, J. M., Alegría, A., & Barberá, R. (2011). Mineral and/or milk supplementation of fruit beverages helps in the prevention of H202-induced oxidative
stress in Caco-2 cells. *Nutr. Hosp.*, 26, 614-621.

[44]   Okello, E. J., Mc Dougall, G. J., Kumar, S., & Seal, C. J. (2011). In vitro protective effects of colon-available extract of Camellia sinensis (tea) against hydrogen peroxide
and beta-amyloid (Aβ(1-42)) induced cytotoxicity in differentiated PC12 cells. *Phyto-
medicine*, 18, 691-696.

[45]   Tavares, L., Figueira, I., Macedo, D., Mc Dougall, G. J., Leitao, M. C., Vieira, H. L. A.,
Stewart, D., Alves, P. M., Ferreira, R. B., & Santos, C. N. (2012). Neuroprotective effect of blackberry (Rubus sp.) polyphenols is potentiated after simulated gastrointestinal digestion. *Food Chem.*, 131, 1443-1452.

[46]   Tavares, L., Figueira, I., Mc Dougall, G. J., Vieira, H. L., Stewart, D., Alves, P. M., Fer-
rerira, R. B., & Santos, C. N. (2012). Neuroprotective effects of digested polyphenols
from wild blackberry species. *Eur. J. Nutr.*, Doi: 10.1007/s00394-012-0307-7.

[47] García-Nebot, M. J., Cilla, A., Alegría, A., & Barberá, R. (2011). Caseinophosphopeptides exert partial and site-specific cytoprotection against H202-induced oxidative stress in Caco-2 cells. *Food Chem.*, 129, 1495-1503.

[48] Elisia, I., Tsopmo, A., Friel, J. K., Diehl-Jones, W., & Kitts, D. D. (2011). Tryptophan from human milk induces oxidative stress and upregulates Nrf-2-mediated stress response in human intestinal cell lines. *J. Nutr.*, 141, 1417-1423.

[49] Ryan, E., Aherne-Bruce, S. A., O'Grady, M. N., Mc Govern, L., Kerry, J. P., & O'Brien, N. M. (2009). Bioactivity of herb-enriched beef patties. *J. Med. Food*, 12, 893-901.

[50] Samaranayaka, A. G. P., Kitts, D. D., & Li-Chan, E. C. Y. (2010). Antioxidative and angiotensin-I-converting enzyme inhibitory potential of Pacific hake (Merluccius productus) fish protein hydrolysate subjected to simulated gastrointestinal digestion and Caco-2 cell permeation. *J. Agric. Food Chem.*, 58, 1535-1542.

[51] Yao, L., Friel, J. K., Suh, M., & Diehl-Jones, W. L. (2010). Antioxidant properties of breast milk in a novel in vitro digestion/enterocyte model. *J. Pediatr. Gastroenterol. Nutr.*, 50, 670-676.

[52] Coates, E. M., Popa, G., Gill, C. I. R., Mc Cann, M. J., Mc Dougall, G. J., Stewart, D., & Rowland, I. (2007). Colon-available raspberry polyphenols exhibit anti-cancer effects on in vitro models of colon cancer. *J. Carcinog.*, 6(4).

[53] Young, D., Nau, F., Pasco, M., & Mine, Y. (2011). Identification of hen egg yolk-derived phosvition phosphopeptides and their effects on gene expression profiling against oxidative-stress induced Caco-2 cells. *J. Agric. Food Chem.*, 59, 9207-9218.

[54] Chohan, M., Naughton, D. P., Jones, L., & Opara, E. I. (2012). An investigation of the relationship between the anti-inflammatory activity, polyphenolic content, and antioxidant activities of cooked and in vitro digested culinary herbs. Oxid. Med. Cell Longev. 627843.

[55] Beyer-Sehlmeyer, G., Glei, M., Hartmann, E., Hughes, R., Persin, C., Böhm, V., Rowland, I., Schubert, R., Jahreis, G., & Pool-Zabel, B. L. (2003). Butyrate is only one of several growth inhibitors produced during gut flora-mediated fermentation of dietary fibre sources. *Br. J. Nutr.*, 90, 1057-1070.

[56] Glei, M., Hofmann, T., Küster, K., Hollmann, J., Lindhauer, M., & Pool-Zabel, B. L. (2006). Both wheat (Triticum aestivum) bran arabinoxylans and gut flora-mediated fermentation products protect human colon cells from genotoxic activities of 4-hydroxynonenal and hydrogen peroxide. *J. Agric. Food Chem.*, 54, 2088-2095.

[57] Veeriah, S., Hofmann, T., Glei, M., Dietrich, H., Will, F., Schreier, P., Knaup, B., & Pool-Zabel, B. L. (2007). Apple polyphenols and products formed in the gut differently inhibit survival of human cell lines derived from colon adenoma (LT97) and carcinoma (HT29). *J. Agric. Food Chem.*, 55, 2892-2900.

[58] Munjal, U., Glei, M., Pool-Zabel, B. L., & Scharlau, D. (2009). Fermentation products of inulin-type fructans reduce proliferation and induce apoptosis in human colon tumour cells of different stages of carcinogenesis. *Br. J. Nutr.*, 102, 663-671.

[59] Borowicki, A., Stein, K., Scharlau, D., & Glei, M. (2010). Fermentation supernatants of wheat (Triticum aestivum L.) aleurone beneficially modulate cancer progression in human colon cells. *J. Agric. Food Chem.*, 58, 2001-2008.

[60] Borowicki, A., Stein, K., Scharlau, D., Scheu, K., Brenner-Weiss, G., Obst, U., Hollmann, J., Lindhauer, M., Wachter, N., & Glei, M. (2010). Fermented wheat aleurone inhibits growth and induces apoptosis in human HT29 colon adenocarcinoma cells. *Br. J. Nutr.*, 103, 360-369.

[61] Lux, S., Scharlau, D., Schlörmann, W., Birringer, M., & Glei, M. (2011). In vitro fermented nuts exhibit chemopreventive effects in HT29 colon cancer cells. *Br J. Nutr.*, 15, 1-10.

[62] Schlörmann, W., Hiller, B., Janhs, F., Zöger, R., Hennemeier, I., Wilheim, A., Lindhauer, M. G., & Glei, M. (2011). Chemopreventive effects of in vitro digested and fermented bread in human colon cells. *Eur. J. Nutr*, 10.1007/s00394-011-0262-8.

[63] Fässler, C., Gill, C. I. R., Arrigoni, E., Rowland, I., & Amadò, R. (2007). Fermentation of resistant starches: influence of in vitro models on colon carcinogenesis. *Nutr. Cancer*, 58, 85-92.

Chapter 5

Geranium Species as Antioxidants

Mirandeli Bautista Ávila,
Juan Antonio Gayosso de Lúcio,
Nancy Vargas Mendoza,
Claudia Velázquez González,
Minarda De la O Arciniega and
Georgina Almaguer Vargas

Additional information is available at the end of the chapter

1. Introduction

Complementary alternative medicine (CAM) has been widely used for a long time for the treatment of multiple diseases, despite the great advances in allopathic medicine. It is estimated that about 80% of the world population use some form of CAM.

CAM encompasses empirical knowledge and medical practice in which use is made of herbal medicinal plants, animals, minerals, manual therapy and exercise, alone or in conjunction for the treatment of diseases. In the early 1980's there emerged a strong interest in their study that has significantly influenced the pharmaceutical industry in developing technologies to identify new chemical entities and structures that are used for the synthesis of drugs. It has been shown that natural products play an important role in the discovery of compounds for drug development to treat multiple diseases.

Also, is important to recognize that use plants and their products have provided proven benefits to humanity, which falls into four areas: (i) food, (ii) essences and flavoring agents, (iii) perfumes and cosmetics, and (iv) biological and pharmaceutical agents [1]. Within the pharmaceutical area, the current outlook for natural products in drug discovery takes a central role, since at the beginning of this new millennium, only about 10% of 350,000 known species have been investigated from a phytochemical or pharmacology point of view [1].

A great examples of molecules that have hit the market as drugs by isolation from natural products metabolites are: taxol (1), an antitumor agent isolated from Taxus species [2] and camptothecin (2), isolated from the Chinese plant Camptotheca acuminate Decne (Nyssaceae), used to treat ovarian, breast and colorectal cancer, another example is ephedrine (3), which is isolated from the plant *Ephedra sinica*[3] and is used as a flu remedy. In drug discovery, researchers around the world use plants as an essential route in the search for new drugs leaders. One of the main objectives of the research laboratories is the preliminary meeting with isolated bioactive natural products, and its uses as anticancer, antiviral, antimalarial, antifungal and anti-inflammatory [3]. The search for active compounds in plants is an essential way for the development of new drugs, a process in which there is now more advanced and specific methodologies for the analysis of biological activities in particular.

Documentary research from 1981 to 2006 showed that natural products have been a source of 5.7% of drugs produced in those years. The derivatives of natural products are most of the times, chemical molecules synthetized from natural products and contributed to the 27.6% of the total of the new molecule.

2. Characterization of *Geranium* genus

Geranium genus is taxonomically classified within the family *Geraniaceae Juss*, which includes five to eleven genuses, and in total near to 750 species. The genus best known are *Geranium* genus, as wild plants (Figure 1) and *Pelargonium* genus, as garden plants. The names of these genuses usually cause confusion because "geranium", is the common name for certain species of *Pelargonium*.

The names come from Greek and refer to the form that its fruits acquire, likes beaks. Thus, the word "Geranium" comes from "geranos" meaning crane, and "Pelargonium" derived from "Pelargos" meaning stork [4].

Figure 1. *Geranium* genus.

Subgenus	Section	Number of Species
Erodioidea	Erodioidea	3
	Aculeolata	1
	Subacaulia	15
	Brasiliensia	3
Geranium	Geranium	339
	Dissecta	4
	Tuberosa	19
	Neurophyllodes	6
	Paramensia	2
	Azorelloida	1
	Polyantha	7
Robertium	Trilopha	5
	Divaricata	2
	Batrachioidea	4
	Ungiculata	5
	Lucida	1
	Ruberta	4
	Anemonifolia	2

Table 1. *Geranium* genus clasification

Within the classification of *Geranium* genus are accepted 423 species, distributed in three subgenuses: Erodioidea, Geranium and Robertium. The following table (Table 1) shows the distribution of this classification.

Currently, in Hidalgo state, in Central Mexico, are classified 8 different species [5] and any-one has chemical or pharmacological studies.

2.1. Biological activities and compounds isolated from Geranium species

Some species of *Geranium* that have been studied has shown biological activity like: hy-potensive, mild astringents, diuretics, hepatoprotection, antioxidants, anti-inflammatory and antiviral. *Geranium* species also are used as a remedy for tonsillitis, cough, whooping cough, urticarial, dysentery, kidney pain and gastrointestinal disorders [6-8]. It is proba-bly that the species of this genus that growing in the State of Hidalgo possess a similar biological activities and metabolites. All phytochemicals studies described for these spe-cies, indicates the presence of polyphenolic compounds called tannins, which have been considered as water-soluble compounds of molecular weight between 500 and 30,000 g/mol with special properties such as the ability to precipitate alkaloids, gelatin and oth-er proteins [9]. Nowadays tannins are well known because of its antioxidant properties. Tannin-protein complexes in the gastrointestinal tract provide persistent antioxidant ac-tivity.

One of the major components in *Geranium* species isgeraniin (4) [10] described by its discov-erer as a crystallizable tannin. This substance first isolated from *Geranium thunbergi* Sieg. Et Zucc. by T. Okuda in 1976, has been evaluated showing an antihypertensive activity, gera-niin inhibits the angiotensin converting enzyme [11,12] and reverse transcriptase of tumoral viruses RNA [13], inhibit HSV-1 and HSV-2 multiplication at different magnitudes of poten-cy and also is an excellent antioxidant [14]. The corilagin (5) [15] is a derivative of geraniin, which has presented antimicrobial activity among other activities [16].

2.2. Different species of geraniums and its relevant compounds

The specie *Geranium* macrorizum presented a significant hypotensive activity in anesthe-tized cats [17], plus antioxidant activity. Of this specie germacrone (6) was isolated which is considered a precursor of pheromones.

Geranium robertianum L. well-known specie and one of the most variable in Britain has been used in conditions where increased diuresis is required, such as cystitis, urethritis, pyelonephritis, gout, hypertension and edema. Nowadays the phytochemistry of this geranium is relatively well known and its most studied active compounds are tannins, volatile oils, flavonoids and polyphenols (hyperoside, ellagic acid, isoquercitrin, querci-trine, kaempferols, caftaric acid, rutoside).Also infusions and decoctions prepared from leaves of this geranium: Robert herb or red Robin, are described as anti-hyperglycae-miant and commonly used in Portuguese herbal medicine [18]. In other hand G. robertia-num extract treatment increased the efficiency of coupling between oxidative and phosphorylative systems, since RCR was considerably higher in GK rats consuming this plant extract [19].

Recently the extracts of G. *sylvaticum* were studied [20] for antioxidant potential and all test-ed extracts had strong antioxidant activity and will be subject for further investigations. From flowers of *Geranium sylvaticum* was isolated 3-O-(6-O-acetyl-•-D-glucopyranoside)-5-O-•-D-glucopyranoside of malvidin (7) [21].*Geranium sanguineum L.* showed significant in-hibitory activity of influenza virus and herpes simplex (8). The methanolic extract of *Geranium pratense* inhibited the action of the amylase enzyme in mouse plasma, isolated for first time the 3-O-(2-O-galloyl) -•-D-glucopyranoside myricetin(9) [22].

Geranium niveum, widely used by the Tarahumara Indians of Mexico. Is a specie rich in proanthocyanidins and other phenolics [23]. Previous in vitro assays have demonstrated that proanthocyanidins exhibited antiinflammatory, antiviral, antibacterial, enzyme-inhib-iting, antioxidant, and radical-scavenging properties, the roots 25 of this species were iso-lated new proanthocyanidins named as geraniins A (9) and B (9a), latterlyin 2001 were found geraniins C (10) and D (10a) [24]. A recently study showed that geraniin A has an-tioxidant activity [25].

Geranium pusillum, commonly known as Small-flowered Cranesbill or (in North America) small Geranium, contains1-O-galloyl-3,6-hexahidroxibifenil-D-galactopyranoside (11) (pusi-lagin) a polyphenolic compound extracted from aerial parts [26]. The aqueous ethanolic ex-tract of *Geranium wallichianum* showed antibacterial activity against *Staphylococcus aureus* [27] and the study of the chemical constituents of the whole plant has resulted in the isola-tion and characterization of six compounds. These six compounds were identified as ursolic acid, β-sitosterol, stigmasterol, β-sitosterol galactoside, herniarin, and 2,4,6-trihydroxyethyl-benzoate which were isolated for the first time from *Geranium wallichianum* [28].

7

8

9: R, R1, R2 = H
9a: R, R2 = H; R1 = OH

10: R, R1, R2 = H   **10a**: R, R2 = H; R1=OH

Geranium caespitosum produces neohesperidoside (12) able to potentiate 10 to 100 times the action of drugs such as ciprofloxacin, norfloxacin, berberine and rhein, against bacterias such as *S. aureus, NorA S. aureus, B.* and *B. megaterium subtilis* [29]. Besides, *Geranium caroli-nianum L.*, isa commonly used traditional Chinese medicine (TCM) with the efficacy of elim-inating wind-damp and treating diarrhea. It is clinically used to treat the arthralgia due to wind-dampness, anaesthetization and muscular constriction. It has been reports that *Gerani-um carolinianum L.* as well of most of the congeneric plants contain significant amounts of tannins, flavonoids, organic acids, and volatile oils [30].Also, has shown that roots contain a substance that is extracted with water and can be a biological mechanism to control bacteria (*Ralstonia solanacearum*) which attacks potatoes [31].

From *Geranium pyrenaicum*, which showed antileishmanial activity [32],a new glycosylate flavonoid: 3-O-(2 ", 3"-di-O-galloyl)- •-D-glucopyranoside of kaempferol (13) was isolated, and anuncommon quercetin derivative: 3-O-(2 ", 3"-di-O-galloyl) - •-D-glucopyranoside of quercetin (13a) too. In *Geranium mexicanum* an antiprotozoal activity was assayed from its roots, where the most active compound founded was the flavan-3-ol-(-)-epicatechin (14), showing moderate activity (+)-catechin (14a), tyramine (15) and 3-O-β-D-glucopyranoside of β-sitosterol [33].

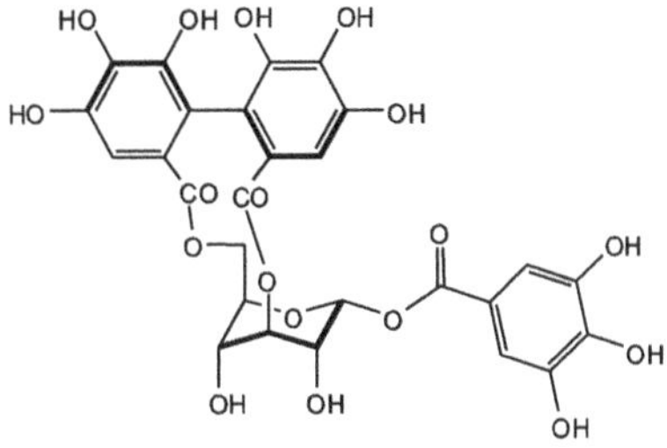

11

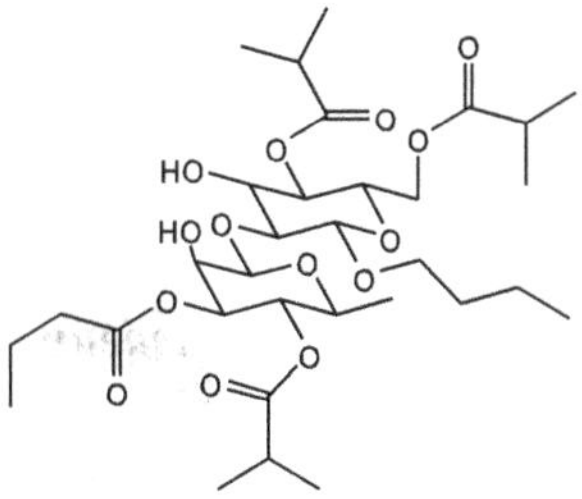

12

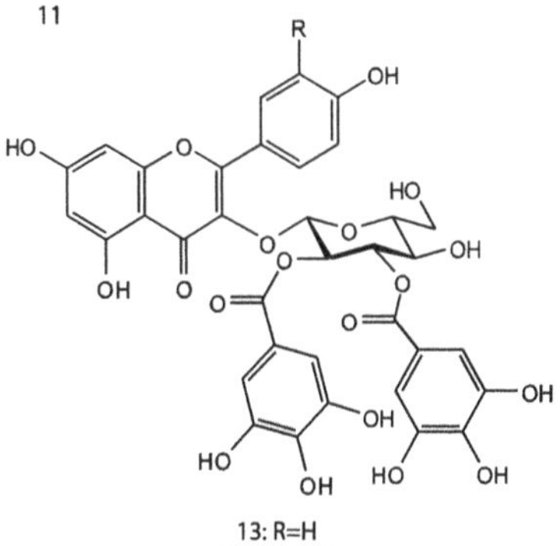

13: R=H
13a: R=OH

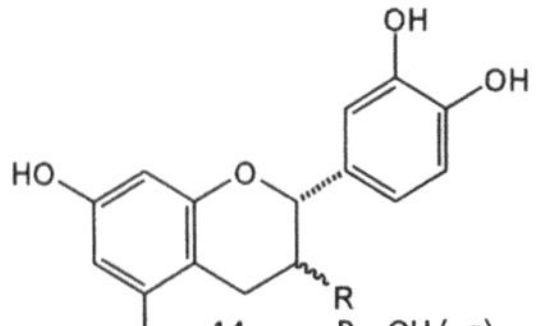

14: R = OH (α)
14a: R = OH (β)

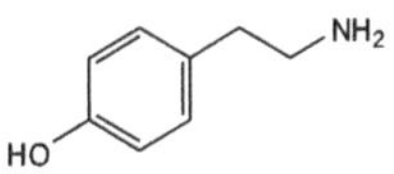

15

Geranium bellum Rose is a perennial plant with long roots, found in the grassy meadows bordering pine/oak forests in the mountains of Hidalgo State, Mexico, where it has the popular name "pata de león" and has been used as a traditional remedy for treatment of fevers, pain, and gastrointestinal disorders. Radical scavenging assay-guided fractionation of the antioxidant EtOAc and MeOH extracts from the aerial parts of *Geranium bellum* resulted in the isolation of bsitosterol 3-O-b-D-glucopiranoside, quercetin 3-O-a-L-(2''-O-acetyl)arabinofuranoside (16), quercetin 3-O-a-L-arabinofuranoside, quercetin, methyl gallate, gallic acid, methyl brevifolin carboxylate (17), and dehydrochebulic acid trimethyl ester (18). Compounds 2, 7 and 8 are iso-lated for the first time from *Geranium* genus [34]. Antioxidant activity of these extracts (both ini-tial fractions and pure compounds), was tested by measuring their capacity to scavenge 2,2'-azino-bis(3-ethylbenzthiazoline-6-sulfonic acid) (ABTS) radicals, an assay widely used for screening of antioxidant activity of natural products [15].

Constituents from the aerial parts of *Geranium potentillaefoium* founded in certain studies were geraniin, corilagin, gallic acid, methyl gallate, methyl brevifolincarboxylate, quercetin, quercetin 3-O-β-Dglucopyranoside, quercetin 3-O-β-D- [6''-O-galloyl)glucopyranoside, kaempferol, β-sitosterol 3-O-β-D-glucopyranoside and β-sitosterol [35].

3. Study of *Geranium schiedeanum*

Geranium schiedeanum (Gs) (Figure 2), species that grows in Central Mexico, has been used as an antipyretic, anti-inflamatory and antiseptic. The use of other geranium species also has been reported a hypoglycemic, antihypertensive and cholesterol-lowering effect. However, scientific evidence does not exist in any literature to corroborate these targets or any other. In the present study the effect of Gs were studied in reference to postnecrotic liver damage induced by thioacetamide (TA).

Figure 2. *Geranium schiedeanum* [5]

3.1. Plant material

Specimen of *Geranium schiedeanum* was collected at Epazoyucan Municipality, in Hidalgo State, México, during June 2009. A voucher specimen (J. A. Gayosoo-de-Lucio) is preserved at the Herbarium of Biological Research Centre, Autonomous University of Hidalgo, Pachuca, Hidalgo, Mexico where Professor Manuel González Ledesma identified the plant material.

3.2. Extraction and purification

Air-dried aerial parts (1 kg) were extracted acetone-H_2O 7:3 (20 L) by maceration for 7 days. Vacuum evaporation of dissolvent give a 5 L residue Filtration give a fatty solid residue (12g) and complete evaporation of water give the acetone-water extract (115 g).

Lots of 3 g of acetone-water extract were purified on a Sephadex LH-20 (25 g) column using H_2O, H_2O-MeOH (9:1, 4:1, 7:3, 3:2, 1:1, 2:3, 3:7, 1:4, 1:9) and MeOH, as eluents. Fractions of 300 mL of each polarity were collected and marked "A–K". They were evaporated and analyzed by TLC and NMR. Fractions "B" gave 75 mg, and were purified over silica gel (10 g), using $CHCl_3$, $CHCl_3$-AcOEt(9:1, 4:1, 7:3, 3:2, 1:1, 2:3, 3:7, 1:4, 1:9) and AcOEt (10 mL of each polarity), as eluents and collecting fractions of 7 mL, fractions 13-16 give I 25 mg. Fractions "C" and "D" gave 56 mg, and were purified over silica gel (10 g), using $CHCl_3$-MeOH (50:7.0, 48:7, 45:7, 40:7, 35:7 and 30:7, 40 mL of each), as eluents and collecting fractions of 7 mL, fractions 33-66 give II 2 mg, (these procedure was repeated ten times to obtain 18 mg of compound), Fractions "F-I" gave 1.8 g, a portion of 500 mg were purified over silica gel C-18 (5 g) using H_2O, H_2O-MeOH (9:1, 4:1, 7:3, 3:2, 1:1, 2:3, 3:7, 1:4, 1:9) and MeOH (20 mL of each polarity), fractions of 10 mL were collected fractions 2-4 gave 325 mg of III, fraction "K" gave 90 mg were added 5 mL of (40°C) pyridine and were placed a room temperature for 72 h, filtrated of mixture give a yellow needles 60 mg of IV (Figure 3).

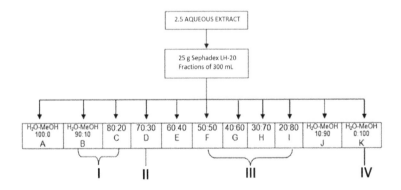

Figure 3. Extract fractions scheme.

3.3. Animals and treatment

Male adult Wistar rats 2 months old (200–220g) were obtained from UAEH Bioterio, and acclimated to our animal room for two weeks before use. Throughout these two weeks rats were supplied with food and water *ad libitum*, exposed to a 12 h light-dark cycle and given intraperitoneally a single necrogenic dose of thioacetamide (6.6 nmol/Kg body weights) freshly dissolved in 0.9% NaCl. The dose of thioacetamide was chosen as the highest dose with survival above 90% [36,37]. Wistar rats were intragastric pre-treated or not with a single dose of Gs extract (300 mg/kg) during 4 days, the fourth day of pretreatment were intraperitoneally injected with a single dose of TA (6.6 mmol/Kg). Samples of blood and liver were obtained from rats at 0, 24, 48,72 and 96 h following TA intoxication. Untreated animals received 0.5 ml of 0.9% NaCl. Experiments were performed on two different groups: rats treated with a single dose of thioacetamide (TA) and rats pre-treated with Gs and treated with a single dose of thioacetamide (Gs + TA). Each experiment was performed in duplicate from four different animals and followed the international criteria for the use and care of experimental animals outlined in *The Guiding Principles in the use of Animals in Toxicology* adopted by the Society of Toxicology in 1989.

3.4. Processing of samples

In order to clarify the sequential changes during the different stages of liver injury and the post-necrotic regenerative response, samples were obtained from control and at 24 and 48 h of TA intoxication in both Gs pre-treated or non pre-treated animals. Rats were sacrificed by cervical dislocation and samples of liver were obtained and processed as previously described. Blood was collected from hearts and kept at 4 °C for 24 h, centrifuged at 3000 rpm for 15 min, and serum was obtained as the supernatant.

3.5. Determination of AST

Enzymatic determination were carried out in serum in optimal conditions of temperature and substrate and cofactor concentrations. Aspartate aminotransferase (AST) activity were determined in serum. AST (EC 2.6.2.1) and was assayed following the method of Rej and Horder [38].

The activity of this enzyme was determined spectrophotometrically, by measuring the decrease in absorbance at 340 nm at 37 ° C, produced by the oxidation of NADH to NAD^+ in the coupled reaction of reduction of oxaloacetate to malate, catalyzed by malate dehydrogenase, according to the following process:

3.6. General

IR spectra measured in MeOH on a Perkin Elmer 2000 FT-IR spectrophotometer. Optical rotations were determined in MeOH on a Perkin Elmer 341 polarimeter. NMR measurements performed at 400 MHz for 1H and 100 MHz for 13C on a VARIAN 400 spectrometer from CDCl3, CD3OH, DMSO-d6 solutions. Column chromatography (CC) was carried out on Merck silica gel 60 (Aldrich, 230-400 mesh ASTM) and sephadex LH-20 Sigma Aldrich.

3.7. Statistical analysis

The results were calculated as the means ± SD of four experimental observations in dupli-cate (four animals). Differences between groups were analyzed by an ANOVA following Snedecor F ($\alpha = 0.05$). Students' test was performed for statistical evaluation as follows: (a) all values against their control; b) differences between two groups Gs + TA versus TA.

3.8. Results

3.8.1. Active compounds of Geranium schiedeanum

One kg of the aerial part of G. schiedeanum was extracted by maceration for 7 days with 20 L acetone-water (7:3), concentrated under reduced pressure to a volume of 3 L, which was ex-tracted with CHCl$_3$ yielding 12.75 g of CHCl$_3$ phase and 125 g of aqueous phase. The phyto-chemical study of Geranium schiedeanum led to the isolation of hydrolysable tannins (I) gallic acid, (III) acetonylgeraniin and (IV) ellagic acid and to a lesser proportion of kaempferol gly-cosideflavonoid (II) (Figure 4 and 5). Is relevant to notice that is the first time discloses the compound II in the Geranium genus and further that the yield of compound III in the crude extract was 40%.

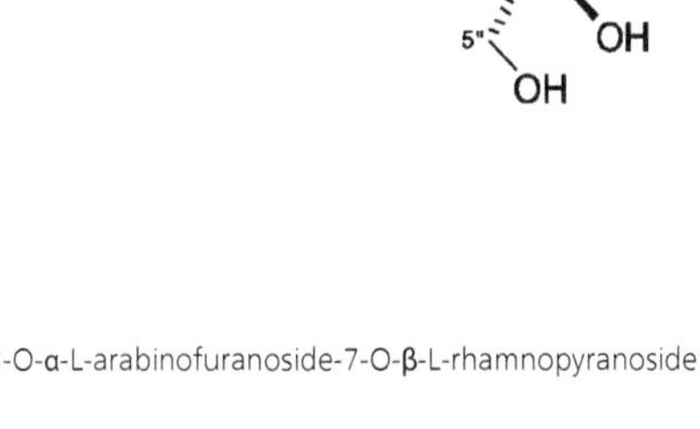

Figure 4. Compound (II) 3-O-α-L-arabinofuranoside-7-O-β-L-rhamnopyranoside de Kaempferol

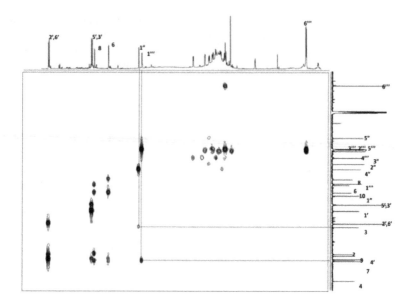

Figure 5. HMBC experiment of compound II

3.8.2. Aspartate aminotransferase

The acute liver injury induced by a necrogenic dose of thioacetamide (TA), a potent hepato-toxic agent, is characterized by a severe perivenous necrosis [39]. The necrosis develops as a consequence of the biotransformation of TA through the microsomal flavin-dependent mon-ooxygenase [40]. The reactive metabolites responsible for TA hepatotoxicity are the radicals derived from thioacetamide-S-oxide and the reactive oxygen species derived as sub prod-ucts in the process of microsomal TA oxidation, both of which can depleted reduced gluta-thione leading to oxidative stress [41, 42].

Liver damage induced by xenobiotics is characterized by the release in serum of hepatic en-zymes due to necrosis of hepatocytes. AST is randomly distributed in the hepatic acinus, and is the enzyme activity used as marker of necrosis. Our results showed that *Gs* extract significantly reduced the level of liver injury. The levels of AST (Figure 6) were significantly lower in the rats pretreated with *Gs*.

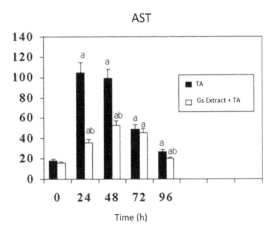

Figure 6. Effect of Gs pre-treatment on aspartate aminotransferase activity in serum of rats intoxicated with one sublethal dose of thioacetamide. Samples were obtained at 0, 24, 48, 72 & 96 h following thioacetamide (TA). The results, expressed as nmol per min per ml of serum, are the mean ± SD of four determinations in duplicate from four rats. Differences against the respective control are expressed as (a) and differences due to Gs extract are expressed as (b) $p<0.05$.

4. Conclusion

There is evidence that free radicals play a critical role in certain pathological conditions such as some cancers, multiple sclerosis, inflammation, arthritis and arterosclerosis [43]. For this reason, some research objectives directed toward the development or discovery of these compounds catchers of these radicals.

A large number of plant species, like *G. schiedeanum* contain chemical compounds that exhibit the ability to trap free radicals. The ability to trap free radicals has been called antioxidant activity. The phytochemical study of *Geranium shiedeanum* led to the isolation of hydrolysable tannins, well known as potent antioxidants: gallic acid, acetonylgeraniin and ellagic acid and a lesser proportion of kaempferol glycoside flavonoid (3-O-α-L-arabinofuranoside-7-O-β-D-rhamnoside de Kaempferol), notably is the first time discloses these compounds in the genus. Further the yield of acetonyl geraniin in the crude extract was 40%.

Also, in the present study TA-induced hepatotoxicity was used to investigate the effect of the pretreatment of *G. schiedeanum* on the events involved in liver regeneration. The results obtained in the present study provide evidence that *Gs*, when administered intravenously prior to TA, significantly reduce liver damage.

The pre-treatment with the crude extract in the model of thioacetamide-induced hepatotoxicity in rats, decreased and delayed liver injury by 66% at 24 h.

The data obtained indicate that the crude *Gs* extract pre-treatment has hepatoprotective and antioxidant effect in damage induced by TA. This result suggests that *Gs* extract may be used as an alternative for reduction of liver damage. However further investigation on the acute toxicity and on the mechanism of the hepatoprotective effect of the plant species needs to be carried out.

Acknowledgements

The authors would like to thank Teresa Vargas for her valuable technical Assistance. Supported by Grant PROMEP-MEXICO UAEHGO-PTC-454.

Author details

Mirandeli Bautista Ávila, Juan Antonio Gayosso de Lúcio, Nancy Vargas Mendoza, Claudia Velázquez González, Minarda De la O Arciniega and Georgina Almaguer Vargas

Universidad Autónoma del Estado de Hidalgo, Mexico

References

[1] K-H Tan, Novel Compounds from Natural Products in the New Millennium: Potential and Challenges. 2004, World Scientific Publishing Company, Singapore.

[2] Dewick P. M., Medicinal Natural Products a Biosynthetic Aproach. 1998, John Wiley & Sons, New York, USA.

[3] Kuo-Hsiung L. *J. Nat. Prod.* 2004, 67(1), 273-283.

[4] Gómez, M. A., Borja y Tomé, López-Lomo V. M. A. M. Biotecnología aplicada a la mejora de *"Pelargonium"*. 2005, Universidad Complutense de Madrid, España.

[5] Pérez Escandón B. E., Villavicencio M. A., Ramirez Aguirre A. Lista Floristica Del Estado De Hidalgo Recopilación Bibliografica, 1998, 1ª edición, Ed. UAEH. México.

[6] Calzada F, Cervantes-Martinez JA, Yepez-Mulia L. In vitro antiprotozoal activity from the roots of *Geranium mexicanum* and its constituents on *Entamoeba histolytica* and *Giardia lamblia*. *J. Ethnopharmacol.* 2005, 98: 191-193.

[7] Ercil D, Kaloga M, Redtke OA, Sakar MK, Kiderlen A, Kolodziej H. O-Galloyl flavonoids from *Geranium pyreniacum* and their *in vitro* antileishmanial activity. *Turk. Chem.* 2005, 29: 437-443.

[8] Küpeli E, Tatl I, Akdemir ZS, Yeflilada E. Estimation of antinociceptive and anti-in-flammatory activity on *Geranium pratense* subsp. *finitimum*. *J. Ethnopharmacol*. 2007, 114: 234-240.

[9] Okuda T., Yoshida T. y Hatano T. *J. Nat. Prod*. 1989, 52(1), 1-31.

[10] Cheng J. T., Chang S. S., Hsu F. L. *J. of Pharm. and Pharmacol*. 1994, 46(1), 469

[11] Kameda K., Takaku T., Okuda H., Kimura Y., Okuda T., Hatano T., Agata I., Arichi S. *J. Nat. Prod*. 1987, 50(4), 680-683.

[12] Ueno H., Hoire S., Nishi Y., Shogawa H., Kawasaki M., Suzuki S., Hayashi, Shimizu A. M., Yoshizaki M., Morita N. *J. Nat. Prod*. 1988, 51(2), 357-359.

[13] Kakiuchi N., Hattori M., Namba T., Nisizahua M., Yamagishi T., Okuda T. *J. Nat. Prod*. 1985, 48(4), 614-621.

[14] Fujiki H., Sagunama M., Kurusu M., Okabe S., Imayoshi. Y., Tanigushi S., Yosida T. *Mutation Research*. 2003; 523-524, 119-25.

[15] Okuda T., Yoshida T., and Mori K. *Phytochemistry* 1975, 14, 1877–1878

[16] Shimizu M., Shiota S., Mizushima T., Ito H., Hatano T., Yoshida T., Tsuchiya T. *Antimicrobial Agents And Chemotherapy* 2001, 45, 3198–3201

[17] Chemical abstracts, vol. 95, 1981, 162140J.

[18] Cunha AP, Silva AP, Roque AR. Plantas e Produtos Vegetais em Fitoterapia. Fundação Calouste Gulbenkian. 2009. Lisboa, Portugal (in Portuguese).

[19] Ferreira FM, Peixoto FP, Nunes E, Sena C, Seiça R, Santos MS. *Vaccinium myrtillus* improves liver mitochondrial oxidative phosphorylation of diabetic Goto-Kakizaki rats. *J Med Plants*. 2010 Res 4: 692–696.

[20] Milena N, Reneta T, and Stephanka I. Evaluation of antioxidant activity in some Geraniacean species *Botanica Serbica*. 2010, 34 (2): 123-125

[21] Andersen M., Viksund R. I., Pedersen A.T. *Phytochemistry* 1995, 38(6), 1513-1517.

[22] Akdemir Z. S., Tatl J. J., Saracoglu J., Ismailoglu U. B., Sahin-ErdemLi I., Calis I., *Phytochemistry*. 2001, 56(2), 189-193.

[23] Maldonado PD, Rivero-Cruz I, Mata R, Pedraza-Chaverrí J. Antioxidant activity of A-type proanthocyanidins from *Geranium niveum* (Geraniaceae). *J Agric Food Chem*. 2005, 23;53(6):1996-200.

[24] Calzada F., García-Rojas C. M., Meches M., Rivera C. R., Bye R., Mata R. *J. Nat. Prod*. 1999, 62, 705-709.

[25] Maldonado P. D., Rivero-Cruz I., Mata R., Pedraza-Chaveri J. *J. Agric. Food Chem*. 2005, 53, 1996-2001.

[26] Kobakhidza K. B., Alaniya M. D. *Chem. of Nat. Comp*. 2004, 39(3), 262-264.

[27] Ahmad B., Ismail M., Iqbal Z., M. Iqbal Chaudhry. *Asian Journal of Plant Sciences* 2003, 2(13), 971-973.

[28] Mohammad I, Zafar Iqbaq, Javid H, Hidayat H, Manzoor Ahmed, Asma Ejaz, Muhammad I. C., Chemical Constituents and Antioxidant Activity of *Geranium wallichianum*. *Rec. Nat. Prod.* 2009, 3:4, 193-197

[29] Oshiro A., Takaesu K., Natsume M., Taba S., Nasu K., Uehara M., Muramoto Y. *Weed Biology and Management* 2004, 4, 187–194

[30] Pharmacopoeia of the People's Republic of China; Chemical Industry Press: Beijing, China, 2010; Vol 1, p.113.

[31] Oshiro A., Takaesu K., Natsume M., Taba S., Nasu K., Uehara M., Muramoto Y. *Weed Biology and Management* 2004, 4, 187–194.

[32] Ercil D., Kaloga M., Ratke O. A., Sakar M. K., Kiderlen F.A., Kolodziej H. *Turk J. Chem.* 2005, 29, 437-443.

[33] Calzada F., Cervantes-Martíneza J. A., Yépez-Muliab L. *Journal of Ethnopharmacology* 2005, 98(1-2), 191-193

[34] Camacho-L A, J Gayosso-De-Lucio, J. Torres-Valencia, J Muñoz-Sánchez, E Alarcón-Hernández, Rogelio L, Blanca L. Barrón. Antioxidant Constituents of *Geranium bellum* Rose. *J. Mex. Chem. Soc.* 2008, 52(2), 103-107

[35] J.A. Gayosso-De-Lucio, J.M. Torres-Valencia, C.M. Cerda-García-Rojas and P. Joseph-Nathan. Ellagitannins from *Geranium potentillaefolium* and *G. bellum*. *Nat. Prod. Comm.*, 2010; 5, 531-534

[36] Sanz N, Diez-Fernández C, Andrés D, Cascales M. Hepatotoxicity and aging: endogenous antioxidant systems in hepatocytes from 2-, 6-, 12-, 18- and 30-month-old rats following a necrogenic dose of thioacetamide. *Biochim Biophys Acta.* 2002; 1587: 12-20.

[37] Zaragoza A, Andrés D, Sarrión D y Cascales M. Potentiation of thioacetamide hepatotoxicity by phenobarbital pretreatment in rats. Inducibility of FAD monooxygenase system and age effect. *Chem Biol Interact.* 2000, 124: 87-101.

[38] Rej R y Horder M. Aspartate aminotransferase. L-aspartate: 2-oxoglutarate aminotranferase, EC 2.6.2.1. Routine U.V. method. En: Bergmeyer HU Editor. Methods of Enzymatic Analysis. 3rd ed., vol III. Weinheim. *Verlag Chemie*, pp. 416-24, (1984).

[39] Cascales M., Martin-Sanz P, Craciunescu DC, Mayo I, Aguilar A, Robles-Chillida EM, Cascales C. Alterations in hepatic peroxidation mechanisms in thioacetamide-induced tumors in rats. Effect of a rhodium complex. *Carcinogenesis* 1991; 12: 233-240.

[40] Dyroff MC y Neal RA. Studies of the mechanism of metabolism of thioacetamide-S-oxide by rat liver microsomes. *Cancer Res.* 1981; 41: 3430-3445.

[41] Sanz N, Diez-Fernandez C, Andres D, Cascales M. Hepatotoxicity and aging: endog-
      enous antioxidant systems in hepatocytes from 2-, 6-, 12-, 18- and 30-month-old rats
      following a necrogenic dose of thioacetamide. *Biochim Biophys Acta* 2002; 1587: 12-20.

[42] Sanz N, Díez-Fernández C, Alvarez AM, Cascales M. Age-dependent modifications
      in rat hepatocyte antioxidant defense systems. *J Hepatol* 1997; 27: 525-534.

[43] Latté P. K., Kolodziej H. *J. Agric. Food Chem*. 2004, 52, 4899-4902.

Cell Damage by Free Radicals - Oxidative Stress in Disease

Chapter 6

Inflammatory Environmental, Oxidative Stress in Tumoral Progression

César Esquivel-Chirino, Jaime Esquivel-Soto,
José Antonio Morales-González,
Delina Montes Sánchez, Jose Luis Ventura-Gallegos,
Luis Enrique Hernández-Mora and
Alejandro Zentella-Dehesa

Additional information is available at the end of the chapter

1. Introduction

The incidence and prevalence of cancer has been increasing in such as degree that it has become the second or third leading cause of death worldwide, depending on ethnicity or country in question and is consequently a major public health, cancer is a leading cause of death in many countries, accounting for 7.6 million deaths (around 13% of all deaths) in 2008. Deaths from cancer worldwide are projected to continue rising, with an estimated 13.1 million deaths in 2030. About 30% of cancer deaths are due to the five leading behavioral and dietary risks: high body mass index, low fruit and vegetable intake, lack of physical activity, tobacco use, alcohol use [1-8].

Cancer is a generic term for a large group of diseases that can affect any part of the body, cancer cells are significantly influenced by the surrounding stromal tissues for the initiation, proliferation, and distant colony formation. When a tumour successfully spreads to other parts of the body and grows, invading and destroying other healthy tissues, this process is referred to as metastasis, and the result is a serious condition that is very difficult to treat, because the progression to metastases is the leading cause of death associated to cancer. Metastatic cells in this process must interact with the endothelium in three stages of tumor progression. In recent years, the interaction between these cell populations has been seen as part of a complex microenvironment tumor-associated. Mantovani et al. have even postulated that this tumor microenvironment inflammatory plays an important role in tumorigenesis and tumor progression [18]. Tumor and normal surrounding cells such as endothelial cells, soluble factors derived from this two cell populations and extracellular matrix [9-12], compose the tumor microenvironment.

There are other factors in the development of cancer such as genetic, environmental, as well as the role of oxidative stress and free radicals in response to damage caused by chemicals, radiation, cellular aging, ischemic lesions and cells immune system [13]. Substances such as antioxidants can protect cells against free radical damage. The damage caused by free radicals can lead to cancer. Antioxidants interact with free radicals to stabilize them so that, being able to avoid some of the damage that free radicals can cause. It is important to analyze the role of antioxidants as an alternative that contributes to cancer treatment and to promote their use and consumption in cancer prevention

2. Tumoral progression

Tumors often become more aggressive in their behavior in more aggressive and their characteristics, although the time course may be quite variable, this phenomenon has been termed tumor progression by Foulds [15].

In the early stages of the tumor progression, there is a detachment of cancer cells from the primary tumor, followed by tumor cell adhesion to endothelial cells of venules in the target organs. After the extravasations occurs extracellular matrix invasion by tumor cells, these cells of primary lesions enter the lymphatics or the bloodstream depending on their anatomical location. In the circulation, many tumor cells are eradicated by physical forces exerted on them to pass through the microvasculature of secondary organs, and immunological mechanisms of action of host defense. Furthermore, once inside the target tissue tumor cells must find favorable conditions for survival and proliferation [16-18]. The biological characteristic that define tumor progression have been extensively described, although the underlying mechanisms are still not completely defined, however there are two theories have been proposed to explain how tumor cells invade secondary sites where metastasis occurs are the following [18-20]. The first is similar to the inflammatory process by cell adhesion and migration, while the second involves the aggregation of circulating tumor cells, and that these cells blocked blood vessels. In this theory in which a cell stably adhered frequently starts a homotypic aggregation, capturing other circulating tumor cells, followed by the formation of multicellular aggregates, these aggregates once grow and emerge from the primary tumor site, is carried out which triggers tumor progression in metastasis, where it requires a coordinated interaction of tumor cells and vascular endothelial cells, play a critical role in most of the events that characterize tumor progression and metastasis [21-23], so it is important to mention the general aspects of endothelial biology:

2.1. Endothelial Biology

The endothelium is the thin layer of cells that lines the interior surface of blood vessels and lymphatic vessels, forming an interface between circulating blood or lymph in the lumen and the rest of the vessel wall. The cells that form the endothelium are called endothelial cells, these cells have very distinct and unique functions that are paramount to vascular biology. These functions include fluid filtration, formation of new blood vessels in the angiogenesis, neutrophil recruitment. The endothelium acts as a semi-selective barrier between the vessel lumen and surrounding tissue, controlling the passage of materials and the transit of white blood cells, hormones into and out of the bloodstream. Excessive or prolonged increases in permeability of the endothelial monolayer, as in cases of chronic inflammation, may lead to tissue edema. It is also important in controlling blood pressure, blood coagulation, vascular tone, degradation of lipoproteins an in the secretion of growth factors and cytokines [24-25]. In recent decades, it has become clear that the endothelium of venules and smaller capillaries, and lymphatic vessels play a central role in the process of tumor growth, dissemination of metastatic cells, which is accompanied by the development of a characteristic tumor vasculature and tumors formed by endothelial cells [26].

There are two phenotypes endothelial (constitutive and activated);

2.1.1. The constitutive phenotype of endothelial cells

Quiescent, resting endothelial cells in the adult form a highly heterogeneous cell population that varies not only in different organs but also in different vessel calibers within an organ. Endothelium in the normal adult male, although being metabolically active, considered quiescent because the turnover of these cells is very low and this called: constitutive phenotype Fig (1).

In this condition, the apical membrane of endothelial cells exhibits a very low amount of intercellular adhesion molecules, so that no adhesion of cellular blood components to the vessel walls [27].

2.1.2. The activated phenotype of endothelial cells

Endothelial cell activation is associated with a number of distinct phenotype changes that, much like differentiation processes of the constitutive phenotype of endothelial cells, serve their need to adapt to functional requirements. [28] The cytokine-induced phenotype of endothelial cells during inflammation has been characterized most extensively in the last few years. When endothelial cells are activated by these cytokines are functional disorders involving immediate responses, for example, some pathological conditions such as sepsis, are associated with endothelial conversion to a phenotype activated [29-30]. The activated phenotype characterized by activation of constitutive nitric oxide synthase (NOS), also accompanied by changes such as increased expression of cell adhesion molecules (CAMs) and E-selectin (CD62E), ICAM-1(CD54), VCAM-1(CD106), P-selectin (CD62P) Fig (1). These changes allow the endothelium to participate in pathological conditions including inflammation, coagulation, cell proliferation, metastasis, tumor angiogenesis. All these cellular interactions are regulated by temporal and spatial presentation of various cell adhesion molecules and chemotactical molecules displaying appropriate specificity and affinity for proper development and functioning of the organism [31-32]. Has been postulated that this phenotype or variants of it, are involved in the processes of metastasis [33].

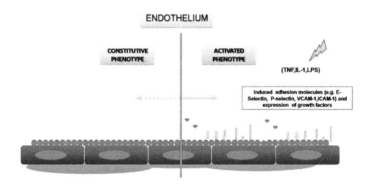

Figure 1. There are two phenotypes endothelial constitutive and activated.

2.2. Metastasis

Metastasis is the result of cancer cell adaptation to a tissue microenvironment at a distance from the primary tumor, is a complex process involving multiple steps: first, when cancer cells break away from the primary tumor, they invade the host stroma, intravasate into lymphatic or blood vessels, spread to the capillary bed of distant organs, where they invade into new surrounding tissues and proliferate to form secondary tumors [34-35]. When cancer is detected at an early stage, before it has spread, it can often be treated successfully by conventional cancer therapies such as surgery, chemotherapy, local irradiation, metastatic diffusion of cancer cells remains the most important clinical problem, because when cancer is detected after known to have metastasized, treatment are much less successful [36]. The metastatic capacity of tumor cells correlates with their ability to exit from the blood circulation, to colonize distant organs, and to grow in distant organs. Metastasis is a complex process that includes local infiltration of tumor cells into the adjacent tissue, transendothelial migration of cancer cells into vessels known as intravasation, survival in the circulatory system, extravasation and subsequent proliferation in competent organs leading to colonization [36-38]. Initially, tumor cell aggregates detachment from the primary tumor, next the cells actively infiltrate the surrounding stroma and enter into the circulatory system, traveling to distinct sites to establish the secondary tumor growth. In the bloodstream, a very small number of tumor cells survive to reach the target organ, indicating that metastasis formation must be regarded as a very ineffective event. Millions of carcinoma cells enter into the circulatory system, but the majority of them die during transportation, and only 1-5% of viable cells are successful in formation of secondary deposits in distinct sites [37-40]. Fig (2).

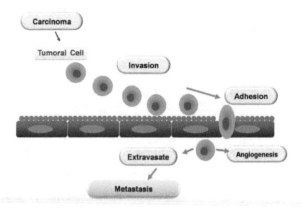

Figure 2. Steps in the Metastasis process.

Metastasis is facilitated by cell-cell interactions between tumor cells and the endothelium in distant tissues and determines the spread. Metastatic cells must act with the endothelium in three different stages of tumor progression: initially during the formation of blood vessels that enable tumor growth (vascularization), during the migration process that allows the passage from tissue into the bloodstream (intravasation), and finally during the process allowing extravasation into the target tissue [41-43]. Metastatic cancer cells require properties that allow them not only to adapt to a foreign microenvironment but also to subvert it in a way that is conducive to their continued proliferation and survival [36-38]. In addition, direct tumor cell interactions with platelets, fibroblasts and monocytes/macrophages, polymorphonuclear cells, soluble components, cytokines, chemokines, proteins of the extracellular matrix, growth factors, and other molecules secreted by host cells, significantly contribute to cancer cell adhesion, extravasation, and the establishment of metastatic lesions [44-47].

2.3. Cellular interactions in the inflammatory reaction and spread tumor

In the early stages of inflammation, neutrophils are cells that migrate to the site of inflammation under the influence of growth factors, cytokines and chemokines, which are produced by macrophages and mast cells residing in the tissue [48]. The process of cell extravasation from the bloodstream can be divided into four stages:

1. bearing

2. activation by stimulation chemoattractant

3. adherence

4. transendothelial migration.

If the inflammatory response is not regulated, the cellular response will become chronic and will be dominated by lymphocytes, plasma cells, macrophages metastasis, which is favored by the microenvironment of the organ target. The installation of tumor cells in blood vessels of the organ target to invade, is related to phenotypic changes in the endothelium allowing vascular extravasation of blood circulation of leukocytes in the inflammatory reaction and, as hypothesized current of tumor cells with metastatic capacity. The phenomenon of extravasation in response to a tumor cell interaction cell endothelial or not allowing the passage of cells whether there are appropriate conditions for the invasion with varied morphology [53-55].

Within the process of inflammation, a phenomenon is well-studied cell migration, which is the entrance of polymorphonuclear neutrophils and the vascular system. This involves a sequential mechanism of recognition, contact formation, and migration mediated by adhesion molecules such as (ICAM-1, VCAM-1, E and P Selectins, Integrins) it has been demonstrated that some of these adhesion molecules, such as E-selectin are not only involved in inflammation, but also in tumor metastasis and play a significant role in cancer progression and metastasis, in some cases of colon cancer. [56-58]. The expression of ICAM-1 has also proven to be a marker associated with an invasive phenotype [59].

Hanahanan et al. suggest that diversity of cancer cell genotypes is a manifestation of six basics alterations in cell physiology that together indicate development of malignant growth Fig (3), these alterations are shared in common by most all types of tumors [19]. In recent years, it has been demonstrated that metastatic dissemination can be influenced by inflammatory-reparative processes [46]. The interaction between these cell populations has been seen as part of a complex inflammatory microenvironment tumor-associated. Mantovani et al. have even postulated that this tumor microenvironment inflammatory plays an important role in tumorigenesis and tumor progression [60].

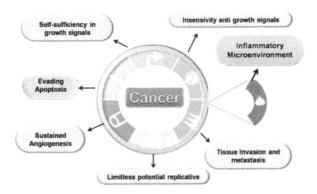

Figure 3. Capabilities of cancer and inflammation [19,60].

2.4. Inflammatory Microenvironment

The tumor microenvironment is composed of stromal fibroblasts, myofibroblasts, myoepithelial cells, macrophages, endothelial cells, leucocytes, and extracellular matrix (ECM) and soluble factors derived from tumor cells. This inflammatory environment surrounding a tumor promotes the breaking of the basal membrane, a process required for the invasion and migration of metastatic cells [60]. Tumor cells are also capable of produce cytokines and chemokines that facilitate evasion of the system immune and help to establishment and development of metastasis (Fig. 4). The increase of tumor-associated macrophages (TAMs) is associated with poor prognosis through various mechanisms: a) release by macrophage IL-10 and prostaglandin E2 which suppress antitumor response, b) easy to release angiogenic factors as VEGF, EGF, endothelin-2 and plasminogen activator promote tumor growth, c) to facilitate cell invasion metastasis by releasing matrix metalloproteinases and induce TNF production and vasodilatation enzyme nitric oxide synthase [61-62].

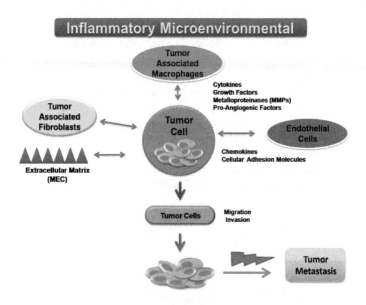

Figure 4. The tumor microenvironment and its role in promoting tumor growth

Cells grow within defined environmental sites and are subject to microenvironmental control. Outside of their sites, normal cells lack appropriate survival signals. During tumor development and progression, malignant cells escape the local tissue control and escape death. Diverse chemoattractant factors promote the recruitment and infiltration of these cells to the tumor microenvironment where they suppress the antitumor immunity or promote tumor angiogenesis and vasculogenesis.

TNF is expressed in low amounts by other cells such as fibroblasts, smooth muscle cells and tumor cells, his target cell are primary endothelial cells, inducing their activation by changing expression levels of some membrane proteins, primarily as adhesion molecules E-selectin, ICAM-1 and VCAM-1 whose expression and synthesis, are regulated by the transcription factor kB nuclear. Activated endothelial phenotype induced by TNF and characteristics of the inflammatory response, have served to comparing the endothelial phenotype has been observed, is produced in response to contact with soluble tumor factors [63-64].

The nuclear factor kappa B (NF-kB) is a transcription factor paramount in regulation of inflammatory response genes Early involved in cell-cell interaction, communication intercellular recruitment or transmigration, amplification of signals pathogenic and acceleration of tumorigenesis [52-53,65].

The study of tumor cells to modify their microenvironment has been a growing area of interest, which has identified the secretion of pro-inflammatory cytokines such as TNF, IL-6) and chemokines such as IL-8, it is interesting to note that these products are known modulators of endothelial function [53].

In recent years, it has been found that tumor cells secrete soluble factors, which modify the endothelial constitutive phenotype, and that exposure to these factors increase to a greater or less extent the capacity to adhere endothelial human tumor cells. It has been recognized that these soluble factors released by tumor cells or non-tumor cells surrounding the tumor play an important role in tumor progression [66].

Our group has shown that breast cancer cells, lymphomas, with high metastatic potentially induces a change in human endothelial cells (HUVECs) that is characterized by promoting a pro-adhesive endothelial phenotype, the expression of intercellular adhesion molecules (ICAM-1, VCAM-1 and E-Selectin) and the activation of NF-kB [66]. These studies include soluble factor leukemias (EUHE, Eusebia), cervical cancer (HeLa) and mammary gland cancer (MCF-7, ZR), oral cancer. In all cases, we have used primary cultures of human endothelial cells (HUVECs), which have generated an *in vitro* model to study the tumor microenvironment. This model is based on induction of a pro-adhesive endothelial phenotype that is associated with expression of adhesion molecules E-selectin, ICAM-1, VCAM dependent activation of NF-kB. These effects are considered essential in the process of adhesion and extravasation during the inflammatory reaction.

Moreover, we have analyzed the biochemical composition of the soluble factors derived from tumor cells. Proteins such as 27, cytokines, chemokines and growth factors associated with the inflammatory reaction; (IL-1 ra, IL-1 beta, IL-2, IL-4, IL-5, IL-6, IL-7, IL-9, IL-10, IL-12, IL-13, IL-15, IL-16, TNF, IFN-gamma, IP-10,RANTES), (IL-8, Eotaxina, MCP-1, MIP-1 beta, MIP-1 alpha), (G-CSF, GM-CSF, PDGF, FGF, VEGF). The molecules identified most significant expression in breast carcinomas, were (VEGF, GM-CSF, IL-1ra, IL6, IL8, IP10, RANTES), which play an important role in endothelial cell activation [52-53,65-66].

VEGF:

An unexpected finding was the abundant presence of the expression of vascular endothelial growth factor (VEGF) in breast carcinoma lines. Whereas in normal cells the expression of VEGF is dependent on a condition of hypoxia, surprisingly VEGF production by tumor cell lines and occurred at a concentration of oxygen partial indicating 20% indicative of impaired VEGF expression. In cancer constitutive secretion of VEGF independent of a condition that favors hypoxic tumor angiogenesis and growth in a clinical setting and is a marker of poor prognosis

[67]. In cancer, constitutive secretion of VEGF-independent hypoxic condition favors tumor angiogenesis and growth in a clinical setting and is a marker of poor prognosis [68].

GM-CSF

The involvement of GM-CSF (Granulocyte Macrophage Colony Stimulating Factor) in cancer is complex, since it seems to require the presence of other cytokines such as IL-4 and IL-6, and there are reports antagonistic to their involvement in tumorigenesis. For example, expression of GM-CSF and interleukin-12, inhibits the immune response and being expressed on B16 murine melanoma cells, decreasing completely tumorigenic capacity of syngenic mice C57B/6. In contrast to the line described breast epithelial MCF7-A benign growth factors secreted able to induce expression of IL-6 and GM-CSF in the tumor line R2T1AS breast cancer that is associated with a higher rate of growth and a higher tumorigenic capacity in vivo. Moreover, continued exposure of GM-CSF plus IL-4 of mesenchymal cells from human bone marrow, resulted in an increase in the morphological transformation and increased rate of growth both *in vitro* and *in vitro* mesenchymal cells, indicating the induction of a transformed phenotype. Whereas previous reports indicate that the involvement of GM-CSF requires other cytokines [69-70].

IL-1ra

It is reported that the soluble form of IL-1 receptor could serve as a functional antagonist of IL-1, the presence of this protein is controversial, since IL-1 by trapping an anti-inflammatory effect would reduce the adhesion of cells added by tumor cells and their extravasation, given this scenario could be proposed that tumor cells not only produce pro-inflammatory factors through the release of soluble factors tumor, but it can regulate the inflammatory process [71].

IL-6

The IL-6 is recognized as a classic inducer of the states of chronic inflammation and that promote the activation of vascular endothelium. It has also been reported that IL-6 is produced by tumors that develop metastases to the liver, as the case of colon and mammary gland cancer. The activity of this cytokine in the soluble factors tumor could be further enhanced by the presence of other co-factors secreted by cells [72-73].

IL-8

Expression of IL-8 in colorectal cancer favors an increase in tumorigenesis and metastasis, this increase is due to IL-8 is associated with expression of MMP-2 and MMP-9, the activity of these metalloproteases have been identified in physiological tissue remodeling processes like normal healing, but also participate in tissue remodeling associated with pathological processes including invasion [74]. Also in melanomas have been identified as IL-8 acts as an angiogenic factor and also promotes mitosis, therefore the IL-8 has been postulated as a potential therapeutic target. In particular, we have sought to interfere with IL-8 with the purpose of reducing tumor growth, an alternative that has developed is the use of an anti-IL-8 in nude mice with liver cancer. The results show that administration of neutralizing anti-IL8 significantly decreased tumor growth, even more interestingly this decrease is associated with decreased expression of MMP-2 and MMP-9. Something similar is observed using the same experimental treatment of melanoma with a decrease in angiogenesis [75]. In our results, we found increases in IL-8 in the soluble factors soluble breast cancer [66]. This indicates that IL-8 could be used as a marker associated with tumor progression, regardless of tumor type. In patients where they identified the production of IL-8, using neutralizing antibodies that interfere with their signaling could serve as a therapeutic alternative.

IP10

IP10 is a protein induced by interferon, has been reported that this protein inhibits proliferation and metastatic tumors, that expression of IP10 in patients with stages II and III colorectal cancer correlate with the development of metastasis and a poor prognosis. The detection of IP10 could be used as a prognostic marker in stage II and III colorectal cancer patients. This has led to propose that IP10 might be used as a prognostic marker in stage II and III colorectal cancer patients [76-77].

RANTES

Regulated upon activation normal T-cell expressed, and secreted is a chemokine that belongs to the CC class, which distinguishes it from IL8, which belongs to the CXC class. RANTES expression in tumor cells has been associated with tumorigenesis and is consistent with our finding of RANTES the products secreted by breast carcinomas. The presence of RANTES in the tumor microenvironment may be chemoattractant to tumor cells. From this point of view is interesting that the tumor-associated endothelial cells, when stimulated with MIF1-alpha can release RANTES [78]. Study that evaluated the expression of RANTES and its receptor CCR5 in 60 patients with metastatic gastric cancer, were identified elevated expression levels, where it is concluded that RANTES and its receptor may contribute to gastric cancer metastasis by promoting responses TH1 and TH2. By comparing a variety of biological markers in a group of biopsies of mammary gland cancer, RANTES was the only marker present in all biopsies [79].

The reported findings strengthen the idea that soluble factors of tumor microenvironment may be relevant in the final stages of the metastatic spread and that these effects may be mediated by cytokines, chemokines, and growth factors present in the soluble factors secreted by tumor cell lines. These elements found in high concentrations are known to be capable of inducing the activated phenotype of endothelial cells to a variety of physiological and pathological cellular responses.

3. Oxidative stress and free radicals: role in cancer development

During the inflammatory process macrophages and endothelial cells, generate a large amount of growth factors, cytokines and reactive oxygen species (ROS) and nitrogen (RNS) that can cause DNA damage. If macrophages and remain on the endothelium may allow the tissue damage continues chronic inflammation predisposes to malignancy [56,80].

3.1. Introduction

In different pathological process the cell injury is induced by free radicals, is an important mechanism of cell damage in many pathological conditions, such as chemical and radiation injury, ischemia injury, cellular aging and some immune system cells such as the phagocytes [82-84]. The free radicals are an example of instability in a biological system; namely, are chemical substances that they have an unpaired electron in its final orbit, this causes that their energy to be unstable and for they become stable they need for molecules which are adjacent to it either organic or inorganic for example: lipids, proteins, carbohydrates and nucleic acids essentials compounds in the cells mainly in the membrane and the core.

These types of chemical species may be either.

1.   Oxygen derived (reactive oxygen species ROS)

2.   Nitrogen derived (reactive nitrogen species RNS)

Reactive oxygen species (ROS) and reactive nitrogen (RNS) (Table 1) are created in the some cells such as the hepatocytes and in different normal physiologic processes, including, oxidative respiration, growth, regeneration, apoptosis and microbial killing by phagocytes [83]. The generation of this species chemical types, is normal in a normal cells; however, when these start to produce in excess and the antioxidant system is deficient, oxidative stress occurs. This causes damage cells: hepatocytes, kupffer cells and endothelial cells, through induction of inflammation, ischemia, fibrosis, apoptosis, necrosis or other atypical transformation in the cell structure and function.

Type of radical	Activating enzyme	Physiological process
Nitrogen derived (nitric oxide) Oxygen derived (ROS)	Nitric oxide synthase NADPH oxidase	Smooth muscle (control or vascular tone) and other cGMP- depended functions. (glycogenolysis, apoptosis, conductance regulator ion channels, vasodilatation and increased blood flow) Oxidation-reduction reactions within the cell, muscle relaxation, control of erythropoietin production, signal transduction from various receptors, enhancement of immunological functions and oxidative

Table 1. *Important physiological process involved with the free radicals.ROS, Reactive oxygen species.*

3.2. Reactive oxygen species

Reactive oxygen species are produced in normal condition them in a living cell during cellular respiration, energy production and various events of growth and cell death, these are degrade by the defensive systems. Therefore, the cells self-regulate their production and degradation of ROS is found transiently in the cell without causing any damage to the cellular level, and for that reason the cell maintains an equilibrium constant but as this production increases oxidative stress is generated, this relates whit different pathological process such as damage in the cell structure and function, degenerative process and cancer, also influence within the inflammation and the immune response, as these are generated by macrophages and neutrophils as mediators for the destruction of pathogens and dead tissue. The generation of free radicals can be made by different pathways [85-86]. (Table 2)

Redox reactions in metabolic processes.	During cellular respiration O_2 is reduced by four electrons to the transport of H_2 for generating two molecules of water through an oxidative enzyme which results is the formation of superoxide anion (electron), hydrogen peroxide (two electron) and hydroxyl ions (three electrons).
Absorption of radiant energy	Electromagnetic radiation (x rays), gamma rays, infrared, UV, microwave etc. These to hydrolyze the water and generate hydroxyl ions and hydrogen
Inflammation and immune response.	This response induced by activated leukocytes, that is caused by a protein complex located at the plasma membrane that employs NADPH oxidase and some intracellular oxidase and this generated a superoxide anion.
Metabolism of drugs	Most chemicals do not show biological activity in its native form these have to become toxic reactive metabolites to act on their target molecules. This is made by oxidases. These metabolites induce the formation of free radicals, such as acetaminophen.
Metals	The transmission metals donate or accept electrons intracellular free during the reactions thus causing the formation of free radicals
Nitric oxide	Nitric oxide is generated by different cells are an example of this are the endothelial cells giving rise to reactive species of nitrogen, such as the peroxynitrite anion ($ONOO^-$) also NO_2 and NO_3^-.

Table 2. The generation of free radicals can be made by different pathways.

3.3. Free radical and carcinogenesis

Free radicals are atoms or groups of atoms that in their atomic structure present one or more unpaired electrons in the outer orbit. These free radicals steal electrons from other molecules in effort to heal themselves, ultimately creating new free radicals in the process by stealing electrons. Free radicals are formed from a number of causes such as cigarette smoke, pollution, exposure to sunlight all cause the formation of free radicals. Other factors include normal daily processes like food digestion and breathing.

When increased production of reactive species and have a deficiency in the antioxidant system they cause significant neoplastic changes. In some diseases, such as Bloom syndrome develops lymphomas, leukemias and carcinomas, in anemia are implicated the production of these and alterations of antioxidant defense mechanisms at the systemic level [82-83]. Some epidemiological information indicates that tumor incidence is lower in populations where the diet is rich in antioxidants like fruits and vegetables [84].

Tumor cells have a high activity of free radical formation in contrast to healthy cells. It is known that tumor cells not only produce oxygen peroxide (H_2O_2) but also decreases the production of antioxidants such as glutathione peroxidase and SOD.

Some pathways by which cancer cells have high amounts of ROS is for multiple factors:

Increasing the metabolic activity of a neoplastic cell, lead to the increased energy requirements (ATP) produced in the mitochondria this in order to enhance the growth and proliferation.

1. The progression of cancer, primarily because of the damage they cause in to the genetic material of a normal cell. Has been shown that the oxidation of guanine to 8-oxo-dG (oxidation product DNA) induces errors in their replication. By dependent of DNA polymerase that generates the nitrogenous base pairing, not complementary, to permit the establishment of hydrogen bonds with (A) adenine and (T) thymine [85].

2. Oncogenic transformation is conditioned by the presence of mutated genes or oncogenes that control essential cellular functions in which the redox state within or outside the cellular microenvironment is very important ROS are potential carcinogens because they facilitate mutagenesis, tumor growth, and metastasis and all those process that has been showed [86].

Cancer is a multifactorial disease where endogenous and exogenous factors are involved but the roles played by free radicals in this disease are very important because, these produce damage in de DNA structure and this produces an important negative effects [86].

4. Antioxidants and Chemoprevention in cancer

Antioxidants are substances that prevent damage to cells caused by free radicals, it can cause damage to DNA, leading to the possible development of cancer [87]. Antioxidants search for these free radicals and lend them an, this stabilizes the molecule, thus preventing damage to other cells. Antioxidants also turn free radicals into waste by products, and they eventually are eliminated from the body. The inability of our body to neutralize free radicals we are exposed daily forces us to rely on foods with antioxidant properties capable of neutralizing them [88].

4.1. Flavonoids

Flavonoids are found in numerous plants and vegetables, with a wide distribution through the plant kingdom. This class compounds numbers more than 4000 members and can be divided into five subcategories: flavones, monomeric flavanols, flavanones, flavonols and anthocyanidines. Are natural compounds chemically derivate from bezo-y-pirone (phenylchromone) or flavone. They are considered important constituents of the human diet. It has been reported that they exert multiple biological effects due to their antioxidant and free radical-scavenging abilities [89].

Flavanoids possess a lot of pharmacological and therapeutically properties; antioxidant, antitumor, antiangiogenic (vascular protective), anti-inflammatory, antiallergic, antihepatotoxic, anticancerigenic, antimutagenic, anticancer effects, antiosteoporotic, and antiviral properties [90]. Many studies emphasize take adequate diets that are active allies against cancer. These diets are based on enzymes and antioxidant substances in certain foods that are rich in components that collect above [91].

They also have the ability to repair previous damage to cells, examples of antioxidants include (beta-carotene, lycopene, vitamins C, E, and A), and other substances. Nutrients such as; green tea, flavonoids, vitamins C, E, and Beta-carotene in the carcinogen process, has been showed that have function in the elimination of carcinogenic factors, inhibition of precarcinogens and reparation of DNA damage. The mechanisms are diverse and range from inhibition to an active reaction of the immune system in general. This has caused the use of multiple antioxidant micronutrients as preventive agents [90]. Several experimental data have demonstrated the antiproliferative and anti-carcinogenic and the role of chemopreventive agent of flavonoids [91-92].

Currently investigations are performed to determine the mechanisms by which act flavonoids, because it has been observed that their effects are greater at high doses, which gives them inducing side effects, so it is important to moderate their consumption by a balanced diet.

5. Conclusions

It is important to analyze the role of tumor-associated inflammatory microenvironment and has been identified that plays an important role in tumor progression. This microenvironment is composed of molecules that play an important role in inflammatory processes and chronic, and favor the invasion and metastasis process that triggers the death of many people with any cancer.

The installation of tumor cells in blood vessels of the target organ to invade, is related to phenotypic changes in the endothelium allowing vascular extravasation of blood circulation of leukocytes in the inflammatory reaction and, as hypothesized current of tumor cells with metastatic capacity. The phenomenon of extravasation in response to a cell interactions between tumor cells and endothelial cells or not allowing the passage of cells whether there are appropriate conditions for the invasion.

Understanding the molecular basis of these interactions between metastatic cells and endothelial cells, will enable us to design strategies to interfere with this inter-cellular communication. It is important to recognize the tumor-associated inflammatory microenvironment and what is the contribution to tumor progression. The importance of these factors on endothelial activation being evaluated by reconstituting the mixture with cytokines, chemokines and growth factors recombinant depleted mixtures of tumor soluble factors of each of these proteins by specific monoclonal antibodies.

Is important mention that during the inflammatory process macrophages, fibroblasts and endothelial cells generate a large amount of growth factors, cytokines, chemokines and reactive oxygen species (ROS) and nitrogen (RNS) that can cause DNA damage. These process allow the tissue damage continues chronic inflammation predisposes to malignancy. Therefore, it is important to note that people with chronic degenerative diseases, which clearly show chronic inflammatory processes, they may promote or contribute to present or develop a tumor lesion.

The use of antioxidants consumed in a balanced diet can be used as an element in the diet that can become a preventive or contributing to diminish the appearance of a tumor lesion.

Author details

César Esquivel-Chirino[1,4,5,6*], Jaime Esquivel-Soto[1,6], José Antonio Morales-González[2], Delina Montes Sánchez[3,4,5], Jose Luis Ventura-Gallegos[4,5], Luis Enrique Hernández-Mora[1] and Alejandro Zentella-Dehesa[4,5]

*Address all correspondence to: cesquivelch@gmail.com

1 Facultad de Odontología, Universidad Nacional Autónoma de México, México

2 Instituto de Ciencias de la Salud, Universidad Autónoma del Estado de Hidalgo (UAEH), México

3 Programa de Genómica Funcional de Procariotes, Centro de Ciencias Genómicas, Universidad Nacional Autónoma de México, Campus Morelos, México

4 Departamento de Medicina y Toxicología Ambiental, Instituto de Investigaciones Biomédicas. Universidad Nacional Autónoma de México, México

5 Departamento de Bioquímica. (INNCMSZ) Instituto Nacional de Ciencias Médicas y Nutrición "Salvador Zubirán" México D.F., México

6 Facultad de Odontología, Universidad Intercontinental, México

References

[1] Ferlay, J., Bray, P., Pisani, P., & Parkin, D. M. (2004). GLOBOCAN 2002: Cancer incidence, mortality and prevalence worldwide. *IARC Cancer Base Version 2.0. Lyon: IARC Press* [5].

[2] World Health Organization. (2008). World cancer report 2008. *Lyon (France): IARC.*

[3] World Health Organization. (2007). Ten statistical highlights in global public health. *World health statistics 2007. Geneva: WHO.*

[4] Jemal, A., Ward, E., & Thun, M. (2005). Cancer statistics. *In: DeVita VJ, Hellman S, Rosenberg S, editors. Cancer principles and practice of oncology. 7th ed. Baltimore (MD): Lipppincott Williams & Wilkins*, 226-240.

[5] International Agency for Cancer Research (IARC). (2012). CANCER Mondial. available from: http://www-dep.iarc.fr/Accessed on June 22

[6] Jemal, A., Bray, F., Center, M. M., Ferlay, J., Ward, E., & Forman, D. (2011). Global cancer statistics. *CA Cancer J Clin*, 61, 69-90.

[7] Kanavos, P. (2006). The rising burden of cancer in the developing world. *Ann Oncol*, 17(8), 15-23.

[8] Kolonel, L., Wilkens, L., Schottenfeld, D., & Fraumeni, J. F. Jr. (2006). Cancer epidemiology and prevention. *3rd ed Oxford: Oxford University Press*, 189-201.

[9] Bacac, M., & Stamenkovic, I. (2008). Metastatic cancer cell. *Annu Rev Pathol*, 3, 221-47.

[10] Jackson, S. P., & Bartek, J. (2009). The DNA-damage response in human biology and disease. *Nature*, 461, 1071-1078.

[11] Pantel, K., Brakenhoff, R. H., & Brandt, B. (2008). Detection, clinical relevance and specific biological properties of disseminating tumor cells. *Nat Rev Cancer*, 8, 329-340.

[12] Hanahan, D., & Weinberg, R. (2000). The hallmarks of cancer. *Cell*, 100, 57-70.

[13] Oldham Hikman, Elizabeth. (2004). Intrinsic oxidative stress in cancer cells a biological basis for therapeutic selectivity" Cancer". *Cancer chemother pharmacol*, 53, 209-19.

[14] Steinmetz, K. A., & Potter, J. D. (1991). Vegetables, fruit, and cancer. *I. Epidemiology. Cancer Causes Control.*, 2, 325-357.

[15] Foulds, L. (1954). The experimental study of tumor progression: a review. *Cancer Res*, 14, 327-339.

[16] Ann, F., Chambers, Alan. C., & Groom, Iand. Mc Donald. (2002). Dissemination and growth cancer cells in metastatic sites. *Nature Reviews*, 2.

[17] Alby, L., & Auerbach, R. (1984). Differential adhesion of tumor cells to capillary endothelial cells in vitro. *Proc Natl Acad Sci USA*, 81, 5739-43.

[18] Balkwill, F., & Mantovani, A. (2001). Inflammation and cancer: back to Virchow? *Lancet*, 357, 539-45.

[19] Hanahan, D., & Weinberg, R. A. (2011). Hallmarks of cancer: the next generation. *Cell.*, 44, 646-674.

[20] Chaffer, C.L, & Weinberg, R. A. (2011). A perspective on cancer cell metastasis. Science. Mar 25; , 331(6024), 1559-64.

[21] Nicolson, G. L. (1993). Cancer progression and growth: relationship of paracrine and autocrine growth mechanisms to organ preference of metastasis. *Exp Cell Res*, 204, 171-80.

[22] Kopfstein, L., & Christofori, G. (2006). Metastasis: cell-autonomous mechanisms versus contributions by the tumor microenvironment. *Cell Mol Life Sci.*, 63(4), 449-68.

[23] Calorini, Lido, & Bianchini, Francesca. (2010). Environmental control of invasiveness and metastatic dissemination of tumor cells: the role of tumor cell-host cell interactions. *Cell Communication and Signaling*, 8, 24.

[24] Tang, D. G., & Conti, C. J. (2004). Endothelial cell development, vasculogenesis, angiogenesis, and tumor neovascularization: an update. *Semin Thromb Hemost.*, 30(1), 109-17.

[25] Aird, W. C. (2009). Cell Tissue Res. Molecular heterogeneity of tumor endothelium. Epub 2008 Aug 23., 335(1), 271-81.

[26] Risau, W. (1995). Differentiation of endothelium. *FASEB J*, 9, 926-33.

[27] Ribatti, Domenico, Nico, Beatrice, Vacca, Angelo, Roncali, Luisa, & Dammacco, Franco. (2002). Journal of Hematotherapy & Stem. *Cell Research.*, 11(1), 81-90.

[28] Augustin, H. G., Kozian, D. H., & Johnson, R. C. (1994). Differentiation of endothelial cells: Analysis of the constitutive and activated endothelial cell phenotypes. *Bioessays*, 16, 901-906.

[29] Geng, J. G. (2003). Interaction of vascular endothelial cells with leukocytes, platelets and cancer cells in inflammation, thrombosis and cancer growth and metastasis. *Acta Pharmacol Sin.*, 24(12), 1297-300.

[30] Pober, J.S. (2002). Arthritis Res. 4(3), 109-16, *Epub May 9. Endothelial activation: intracellular signaling pathways.*

[31] Riscoe, D. M., Cotran, R. S., & Pober, J. S. (1992). Effects of tumor necrosis factor, lipopolysaccharide, and IL-4 on the expression of vascular cell adhesion molecule-1 in vivo. *Correlation with CD3+ T cell infiltration. J Immunol.*, 149, 2954-2960.

[32] Joan, M., & Cook-Mills, Tracy. L. Deem. (2005). Active participation of endothelial cells in inflammation. *J Leukoc Biol.*, 77(4), 487-495.

[33] Wagner, M., Bjerkvig, R., Wiig, H., Melero-Martin, J. M., Lin, R. Z., Klagsbrun, M., & Dudley, A. C. (2012). Inflamed tumor-associated adipose tissue is a depot for macrophages that stimulate tumor growth and angiogenesis. Angiogenesis. May 22.

[34] Bendas, G, & Borsig, L. (2012). Cancer cell adhesion and metastasis: selectins, integrins, and the inhibitory potential of heparins. *Int J Cell Biol*, 676-731.

[35] Chambers, A. F., Groom, A. C., & Mac Donald, I. C. (2002). Dissemination and growth of cancer cells in metastatic sites. *Nature Reviews Cancer*, 2(8), 563-572.

[36] Calorini, Lido, & Bianchini, Francesca. (2010). Environmental control of invasiveness and metastatic dissemination of tumor cells: the role of tumor cell-host cell interactions. Cell Communication and Signaling, , 8, 24.

[37] Rubin, H. (2008). Contact interactions between cells that suppress neoplastic development: can they also explain metastatic dormancy? *Adv Cancer Res*, 100, 159-202.

[38] Aguirre-Ghiso, J. A. (2007). Models, mechanisms and clinical evidence for cancer dormancy. *Nat Rev Cancer*, 7, 834-846.

[39] Nicolson, G. L. (1993). Cancer progression and growth: relationship of paracrine and autocrine growth mechanisms to organ preference of metastasis. *Exp Cell Res*, 204, 171-80.

[40] Gasic, G. J. (1986). Role of plasma, platelets and endothelial cells in tumor metastasis. *Cancer Metastasis Rev*, 3, 99-116.

[41] De Visser, K. E., Korets, L. V., & Coussens, L. M. (2005). De novo carcinogenesis promoted by chronic inflammation is B lymphocyte dependent. *Cancer Cell*, 7, 411-423.

[42] Di Carlo, E., Forni, G., Lollini, P., Colombo, M. P., Modesti, A., & Musiani, P. (2001). The intriguing role of polymorphonuclear neutrophils in antitumor reactions. *Blood*, 97, 339-345.

[43] De Larco, J. E., Wuertz, B. R., & Furcht, L. T. (2004). The potential role of neutrophils in promoting the metastatic phenotype of tumors releasing interleukin-8. *Clin Cancer Res*, 10, 4895-4900.

[44] Silzle, T., Randolph, G. J., Kreutz, M., & Kunz-Schughart, L. A. (2004). The fibroblast: sentinel cell and local immune modulator in tumor tissue. *Int J Cancer*, 108, 173-180.

[45] Kalluri, R., & Zeisberg, M. (2006). Fibroblasts in cancer. *Nat Rev Cancer*, 6, 392-401.

[46] Nicolson, G. L. (1993). Cancer progression and growth: relationship of paracrine and autocrine growth mechanisms to organ preference of metastasis. *Exp Cell Res*, 204, 171-80.

[47] Mehlen, P., & Puisieux, A. (2006). Metastasis: a question of life or death Nat. *Rev. Cancer*, 6, 449-458.

[48] Fidler, I. J. (2005). Cancer biology is the foundation for therapy. *Cancer Biol. Ther.*, 4, 1036-1039.

[49] Almog, N. (2010). Molecular mechanisms underlying tumor dormancy. *Cancer Lett*, 294, 139-146.

[50] Klein, C. A. (2009). Parallel progression of primary tumours and metastases. *Nat. Rev. Cancer*, 9, 302-312.

[51] Lu, H., Ouyang, W., & Huang, C. (2006). Inflammation, a key event in cancer development. *Mol Cancer Res*, 4, 221-233.

[52] Lopez-Bojorquez, L. N. (2004). Regulation of NF-kappaB transcription factor. *A molecular mediator in inflammatory process. Rev Invest Clin.*, 56, 83-92.

[53] Lopez-Bojorquez, L. N., Arechavaleta-Velasco, F., Vadillo-Ortega, F., Montes-Sanchez, D., Ventura-Gallegos, J. L., & Zentella-Dehesa, A. (2004). NF-kappaB translocation and endothelial cell activation is potentiated by macrophage released signals cosecreted with TNF-alpha and IL-1beta. *Inflamm Res.*, 53, 567-575.

[54] Smid, M., Wang, Y., Zhang, Y., Sieuwerts, A. M., Yu, J., Klijn, J. G., Foekens, J. A., & Martens, J. W. (2008). Subtypes of breast cancer show preferential site of relapse. *Cancer Res*, 68, 3108-3114.

[55] Wu, J. M., et al. (2008). Heterogeneity of breast cancer metastases: comparison of therapeutic target expression and promoter methylation between primary tumors and their multifocal metastases. *Clin Cancer Res*, 14, 1938-1946.

[56] O'Hanlon, D. M, Fitzsimons, H, Lynch, J, Tormey, S, Malone, C, & Given, H. F. (2002). Soluble adhesion molecules (E-selectin, ICAM-1 and VCAM-1) in breast carcinoma. *Eur J Cancer.*, 38, 2252-2257.

[57] Kim, I., Moon, S. O., Kim, S. H., Kim, H. J., Koh, Y. S., & Koh, G. Y. (2001). Vascular endothelial growth factor expression of intercellular adhesion molecule 1 (ICAM-1), vascular cell adhesion molecule 1 (VCAM-1), and E-selectin through nuclear factor-kappa B activation in endothelial cells. *J Biol Chem.*, 276, 7614-7620.

[58] Lieder, A. M., Prior, T. G., Wood, K. J., & Werner, J. A. (2005). The relevance of adhesion molecules in the classification of 72 squamous cell carcinoma of the head and neck. *Anticancer Res*, 25, 4141-4147.

[59] Okegawa, T, Li, Y, Pong, R. C., & Hsieh, J. T. (2002). Cell adhesion proteins as tumor suppressors. *J Urol.*, 167, 1836-1843.

[60] Colotta, F., Allavena, P., Sica, A., Garlanda, C., & Mantovani, A. (2009). Cancer-related inflammation, the seventh hallmark of cancer: links to genetic instability. *Carcinogenesis.*, 30, 1073-81.

[61] Kawaguchi, T. (2005). Cancer metastasis: characterization and identification of the behavior of metastatic tumor cells and the cell adhesion molecules, including carbohydrates. *Curr Drug Targets Cardiovasc Haematol Disord*, 39-64.

[62] Thorne, R. F., Legg, J. W., & Isacke, C. M. (2004). The role of the CD44 transmembrane and cytoplasmic domains in coordinating adhesive and signalling events. *J Cell Sci.*, 117, 373-380.

[63] Li, A., Li, H., Jin, G., & Xiu, R. (2003). A proteomic study on cell cycle progression of endothelium exposed to tumor conditioned medium and the possible role of cyclin D1/E. *Clin Hemorheol Microcirc.*, 29, 383-390.

[64] Watts, M. E., Parkins, C. S., & Chaplin, D. J. (2002). Influence of hypoxia and tumour-conditioned medium on endothelial cell adhesion molecule expression in vitro. *Anticancer Res.*, 22, 953-958.

[65] Estrada-Bernal, A., Mendoza-Milla, C., Ventura-Gallegos, J. L., Lopez-Bojorquez, L. N., Miranda-Peralta, E., Arechavaleta-Velasco, F., Vadillo-Ortega, F., Sanchez-Sanchez, L., & Zentella-Dehesa, A. (2003). NF-kappaB dependent activation of human endothelial cells treated with soluble products derived from human lymphomas. *Cancer Lett*, 191, 239-48.

[66] Montes-Sanchez, D., Ventura, J. L., Mitre, I., Frias, S., Michan, L., Espejel-Nunez, A., Vadillo-Ortega, F., & Zentella, A. (2009). Glycosylated VCAM-1 isoforms revealed in 2D western blots of HUVECs treated with tumoral soluble factors of breast cancer cells. *BMC Chem Biol*, 9, 7.

[67] Baldewijns , M. M., van Vlodrop, I. J., Vermeulen, P. B., Soetekouw, P. M., van Engeland, M., & de Bruïne, A. P. (2010). VHL and HIF signalling in renal cell carcinogenesis. *J Pathol.*, 221(2), 125-38.

[68] Kamada, H., Tsutsumi, Y., Kihira, T., Tsunoda, S., Yamamoto, Y., & Mayumi, T. (2000). In vitro remodeling of tumor vascular endothelial cells using conditioned medium from various tumor cells and their sensitivity to TNF-alpha. *Biochemical and Biophysical Research Communications*, 268, 809-813.

[69] Edeline, J., Vigneau, C., Patard, J. J., & Rioux-Leclercq, N. (2010). Signalling pathways in renal-cell carcinoma: from the molecular biology to the future therapy]. *Bull Cancer*, 97, 5-15.

[70] Kulbe, H., Chakravarty, P., Leinster, D. A., Charles, K. A., Kwong, J., Thompson, R. G., Gallagher, W. M., Galletta, L., Salako, M. A., Smyth, J. F., Hagemann, T., Brennan, D. J., Bowtell, D. D., & Balkwill, F. R. (2011). A dynamic inflammatory cytokine network in the human ovarian cancer microenvironment. *Cancer Res.*

[71] Friberg, E., Orsini, N., Mantzoros, C. S., & Wolk, A. (2007). Diabetes mellitus and risk of endometrial cancer: a meta-analysis. *Diabetologia.*, 50(7), 1365-74.

[72] Mc Lean, M. H., Murray, G. I., Stewart, K. N., Norrie, G., Mayer, C., Hold, G. L., Thomson, J., & El -Omar, E. M. (2011). The inflammatory microenvironment in colorectal neoplasia. PLoS One. Jan 7 , 6(1), e15366.

[73] Rajkumar, T., Shirley, S., Raja, U. M., & Ramakrishnan, S. A. Identification and validation of genes involved in gastric tumorigenesis. *Cancer Cell Int.*, 10, 45.

[74] Huang, S., Mills, L., Mian, B., Tellez, C., Mc Carty, M., Yang, X. D., Gudas, J. M., & Bar-Eli, M. (2002). Fully humanized neutralizing antibodies to interleukin-8 (ABX-IL8) inhibit angiogenesis, tumor growth, and metastasis of human melanoma. *Am J Pathol.*, 161(1), 125-34.

[75] Mc Conkey, D. J., & Bar-Eli, M. (2003). Fully human anti-interleukin 8 antibody inhibits tumor growth in orthotopic bladder cancer xenografts via down-regulation of matrix metalloproteases and nuclear factor-kappaB. *Clin Cancer Res.*, 9(8), 3167-75.

[76] Jiang, Z., Xu, Y., & Cai, S. (2010). CXCL10 expression and prognostic significance in stage II and III colorectal cancer. *Mol Biol Rep.*, 37(6), 3029-36.

[77] Utoguchi, N. H., Makimoto, Y., Wakai, Y., Tsutsumi, S., Nakagawa, , & Mayumi, T. (1996). Effect of tumour cell-conditioned medium on endothelial macromolecular permeability and its correlation with collagen. *British Journal of Cancer*, 73, 24-28.

[78] Cao, Z., Xu, X., Luo, X., Li, L., Huang, B., Li, X., Tao, D., Hu, J., & Gong, J. J. (2011). Huazhong Role of RANTES and its receptor in gastric cancer metastasis. Univ Sci Technolog Med Sci. Epub Jun, 31(3), 342-7.

[79] The high level of RANTES in the ectopic milieu recruits macrophages and induces their tolerance in progression of endometriosis. (2010). *J Mol Endocrinol*, 45, 291-299.

[80] Lu, H., Ouyang, W., & Huang, C. (2006). Inflammation, a key event in cancer development. *Mol Cancer Res*, 4, 221-233.

[81] Yan, B., Wang, H., Rabbani, Z. N., Zhao, Y., Li, W., Yuan, Y., Li, F., Dewhirst, M. W., & Li, C. Y. (2006). Tumor necrosis factor-alpha is a potent endogenous mutagen that promotes cellular transformation. *Cancer Res*, 66, 11565-11570.

[82] Kumar et.al. (2010). Robbins and Cortan structural and functional pathology. *ed. the eighth edition. Elsevier sounders Barcelona Spain.*, 19-20.

[83] Hikman, Elizabeth Oldham. (2004). Intrinsic oxidative stress in cancer cells a biological basis for therapeutic selectivity" Cancer". *Cancer chemother pharmacol*, 53, 209-19.

[84] Cerutti, P. (1985). Pro-oxidant states and tumor promotion. *Science*, 227, 375-80.

[85] Migliori, L., et al. (1991). Genetic and environmental factors in cancer an neurodegenerative disease. *Mut Res*, 202(512), 135-153.

[86] Kouchakgjian, M., et al. (1991). MR structural studies of the ionizing radiation adduct 7-hydro-8oxodeoxyguanosine (8-oxo-7H-dG) opposites deoxyadenisine in a D A duplex 8-oxo-7H-7dG(syn)-dA(anti)aligment a lesion site. *Biochem*, 30, 1403.

[87] Elejalde Guerra, J. I. (2001). Oxidative Stress, diseases and antioxidants treatments. *An Med Int (Madrid-Spain)*, 18, 326-335.

[88] Aherne, S. A., & y O'Brien, N. M. (2002). Dietary flavonols: chemistry, food content, and metabolism. *Nutrition*, 18, 75-81.

[89] Jovanovic, S. V., Steenken, S., Simic, M. G. y., & Hara, Y. (1998). Antioxidant properties of flavonoids: reduction potentials and electron transfer reactions of flavonoid radicals. En: Rice Evans C, Parker L (eds.): Flavonoids in health and disease. *Marcel Dekker, Nueva York*, 137-161.

[90] Letan, A. (1966). The relation of structure to antioxidant activity of quercitin and some of its derivates. *J Food Sci*, 31, 518-523.

[91] Stahl, W., Ale-Agha, N. Y., & Polidori, M. C. (2002). Non-antioxidant properties of carotenoids. *Biol Chem*, 383, 553-558.

[92] Stacvric, B. (1994). Quercitin in our diet: From potent mutagen to probable anticarcinogen. *Clinical Biochemistry*, 27, 245-248.

[93] Da Silva, J., Herrmann, S. M., Peres, W., Possa, Marroni. N., Gonzalez Gallego, J. Y., & Erdtmann, B. (2002). Evaluation of the genotoxic effect of rutin and quercetin by comet assay and micronucleus test. *Food Chem Toxicol*, 40, 941-947.

Chapter 7

The Chemoprevention of Chronic Degenerative Disease Through Dietary Antioxidants: Progress, Promise and Evidences

Eduardo Madrigal-Santillán,
Eduardo Madrigal-Bujaidar, Sandra Cruz-Jaime,
María del Carmen Valadez-Vega,
María Teresa Sumaya-Martínez,
Karla Guadalupe Pérez-Ávila and
José Antonio Morales-González

Additional information is available at the end of the chapter

1. Introduction

1.1. Epidemiology of chronic degenerative diseases in Mexico and the world

During the last 30 years relevant changes in the public health field have arisen worldwide, among which the most representative are observed in developed countries where a big deal of infectious diseases have been reduced and controlled as a result of the creation and intro-duction of powerful antibiotics [1].

In countries such as Australia, Austria, Belgium, Canada, Denmark, Finland, France, Ger-many, Greece, Ireland, Italy, Japan, Luxembourg, Netherlands, New Zealand, Norway, Portugal, Spain, Sweden, Switzerland, the United Kingdom, and the United States, incor-porated to the OECD (Organization for Economic Cooperation and Development), mortal-ity due to those diseases has diminished up to 38% in people between 35 and 69 years old. Likewise, the risk of mortality before age 70 has diminished up to the 23%. Those re-ductions have been the result of social changes and of the improvement of preventive methods of infectious diseases. However, in recent years the prevalence of chronic degen-erative diseases has increased [1].

Chronic degenerative diseases (CDDs) represent a problem of public health for they have become the cause of death worldwide both in adolescents and adults. Among the most prevalent CDDs worldwide is obesity, the cardiovascular diseases (such as hypertension, atherosclerosis), heart diseases, diabetes, chronic respiratory diseases, and cancer; which have caused the 60% of the 58 million yearly deaths, which are approximately 35 million people death for these diseases between 2005 and 2007 [2].

Prevalence in chronic degenerative diseases results from different factors, among which the technologic advance and modernization affect life styles where an increase in process-ed foods consumption with a high level of fat content, a sedentary lifestyle since child-hood, alcohol and tobacco, stress and a lack of culture in terms of damage prevention and health risks [1,2].

Mexico does not escape this situation as a result of specific factors to our country such as economic development, concentration of population in urban areas, lack of support to im-prove the health services and the limitations in preventive programs, particularly in the population under 10 years. Besides, there is a transformation of the population pyramid due to a reduction in mortality and a decrease in birth rate; both phenomena are identi-fied as epidemiologic and demographic transitions [2]. In México, the morbidity data pro-duced by the CDDs are taken from the statistics of the healthcare sector and published by the healthcare ministry. Although in those reports not all the existing cases are included (not all patients request healthcare services), they are a good help to understand the dam-age behavior along with other indicators of prevalence that estimate the number of cases in the population within a specific period of time. Such indicators are obtained from the national healthcare survey and from the national healthcare and nutrition survey 2006 [2]. On the other hand, the mortality statistics are considered as more reliable due to the per-manent job in updating the database. The information is obtained from the records of the national institute of statistics, geography and informatics (INEGI) and the general bureau of health information, in conjunction with the epidemiological AVAD index, which is a measure that combines years of healthy life lost due to premature mortality and years of life lost due to disability [3].

As mentioned above, the epidemiological and demographic transitions are important factors for the prevalence of chronic degenerative diseases and indicate changes in the behavior of population dynamics, as well as damage to health which are the result of the low socioeco-nomic development and the impact of government policies on public health. The demo-graphic transition shows the change in a steady state population with high fertility and mortality associated with the low socioeconomic development process and/or moderniza-tion. This process is irreversible and was constructed from the first countries reaching socio-economic development in Europe such as France and England. In recent years it has made rapid changes affecting the world population [2].

According to data from INEGI and the national population council (CONAPO), Mexico has experienced an accelerated process of demographic transition, which has influenced the eco-nomic development and migration, leading to a reduction in mortality and a parallel high birth rates, as well as the consequent population growth, so it is estimated that between 2010 and 2050 the proportion of elderly people in Mexico will grow from 7% to 28% and with it the possibility of an chronic degenerative disease is greater [2,3].

In the case of the epidemiological transition, this is characterized by a reduction of morbidity and mortality from transmissible diseases and an increase in chronic degenerative diseases. In recent years, this parameter has shown that in both developed and developing countries, the proportion of infectious diseases in individuals over age 15 is stable, but unfortunately the CDDs are increasing, showing that they occupy almost half of value of morbility globally. A relevant fact is observed in developed countries (like France, Germany, Japan, United Kingdom, and United States) where the greatest impact of transmissible diseases remains the HIV/AIDS; but the cerebrovascular diseases and the ischemic heart disease are among the main causes of morbidity and mortality (Table 1) in individuals over age 15, both diseases represent more than 36% of deaths worldwide [2].

In the specific case of Mexico, it is well-known that infectious diseases made up the profile of mortality in the fifties, since half of the deaths were caused by diarrhea and respiratory infections, for reproductive problems and associated malnutrition conditions. Nowadays, these diseases (classified as lag diseases) are concentrated in less than 15% of deaths [2].

In the last 10 years, there has been an overlap between lag diseases and the so-called emerging diseases. Thus, the epidemiological transition has ranked the chronic degenerative diseases among the 10 leading causes of death, highlighting the type 2 diabetes, obesity, cardiovascular diseases, malignant neoplasms and cerebrovascular diseases [4].

Mortality (individuals between 15 and 50 years)		Deaths (thousands)	Mortality (over 60 years)		Deaths (thousands)
	Cause			Cause	
1	HIV / AIDS	2279	1	Ischemic heart disease	5825
2	Ischemic heart disease	1332	2	Cerebrovascular diseases	4689
3	Tuberculosis	1036	3	Chronic obstructive pulmonary disease	2399
4	Injuries from traffic accidents	814	4	Infections of lower respiratory	1396
5	Cerebrovascular diseases	783	5	Trachea and lung cancer	928
6	Self-harm	672	6	Diabetes	754
7	Violence	473	7	Hypertensive heart disease	735
8	Liver cirrhosis	382	8	Stomach Cancer	605
9	Infections of lower respiratory	352	9	Tuberculosis	495
10	Chronic obstructive pulmonary disease	343	10	Colorectal cancer	477

Table 1. Leading causes of death in people over 15 years in the world (as a function of AVAD index)

2. Definition, importance and control of oxidative stress

The term "oxidative stress" was first introduced in the eighties by Helmut Sies (1985), defining it as a disturbance in the prooxidant-oxidant balance in favor of the first. From that time, a great number of researchers have studied this phenomenon; so, the concept has evolved and now, has been defined as "A situation when steady-state ROS concentration is transiently or chronically enhanced, disturbing cellular metabolism and its regulation and damaging cellular constituents" [5,6].

However, oxidative stress is a phenomenon not entirely detrimental for the organism; also, free radicals (FR) have an important function in several homeostatic processes. They act as intermediate agents in essential oxidation-reduction (redox) reactions. Some examples are the destruction of microorganisms through phagocytosis, synthesis of inflammatory mediators and detoxification. Therefore, FR in low concentrations are useful and even essential [7].

FR represents any chemical species that exists independently and has one or more unmatched (odd) electrons rotating in its external atomic orbits. This highly unstable configuration causes this chemical species to be very aggressive and to have a short life span. Once generated, FR interact with other molecules through redox reactions to obtain a stable electronic configuration [8-10].

Several authors have classified FR according to the functional group in their molecule, being the most frequent the reactive oxygen species (ROS) and reactive nitrogen species (RNS). Thiol radicals are less important, their reactive group contains sulfur; well as those containing carbon or phosphorus in their reactive center. ROS are constituted by superoxide anion ($O_2 \bullet -$), hydroxyl radical ($\bullet OH$), hydrogen peroxide (H_2O_2), and singlet oxygen. While, that the RNS are nitric oxide (NO), nitrogen dioxide ($NO_2 \bullet -$) and peroxynitrite ($OONO-$) [11,12].

Due to the constant production of ROS and RNS during metabolic processes, the organism has developed a powerful, complex defense system that limits its exposure to these agents these are the so-called antioxidants (AO). Several antioxidants are enzymes or essential nutrients, or include these in their molecular structure. An essential nutrient is a compound that must be eaten because the organism is unable to synthesize it. Based on this characteristic, some authors classify AO as non-enzymatic and enzymatic [12-14].

2.1. Enzimatic antioxidants

Some researchers state that the AO function performed by enzymes has advantages compared to AO compounds, for this activity is regulated according to cellular requirements: they can be induced, inhibited or activated by endogenous effectors [15]. Ho and colleagues (1998) showed evidence of the importance of AO enzymes in protection against oxidant agents. When using transgenic mice designed to overexpress the activity of some AO enzymes, it was noticed that there is a notorious tolerance of certain tissues when they are exposed to toxics and pathologic conditions that would promote ROS action [9,16].

Enzymatic AO catalyze electron transference from a substrate towards FR. Later, the substrates or reducing agents used in these reactions are regenerated to be used again, they achieve this by using the NADPH produced in different metabolic pathways [14]. A prolonged exposure to ROS can result in diminished NADPH concentration, which is needed in other important physiologic processes, even though some enzymatic AO do not consume cofactors. Superoxide dismutase (SOD), catalase (CAT) and glutathione peroxidase (GSH-Px) belong to this group [12,14,17].

2.2. Non-enzymatic antioxidants

Non-enzymatic antioxidants constitute a heterogeneous group of hydrophobic and hydrophilic molecules that trap FR and create chemical species that are less noxious to cell integrity [18]. Essentially, they give an electron to a FR to stabilize it. Hydrophilic non-enzymatic antioxidants are located mainly in the cytosol, mitochondrial and nuclear matrixes and in extracellular fluids. They are vitamin C, glutathione, uric acid, ergothioneine and polyphenolic flavonoids [9,18].

3. Role of oxidative stress in the development and pathogenesis of the chronic degenerative diseases

Currently, studies related to reactive oxygen species (ROS) and reactive nitrogen species (RNS) have become a relevant issue in research with the main purpose of understanding their functions and effects in the organism. Studies developed throughout the 20th century have explained the action mechanisms of ROS and the operation of the systems responsible for their elimination. These evidences have shown the existence of enzyme systems that produce ROS (cytochrome P450, xanthine oxidase, respiratory chain) of the Fenton reaction, catalase, peroxidase, and superoxide dismutase [19].

All researches lead to the same conclusions so far: a) the evidence that the cells have specialized systems to convert ROS into less reactive compounds, and b) if those systems would fail, the ROS could be preexisting compounds for the development of diseases. Thus, several researchers agree in the relevance of "oxidative stress" in medical problems, specifically in pathogenesis and/or complications of chronic degenerative diseases [9,12,20,21].

Different observations suggest that these pathologies could be originated when reactive species are formed and suffer alterations, or when they are eliminated, or both. However, the situation is real and much more complicated, for it is difficult to determine the crucial event that originates this disease due to the diversity of forms of oxidative stress (Figure 1). Different researches indicate that mutations produced in genes are responsible of the metabolic unbalance of ROS, while others suggest that environmental changes and common habits weigh on human metabolic processes. However, doubt remains, if oxidative stress is the primary event that leads to the disease or the oxidative phenomenon is developed throughout the disease [22].

Whatever the means by which oxidative stress is induced and pathology is developed, the majority of evidences coincide in the relevance of alterations or enzyme deficiencies. These deficiencies are often caused by mutations in genes coding antioxidant or related enzymes, for example, by genetic polymorphism.

This concept is frequently related to large number of pathologies. Enzymes involved in defence against ROS are not an exception. All enzymes contributing to antioxidant defence can be classified to really antioxidant ones, dealing directly with ROS as substrates, and auxiliary ones. The latter enzymes respond for reparation or degradation of oxidatively modified molecules, maturation and posttranslational modification of antioxidant enzymes and metabolism of low molecular mass antioxidants. As a rule, genetic polymorphisms of enzymes may lead to oxidative stress and consequent diseases, among which cancer, neurodegeneration, cardiovascular disorders, and diabetes are most frequently mentioned. Among the most studied enzymes with genetic polymorphism is the glucose-6-phosphate dehydrogenase, catalase, superoxide dismutase, glutathione peroxidase and those involved in reparation of oxidized molecules and the disease progression [22].

3.1. Glucose-6-phosphate dehydrogenase deficiency

The most striking example among polymorphisms of genes coding enzymes related to antioxidant defence is well-known deficiency in glucose-6-phosphate dehydrogenase (G6PDH) which leads to favism; genetic disease characterized by the lysis of erythrocytes when consumed broad beans and other substances which are harmless to the general population [23].

Other pathologies which are related to the same deficiency are diabetes [24,25], vascular diseases [24], and cancer [26]. In these cases, oxidative stress is induced in specific cells; it was shown that GSH may react with superoxide anion radical providing partial defence against this ROS [27]. When decreasing GSH concentration in G6PDH-deficient individuals enhances their sensitivity to redox-active compounds, producing superoxide. Superoxide is able to react also with nitric oxide, leading to the formation of rather harmful oxidant peroxynitrite. However, relation of this reaction to diabetes and vascular diseases is not because of peroxynitrite production and subsequent oxidative damage, but rather because of decrease in nitric oxide level [28].

The latter is an important second messenger in certain signalling pathways particularly related to vasodilation [29]. There is some probability also that individuals with G6PDH-deficiency may fail to regulate properly blood pressure [30]. Despite possible impairment in nitric oxide production, there is also other way to connect G6PDH deficiency with vascular diseases. It is known, that development of vascular diseases depends on the levels of homocysteine and folate, intermediates in metabolism of sulfur-containing amino acids [31]. Production of two these metabolites depends on GSH and NADPH levels in cells [32].

Data regarding association of G6PDH deficiency with cancer are controversial, because some studies demonstrated that G6PDH-deficient patients may additionally suffer from cancer [33], while others state opposite [34]. Nevertheless, both situations are possible. In particular, there is a large data body indicating that different cancer types are developed at increased DNA damage. It often happens under polymorphism in enzymes contributing to DNA repair, what will be discussed below.

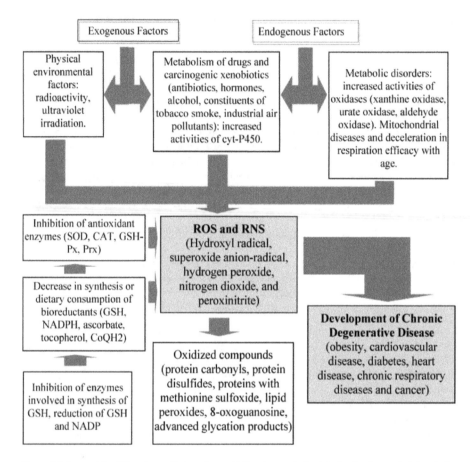

Figure 1. Scheme of the different ways that produce oxidative stress and stimulate the development of chronic degenerative diseases. *Abbreviations*: ROS – reactive oxygen species, RNS – reactive nitrogen species, SOD – superoxide dismutase, CAT - catalase, GSH-Px – glutathione-peroxidase, Prx - peroxyredoxin, GSH – reduced glutathione, CoQH2 – ubiquinol.

On the other hand, NADPH supply at certain conditions may be even harmful leading to enhanced oxidative damage and cancer development. Indeed, it was shown that G6PDH was particularly responsible for cell growth and frequently correlated with cell growth [26]. Tian and colleagues (1998) found that cancer cells possessed several times higher G6PDH activity. The positive correlation between tumour progression and G6PDH activity was found also for humans [35,36].

Increased NADPH supply resulting from G6PDH overexpression can lead to so-called "reductive stress" [37]. Enhanced activity of G6PDH, a lipogenic enzyme, was found at diabetes and obesity [38]. In humans, G6PDH is regulated by many transcription factors, in particular, SREBP-1a (sterol regulatory element binding protein) [39], AP-1 [40] and Sp1 [41]. It was shown that elevation of G6PDH activity might lead to enhanced lipid synthesis [42] and to possible reductive stress [43].

3.2. Catalase deficiency

The first case of catalase deficiency was described by Shigeo Takahara (1947) in a child with cold sores and called acatalasemia to the patology [44]. The cause of this patology is related to ability of oral Streptococci to produce hydrogen peroxide which may promote death of mouth mucosa cells in acatalasemic patients [45]. Catalase deficiency is also associated with diabetes mellitus [46]. This association is attributed for Hungarian hypocatalasemic patients. They were shown to possess higher levels of homocysteine and lower levels of folate [32]. It hints, on one hand, to abnormalities of sulfur metabolism, but on the other hand, it is commonly known that higher homocysteine levels are related to cardiovascular diseases [47], the fact we mentioned above in the context of G6PDH deficiency.

3.3. Polymorphism of Cu,Zn-SOD and protein aggregation

In recent years, the main attention has focused on the polymorphism of genes coding the enzyme superoxide dismutase. More than 100 nucleotide substitutions for the gene SOD1 coding human cytosolic copper- and zinc containing SOD (Cu,Zn-SOD) were described [48]. It is known that several mutations in SOD1 gene are associated with cases of familial amyotrophic lateral sclerosis (ALS), a neurodegenerative disease which is characterized by paralysis and subsequent death [49]. Mechanisms of the disease development are still unknown, but there are many evidences that oxidative stress, developed in neurons, is rather caused by unexpected pro-oxidative activity of SOD than by the loss of the activity at all [50]. It was found that the aggregates cause harm to the cells not only via oxidative stress, but also via inhibition of glutamate receptors [51] and induction of apoptosis [52].

Irwin Fridovich presented some examples of unusual activities of SOD, such as oxidase-like or reductase-like ones [53]. His works and data of other authors suggest that SOD, being mutated or placed in specific conditions, may produce more harmful ROS tan hydrogen peroxide, i.e. hydroxyl radical [54, 55]. Some studies suggested that SOD aggregation can be triggered by higher susceptibility to oxidation of mutated protein [56,57]. Indeed, Cu,Zn-SOD is considered to be rather stable, resistant to many, deleterious to other proteins, compounds [48]. These evidences suggest that Alzheimer, Huntington, and Parkinson diseases are other pathologies related to this enzymatic alteration [22].

3.4. Polymorphism of Mn-SOD, extracellular SOD and glutathione peroxidase

Unlike Cu,Zn-SOD, less mutations were found in the gene coding human manganese containing superoxide dismutase (SOD2). Substitution of alanine-16 to valine (so called "Ala variant") is the most known mutation [58]. This mutation has recently been associated with cancers of breast, prostate, ovaries and bladder, as well as non-Hodgkin lymphoma, mesothelioma and hepatic carcinoma [58]. Mammals possess also extracellular Cu,Zn-SOD (EC-

SOD) encoded in humans by gene SOD3. The enzyme is a homotetramer presenting in plasma, lymph, and synovial fluid [59]. Extracellular SOD is abundant particularly in the lung, blood vessels, and the heart. Consequently, polymorphism of SOD3 gene is associated with pulmonary and cardiovascular diseases [60].

Polymorphism of glutathione peroxidase (GSH-Px) was found to be associated with some cancers. Four GSH-Px isoforms have been described in humans. It was found that mutations in exon 1 of human GSH-Px-1 gene lead to appearance of polyalanine tract at N-terminus of the protein [59]. These tracts themselves are not connected with diminished enzyme activity. Another polymorphism, substitution of proline-198 to leucine, was found in Japanese diabetic patients and associated with intima-media thickness of carotid arteries [61]. The same substitution for adjacent proline-197 was associated with lung and breast cancers, as well as with cardiovascular diseases [59].

3.5. Polymorphism of enzymes involved in reparation of oxidized molecules

Mutations may also affect enzymes involved in DNA reparation. The enzyme 8-hydroxy-2'-deoxyguanosine glycosylase (hOGG) encoded in human genome by the gene hOGG1 is probably the most known example. Recent studies associate mutations in hOGG1 with different cancer types, such as lung, stomach and bladder cancers [62]. Most of the mutations in this gene affect exon 7 and cause serine-to-cysteine substitution. It was demonstrated that substitution S326C in hOGG1 protein confers susceptibility to oxidation and makes the enzyme prone to form disulfide bond between different polypeptide chains [63].

Hydrolase MTH1 is other important enzyme preventing incorporation of oxidized purine nucleotide triphosphates in DNA [64]. Knockout of this enzyme in mice resulted in increased frequency of lung, stomach and liver tumours with age [65].

Other important antioxidant enzymes are glutathione S-transferases (GSTs). Its main function is to conjugation of different electrophilic compounds with glutathione [66]. Oxidatively modified compounds as well as lipid oxidation products, like 4-hydroxy-2-nonenal, are subjected to conjugation with glutathione. In general, GSTs are belong to xenobiotic-elimitating system. Some of them, namely GSTs of μ class, are known well by their ability to eliminate polycyclic aromatic hydrocarbons, oxidized previously by cytochrome P450 monooxygenases. To date, eight classes of GSTs have been described: α, κ, μ,σ, ξ, π, θ, and ω. Cytosolic enzymes belong to classes α, μ, π and θ [67]. The gene coding GSTM1 (GST of μ class, isoform 1) is appeared to be highly polymorphic and found inactivated in half of human population.

Some studies associate polymorphism of GSTM1 with lung cancer [59,68], although reports are controversial. For example, meta-analysis conducted by [69] found no association of GSTM1 null genotype with lung cancers as well as with smoking. Other authors found such association and reported increased susceptibility to cancerogens among Caucasian and African-American populations [70]. Polymorphism of GSTM1 was also found to be associated with head and neck carcinomas [67]. The need in GSTM1 and its role in prevention of lung cancer are explained by the ability of the enzyme to detoxify constituents of cigarette smoke, such as mentioned above polycyclic aromatic hydrocarbons. Some studies also associate lung cancer with polymorphism of GSTT1 (GST of θ class) which participates in catabolism of tobacco smoke constituents, such as halomethanes and butadione [70].

3.6. Role of oxidative modifications of antioxidant and related enzymes in disease progression

Many disorders related to the metabolism of transition metals, amino acids or low molecular mass reductants are known to be connected with activities of antioxidant enzymes. Particularly, impairement in selenium uptake or synthesis of selenocysteine needed for glutathione peroxidase may lead to GSH-Px deficiency and subsequent disorders such as cardiovascular ones [47]. Disruption of iron-sulfur clusters by superoxide anion radicals or peroxynitrite leads frequently to impairment of many metabolic pathways. Indeed, aconitase, NADH-ubiquinone-oxidoreductase (complex I of mitochondrial electron transport chain), ubiquinol-cytochrome c oxidoreductase (complex III), ribonucleotide reductase, ferredoxins possess iron-sulfur clusters, susceptible to oxidation. Owing to this, aconitase is used as one of oxidative stress markers [71]. On the other hand, iron is a component of haem, a prosthetic group in catalase holoenzyme. Susceptibility to oxidative modification is described for catalase, glutathione peroxidase, Cu,Zn-SOD, and G6PDH. The latter is believed to be one of the most susceptible to oxidation enzymes [22]. Thus, oxidative stress induced by exogenous factors, like carcinogens, certain drugs, ions of transition metals, etc., or by metabolic disorders, like diabetes, can be exacerbated by oxidative modification of antioxidant enzymes. These assumptions demonstrate the potential of antioxidant therapy in particular cases. At some pathological states, whatever the cause of the disease, oxidative stress is seen to be a powerful exacerbating factor. Type II diabetes, cardiovascular diseases and neurodegenerative diseases, associated with protein aggregation are among such pathologies. Indeed, enhanced level of glucose results in higher probability of protein glycation [72].

4. Impact of chemopreventive agents in the chronic degenerative diseases

The available evidences indicates that individuals with chronic degenerative diseases are more susceptible to oxidative stress and damage because they have elevated levels of oxidants and/or reduced antioxidants. Therefore, it has been posited that antioxidant supplementation in such individuals may be beneficial. Different research has confirmed that many common foods contain nonnutritive components that may provide protection against chronic degenerative diseases, however, the most studies have had impact on the cancer [20,21].

The "chemoprevention" seeks to eliminate precancerous cells in order to avoid the necessity of chemotherapy. It can be further classified as primary, secondary, or tertiary prevention. Primary chemoprevention focuses on preventing the development of precancerous lesions, secondary chemoprevention focuses on preventing the progression of these lesions to cancer, and tertiary chemoprevention aims to prevent the recurrence or spread of a primary cancer [73].

It has been known for some time that dietary factors play a role in the development of some human cancers [73,74] and that some foods contain mutagens and carcinogens [74,75]. Investigations of last decades, has focused on the existence of a number of non nutritional components in our regular diet that possess antimutagenic and anticarcinogenic properties, these compounds have been called as chemopreventers [76,77].

The chemopreventers are classified as food entities that can prevent the appearance of some long-term diseases like cancer or cardiovascular disorders. It has been suggested that chemoprevention should be considered as an inexpensive, easily applicable approach to cancer control and "may become a major weapon in the anticancer arsenal" [76,78,79]. These compounds can be found in all food categories, but mainly in fruits, vegetables, grains and tea [78,79]. Chemopreventers belong to different classes of chemicals but the most recognized are some vitamins, food polyphenols, flavonoids, catechins, and some components in spices [78, 79].

The mechanisms of action of the chemopreventers are complex and can be categorized according to the site of action or by the specific type of action. It appears that most chemopreventers act primarily as antioxidants. As such, they may scavenge free radicals formed during the preparation of food or as a normal biological process in the body. Recall, that the free radicals can react with DNA, lipids, or cell membranes, leading to aging, injuries of the organ, and greater susceptibility to develop the chronic degenerative disease. Therefore, any event that removes free radicals in the human body is considered beneficial for human health. In addition to their antioxidative activities, there are other mechanisms that show in the Table 2 [80-82].

5. Chemopreventive evidence of some fruits and food supplements evaluated by our research group

5.1. Cactus pears

Plants from the genus *Opuntia* are the most abundant of the Cactaceae family, grown throughout the Americas as well as the central area of the Mediterranean, Europe, Asia, Africa, and Australia. *Opuntia* species display flattened stems called "pencas" or cladodes. The cactus pear (also called prickly pears) is the fruit of this plant (*Opuntia* spp.). The fruit is an oval berry with a large number of seeds and a semi-hard rind with thorns, which may be grouped by fruit colors: red, purple, orange-yellow and white. The fruits with white pulp and green rind are preferred for consumption as food, and their domestic production corresponds to almost 95% of the total production. Mexico is the main producer of cactus pears and accounts for more than 45% of the worldwide production; however, only 1.5% of this production is exported [83,84].

Mechanism	Action	Examples
Inhibition of carcinogen formation	Agents that block or inhibit to the enzymes responsible for the biotransformation of procarcinogens to carcinogen form	Dithiocarbamates, isothiocyanates, diallyl sulfide, and ellagic acid
Inducing agents	Agents that induce or enhance enzyme activity (e.g., glutathione S-transferase, GST) for detoxify and reduce the level of mutagenic/carcinogenic species of the body	Isothiocyanates, sulfaraphane, d-limonene, terpinoids, turmeric, and curcurains
Suppressing agents	Agents that may react on different processes (e.g., inhibition of arachidonic acid metabolism, activity of protease or protein kinase C) involved in tumor promotion/progression	isoflavones, phytoestrogens, selenium
Immune activity and modulation	Since the immune system can influence on growth either via effects on the inflammation status or by causing apoptosis. Some chemopreventers can act on the early stages in neoplasia or have effects on frank malignancies	Carotenoid, flavonoids, lactoferrin
Signal transduction pathways and their regulation	Some chemopreventers may alter signaling pathways of receptors for hormones and others factors responsible for cell regulation and can be modified the potential for growth, either by acting to increase mitosis or alter the level of apoptosis.	d-limonene, sulfur compounds, lactoferrin, retinoids

Table 2. Other mechanisms of chemoprevention

A viable strategy to increase the competitiveness of the Mexican cactus pear in national and international markets is the innovation and creation of new high value-added products. This could be achieved by determining the nutritional and functional properties that differentiate the Mexican cactus pear from analogous products. In addition, providing functional products for a market in constant growth would offer a key competitive advantage and would allow the producers to diversify its commercialization, not as fresh fruit only, but also as an ingredient or high-value additive for the food industry. A commercialization of the cactus-pear based on its antioxidant properties could generate competitive advantages that may turn into business opportunities and the development of new products [85].

Different studies with the varieties of European and Asian cactus pears have shown notable antioxidant activities that significantly reduce oxidative stress in patients and may help in preventing chronic pathologies (as diabetes and cancer) [85-87]. For this reason, the cactus pear is considered a functional food; this feature is attributed to its bioactive compounds such as vitamin C and vitamin E, polyphenols, carotenoids, flavonoid compounds (e.g., kaempferol, quercetin, and isorhamnetin), taurine and pigments [88,89].

Betalains are water-soluble pigments. Two betalain derivatives are present in cactus-pears: betacyanin, which gives the red-purple color, and betaxanthin, which gives a yellow-orange color. These pigments have shown beneficial effects on the redox-regulated pathways involved in cell growth and inflammation, and have not shown toxic effects in humans [90,91].

In addition, a neuroprotector activity against oxidative damage induced in cultures of rat cortical cells has been attributed to the cactus pear flavonoids [92]. Another beneficial effect of the fruit was observed in the prevention of stomach ulcers through the stimulation of prostaglandin production: cactus pear promoted mucous secretion of bicarbonate, involved in the protection of gastric mucosa [93]. On the other hand, their contents of natural antioxidants has raised interest in the use of cactus pears as substitute for synthetic antioxidants, such as butylhydroxytoluene (BHT), butylhydroxyanisole (BHA) [88].

In the Institute of Health Sciences (Autonomous University of Hidalgo State) have been performed studies to demostrate the chemopreventive capacity of the cactus pear. The first studies were developed by Hernández-Ceruelos et al. (2009) with the main objective to evaluate the antioxidant effect of three varieties of prickly pear juice (red-purple, white-green and yellow-orange) in four different concentrations (25, 50, 75 and 100%) by the technique of DPPH (1,1-diphenyl-2-picrylhydrazyl). Their results indicated that the juice of princkly pear variety red-purple (PPRP) had the highest antioxidant capacity in all concentrations in comparison with the positive control (vitamin E). Subsequently, researchers evaluated the anticlastogenic potential of PPRP by micronucleus assay against of methyl methane sulfonate (MMS) in mice. This experiment had a duration of 2 weeks, was included a negative control (animals treated with water), a positive control of MMS (40 mg/kg), a group of mice treated with princkly pear variety red-purple (25mL/Kg), and three groups with PPRP (in doses of 25, 16.5 and 8.3 mL/Kg) plus the mutagen. The PPRP was administered by oral gavage and the mutagen was injected intraperitoneally 5 days before the end of the experiment (single dose). Finally, blood samples were obtained in four times (0, 24, 48 and 72 hours) to determine the frequency of micronucleated polychromatic erythrocytes (MNPE). The result indicated that PPRP is not a genotoxic agent, on the contrary, may reduce the number of micronucleated polychromatic erythrocytes. In this regard, the princkly pear variety red-purple showed an anticlastogenic effect directly proportional to the concentrations. The highest protection was obtained with the concentration of 25 mL/Kg (approximately, 80%) after 48 hours of treatment [94].

In the second study was evaluated the antioxidant activities [with three assays: a)1,1-diphenyl-2-picrylhydrazyl radical-scavenging, b) protection against oxidation of a β-carotene-linoleic acid emulsion, and c) iron(II) chelation], the content of total phenolic compounds, ascorbic acid, betacyanin, betaxanthins and the stability of betacyanin pigments in presence of Cu(II)-dependent hydroxyl radicals (OH•), in 18 cultivars of purple, red, yellow and white cactus pear from six Mexican states (Hidalgo, Puebla, Guanajuato, Jalisco, Zacatecas and the State of Mexico). The results indicated that the antiradical activities from yellow and white cactus pear cultivars were not significantly different and were lower than the average antiradical activities in red and purple cultivars. The red cactus pear from the state of Zacatecas showed the highest antioxidant activity. The free radical scavenging activity for red cactus pears was significantly correlated to the concentration of total phenolic compounds ($R2 = 0.90$) and ascorbic acid ($R2 = 0.86$). All 18 cultivars of cactus pears studied showed significant chelating activity of ferrous ions. The red and purple cactus pears showed a great stability when exposed to OH• [88].

5.2. Cranberries

Among small soft-fleshed colorful fruits, berries make up the largest proportion consumed in our diet. Berry fruits are popularly consumed not only in fresh and frozen forms, but also as processed and derived products including canned fruits, yogurts, beverages, jams, and jellies. In addition, there has been a growing trend in the intake of berry extracts as ingredients in functional foods and dietary supplements, which may or may not be combined with other colorful fruits, vegetables, and herbal extracts [95]. Berry fruits commonly consumed in America include blackberries, black raspberries, red raspberries and strawberries, blueberries, and cranberries.

Other "niche-cultivated" berries and forest/wild berries, for example, bilberries, black currant, lingonberry, and cloudberry, are also popularly consumed in other regions of the World [95]. The North American cranberry (*Vaccinium macrocarpon*) is of a growing public interest as a functional food because of potential health benefits linked to phytochemicals of the fruit. Cranberry juice has long been consumed for the prevention of urinary tract infections, and research linked this property to the ability of cranberry proanthocyanidins to inhibit the adhesion of *Escherichia coli* bacteria responsible for these infections [96]. These studies, which brought to light the unique structural features of cranberry proanthocyanidins [97], have sparked numerous clinical studies probing a cranberry's role in the prevention of urinary tract infections and targeted the nature of the active metabolites. Further antibacterial adhesion studies demonstrated that cranberry constituents also inhibit the adhesion of *Helicobacter pylori*, a major cause of gastric cancer, to human gastric mucus [98]. The earliest report of potential anti-carcinogenic activity appeared in 1996 in the University of Illinois [99].

Extracts of cranberry and bilberry were observed to inhibit ornithine decarboxylase (ODC) expression and induce the xenobiotic detoxification enzyme quinonereductase *in vitro* [99]. Subsequent studies with cranberry and other berries in cellular models have focused on some cancers such as breast, colon, liver, prostate and lung [100-102]. This biological activity of berries are partially attributed to their high content of a diverse range of phytochemicals such as flavonoids (anthocyanins, flavonols, and flavanols), tannins (proanthocyanidins, ellagitannins, and gallotannins), quercetin, phenolic acids, lignans, and stilbenoids (e.g., resveratrol) [100]. With respect to his genotoxic and/or antigenotoxic potential, there are few reports in the literature that demonstrate this effect and the majority of studies were performed *in vitro* cell culture models [101,103,104]. Boateng *et al.* demonstrated that consumption of some juices of berries (as blueberries, blackberries, and cranberry) can reduce the formation of aberrant crypt foci (ACF) induced by azoxymethane in Fisher male rats [105]. Another study, in which it was administrated a lyophilized extract of *Vaccinium ashei* berries in male Swiss mice during 30 days, showed to have improved the performance on memory tasks and has a protective effect on the DNA damage in brain tissue evaluated with the comet assay [106].

Although the types of berry fruits consumed worldwide are many, the experiment executed in our laboratory is focuses on cranberries that are commonly consumed in Mexico, especially in the states of Tlaxcala, Hidalgo, and Puebla. The purpose of our study was to determine whether cranberry ethanolic extract (CEE) can prevent the DNA damage produced by benzo[a]pyrene (B[a]P) using an in vivo mouse peripheral blood micronucleus assay. The experimental groups were organized as follows: a negative control group (without treatment), a positive group treated with B[a]P (200 mg/kg), a group administered with 800 mg/kg of cranberry ethanolic extract, and three groups treated with B[a]P and cranberry ethanolic extract (200, 400, and 800 mg/kg) respectively. The CEE and benzo[a]pyrene were administered orally for a week, on a daily basis. During this period the body weight, the feed intake, and the determination of antigenotoxic potential were quantified. At the end of this period, we continued with the same determinations for one week more (recovery period) but anymore administration of the substances. The animals treated with B[a]P showed a weight increase after the first week of administration. The same phenomenon was observed in the lots combined with B[a]P and CEE (low and medium doses). The dose of 800 mg/kg of CEE showed similar values to the control group at the end of the treatment period. In the second part of the assay, when the substances were not administered, these experimental groups regained their normal weight. The dose of CEE (800 mg/kg) was not genotoxic nor cytotoxic. On the contrary, the B[a]P increases the frequency of micronucleated normochromatic erythrocytes (MNNE) and reduces the rate of polychromatic erythrocytes (PE) at the end of the treatment period. With respect to the combined lots, a significant decrease in the MN rate was observed from the sixth to the eighth day of treatment with the two high doses applied; the highest protection (60%) was obtained with 800 mg/kg of CEE. The same dose showed an anticytotoxic effect which corresponded to an improvement of 62.5% in relation to the animals administered with the B[a]P. In the second period, all groups reached values that have been seen in the control group animals. Our results suggest that the inhibition of clastogenicity of the cranberry ethanolic extract against B[a]P is related to the antioxidant capacity of the combination of phytochemicals present in its chemical composition [107].

5.3. Grapefruit juice and naringin

The grapefruit is a subtropical citrus tree known for its bitter fruit. These evergreen trees usually grow around 6 meters tall. The leaves are dark green, long and thin. His fruit (called toronja in Spanish) has become popular since the late 19th century, is yellow-orange skinned and largely an oblate spheroid and generally, is consumed in form of juice [108].

The grapefruit juice is an excellent source of many nutrients and phytochemicals that contribute to a healthy diet. Is a good source of vitamin C, contains the fiber pectin, and the varieties pink and red contain the beneficial antioxidant lycopene [108]. But, the main flavonoid, existing in highest concentration in grapefruit juice is naringin, which in humans is metabolized to naringenin [109].

Since grapefruit juice is known to inhibit enzymes necessary for the clearance of some drugs and hormones, some researchers have hypothesized that grapefruit juice and the naringin may play an indirect role in the development of hormone-dependent cancers. A study found a correlation between eating a quarter of grapefruit daily and a 30% increase in risk for breast cancer in post-menopausal women. The study points to the inhibition of CYP3A4 enzyme by grapefruit, which metabolizes estrogen [110]. However, an investigation conducted in 2008 has shown that grapefruit consumption does not increase breast cancer risk and found a significant decrease in breast cancer risk with greater intake of grapefruit in women who never used hormone therapy [111].

In the case of naringin, this compound exerts a variety of pharmacological effects such as antioxidant activity, blood lipid lowering, anticancer activity, and inhibition of selected drug-metabolizing cytochrome P450 enzymes, including CYP3A4 and CYP1A2, which may result in drug-drug interactions in vivo. Ingestion of naringin and related flavonoids can also affect the intestinal absorption of certain drugs, leading to either an increase or decrease in circulating drug levels [112].

This evidence has motivated to our research group to develop various studies with grapefruit juice (GJ) and the naringin (Nar) to assess his chemoprotective ability.

Our first experience was with naringin in 2001. On that occasion, the study was designed for three main purposes: (1) to determine whether Nar has a genotoxic effect in mouse in vivo. This was evaluated by measuring the rate of micronucleated polychromatic erythrocytes (MNPE); (2) to determine its antigenotoxic and its anticytotoxic potential on the damage produced by ifosfamide. The first study was done by scoring the rate of MNPE, and the second one by establishing the index polychromatic erythrocytes/normochromatic erythrocytes (PE/NE); and (3) to explore whether its antigenotoxic mechanism of action is related to an inhibitory effect of Nar on the expression of the CYP3A enzyme, an effect which could avoid the biotransformation of ifosfamide.

A single oral administration was used for all groups in the experiment: three groups were given different doses of Nar (50, 250, and 500 mg/kg), other groups received the same doses of Nar plus an administration of ifosfamide (60 mg/kg), another group treated with distilled water and another with ifosfamide (60 mg/kg) were used as negative and positive controls, respectively. The micronuclei and the cell scoring were made in blood samples taken from the tail of the animals at 0, 24, 48, 72, and 96 h. The results showed that Nar was neither genotoxic nor cytotoxic with the doses tested, but ifosfamide produced an increase in the rate of MNPE at 24 and 48 h. The highest value was 24+/-1.57 MNPE per thousand cells at 48 h. The index PE/NE was significantly reduced by ifosfamide at 24 and 48 h. Concerning the antigenotoxic capacity of Nar, a significant decrease was observed in the MNPE produced by ifosfamide at the three tested doses. This effect was dose-dependent, showing the highest reduction in MNPE frequency (54.2%) at 48 h with 500 mg/kg of Nar. However, no protection on the cytotoxicity produced by ifosfamide was observed. Immunoblot analysis was used to assess the CYP3A expression in liver and intestinal microsomes from mouse exposed orally to Nar. An induction in the CYP3A protein was observed in both intestinal and hepatic microsomes from treated mice. This induction correlated with an increase in erythromycin N-demethylase activity. These data suggest that other mechanism(s) are involved in the antigenotoxic action of naringin [113].

With regard to grapefruit juice (GJ), we performed two experiments which are summarized below. The first evaluated the capacity of GJ to inhibit the micronucleated polychromatic erythrocytes (MNPE) produced by daunorubicin in an acute assay in mice, as well as to determine its antioxidant potential in mouse hepatic microsomes, and its capacity to trap free radicals in vitro.

The results showed that GJ is not toxic or genotoxic damage; on the contrary, it generated a significant reduction of the MNPE formed by daunorubicin. The effect was found throughout the examined schedule (from 24 to 96 h). The two high doses produced inhibition of about 60% at 48 h, 86% at 72 h and 100% at 96 h after the treatment. With respect to the grapefruit juice antioxidant potential, a 50% decrease in liver microsomal lipid peroxidation produced by daunorubicin was found by quantifying malondialdehyde formation. Finally, a strong GJ scavenging activity evaluated with the 1,1-diphenyl-2-picryl-hydrazyl (DPPH) was observed, giving rise to a concentration-dependent curve with a correlation coefficient of 0.98. Overall, our results established an efficient anticlastogenic potential of grapefruit juice, probably related to its antioxidant capacity, or to alterations of daunorubicin metabolism [114].

Based on this background; recently, we finished another study in which using the comet assay was demonstrated a strong effect by hydrogen peroxide (HP) and no damage by grapefruit juice (GJ) in human lymphocytes. Cells exposed to HP and treated with GJ was shown an increase of DNA damage by HP over the control level, and a decrease of such damage by GJ. With the comet assay plus formamidopyrimidine-DNA-glycosylase we found the strongest increase of DNA damage by HP over the control level, and the strongest reduction of such damage by GJ. By applying the comet/FISH method we determined 98% of the p53 gene signals in the comet head of control cells along the experiment, in contrast with about 90% signals in the comet tail of cells exposed to HP. Cells treated with both agents showed a significant, concentration/time dependent return of p53 signals to the head, suggesting enhancement of the gene repair. Finally, with the annexin V assay we found an increase in apoptosis and necrosis by HP, and no effect by GJ; when GJ was added to HP treated cells no modification was observed in regard to apoptosis, although a decrease of necrosis was observed [115].

5.4. Chamomile

Chamomile (*Matricaria chamomilla* or *Chamomilla recutita*) is an asteraceae plant native to Europe and distributed around the world, except in tropical and polar regions. This plant has been used for its curative properties since ancient Egyptian and Greek times, and at present is frequently used as an antiseptic, antiflogistic, diuretic, expectorant, febrifuge, sedative, anti-inflammatory and anticarcinogen [116]. Pharmacological activities of various components of the plant have been reported, for example, the anti-inflammatory capacity and the modulating effects of the heat shock protein on apigenin and quercetin flavonoids, as well as the anti-inflammatory, antioxidant, and antiseptic activities detected on α-bisabolol, guargazulene, and chamazulene [117, 118]. The essential oil extracted from the chamomile flower varies from 0.42 to 2%, and consists of compounds such as bisabolol, chamazulene, cyclic sesquiterpenes, bisabolol oxides, and other azulenes and terpenes [119].

With respect to his genotoxic and/or antigenotoxic potential, there are few reports in the literature that demonstrate this effect. Therefore, our laboratory performed two investigations with the main purpose to evaluate the chemoprotection capacity of chamomile. Initially, we obtained the chamomile essential oil (CEO) from flowers of *Chamomilla recutita* by steam distillation, and then it was analyzed by gas chromatography to identify the chemical species. Thirteen compounds were determined with this assay, including bisabolol and its oxides, β-farnecene, chamazulene, germacrene, and sesquiterpenes (Table 3).

Compound	RT[a]	Area (%)
(E)-β-Farnecene	38.46	28.17
Germacrene-D	39.23	2.19
Unidentified sesquiterpene	40.07	1.40
Unidentified sesquiterpene	41.17	0.78
(Z,E)–α–Farnecene	41.35	1.59
Unidentifiedsesquiterpene	48.52	0.71
α–Bisabolol oxide A	54.46	41.77
α–Bisabolol oxide B	49.28	4.31
α-Bisabolol oxide	50.65	5.30
α–Bisabolol	51.18	2.31
Chamazulene	52.80	2.39
1,6-Dioxaspiro[4,4]non-3-ene,2-(2,4hexadyn-1-ylidene)	60.73	2.19
Hexatriacontane	67.49	0.50

RT[a], Retention time obtained with gas chromatography.

Table 3. Components of the tested chamomile essential oil

The first work was to determine the inhibitory effect of the CEO, on the sister chromatid exchanges (SCEs) produced by daunorubicin and methyl methanesulfonate (MMS) in mouse bone marrow cells.

The authors performed a toxic and genotoxic assay of chamomile essential oil; both showed negative results. To determine whether CEO can inhibit the mutagenic effects induced by daunorubicin, one group of mice was administered corn oil, another group was treated with the mutagen (10 mg/kg), a third group was treated with 500 mg/kg of CEO; three other groups were treated first with CEO (5, 50 and 500 mg/kg) and then with 10 mg/kg of daunorubicin. In the case of MMS, the experimental groups consisted of the following: the negative control group which was administered corn oil, a group treated with 25 mg/kg of MMS, a group treated with 1000 mg/kg of CEO, and three groups treated first with CEO (250, 500 and 1000 mg/kg) and then with MMS (25 mg/kg). The results indicated a dose-dependent inhibitory effect on the SCEs formed by both mutagens. In the case of daunorubicin, a statistically significant result was observed in the three tested doses: from the lowest to the highest dose, the inhibitory values corresponded to 25.7, 63.1 and 75.5%. No alterations were found with respect to the cellular proliferation kinetics, but a reduction in the mitotic index was detected. As regards MMS, the inhibitory values were 24.8, 45.8 and 60.6%; no alterations were found in either the cellular proliferation kinetics or in the mitotic indices [120]. This results suggested that CEO may be an effective antimutagen and was the reason for develop the second study.

The aim of the second investigation was to determine the inhibitory potential of CEO on the genotoxic damage produced by daunorubicin (DAU) in mice germ cells. We evaluated the effect of 5, 50, and 500 mg/kg of essential oil on the rate of sister chromatid exchange (SCE) induced in spermatogonia by 10 mg/kg of the mutagen. We found no genotoxicity of CEO, but detected an inhibition of SCE after the damage induced by DAU; from the lowest to the highest dose of CEO we found an inhibition of 47.5%, 61.9%, and 93.5%, respectively. As a possible mechanism of action, the antioxidant capacity of CEO was determined using the 1,1-diphenyl-2-picrylhydrazyl (DPPH) free radical scavenging method and ferric thiocyanate assays. In the first test we observed a moderate scavenging potential of the oil; nevertheless, the second assay showed an antioxidant capacity similar to that observed with vitamin E. In conclusion, we found that CEO is an efficient chemoprotective agent against the damage induced by DAU in the precursor cells of the germinal line of mice, and that its antioxidant capacity may induce this effect [116].

5.5. Silymarin

Silybum marianum is the scientific name of milk thistle or St. Mary's thistle. It is a Mediterranean native plant belonging to the Asteraceae family. It is characterized by thorny branches, a milky sap, with oval leaves that reach up to 30 centimeters, its flowers are bright pink and can measure up to 8 cm to diameter [121].

Milk thistle (Mt) grows of wild form in the southern Europe, the northern Africa and the Middle East but is cultivated in Hungary, China and South American countries as Argentina, Venezuela and Ecuador. In México, is consumed as supplement food for many years ago [122].

In the sixties years, German scientists performed a chemical investigation of his fruits, isolating a crude extract formed by active compounds with hepatoprotective capacity; this group of compounds was called silymarin. In 1975, it was found that the principal components of silymarin are silybin A, silybin B, isosilybin A, isosilybin B, silychristin A, silychristin B and silydianin [123]. Currently it is known that the chemical constituents of silymarin are flavonolignans, ie, a combination conformed by flavonoids and lignins structures [124].

Mt is one of the most investigated plant extracts with known mechanisms of action for oral treatment of toxic liver damage. Silymarin is used as a protective treatment in acute and chronic liver diseases [125]. His protective capacity is related with different mechanisms as suppress toxin penetration into the hepatic cells, increasing superoxide dismutase activity, increasing glutathione tissue level, inhibition of lipid peroxidation and enhancing hepatocyte protein synthesis. The hepatoprotective activity of silymarin can be explained based on antioxidant properties due to the phenolic nature of flavonolignans. It also acts through stimulating liver cells regeneration and cell membrane stabilization to prevent hepatotoxic agents from entering hepatocytes [126].

Silymarin is also beneficial for reducing the chances for developing certain cancers [127]. The molecular targets of silymarin for cancer prevention have been studied. Milk thistle interfere with the expressions of cell cycle regulators and proteins involved in apoptosis to modulate the imbalance between cell survival and apoptosis. Sy-Cordero et al. (2010) isolated four key flavonolign and diastereoisomers (silybin A, silybin B, isosilybin A and isosilybin B) from *S. marianum* in gram scale. These compounds and other two related analogues, present in extremely minute quantities, were evaluated for antiproliferative/cytotoxic activity against human prostate cancer cell lines. Isosilybin B showed the most potent activity [126]. The isolation of six isomers afforded a preliminary analysis of structure-activity relationship toward prostate cancer prevention. The results suggested that an *ortho* relationship for the hydroxyl and methoxy substituents in silybin A, silybin B, isosilybin A and isosilybin B was more favorable than the *meta* relationship for the same substituents in the minor flavonolignans. Silymarin suppressed UVA-induced oxidative stress that can induce skin damage. Therefore, topical application of silymarin can be a useful strategy for protecting against skin cancer [128].

In our laboratory, we evaluated the antigenotoxic effect of two doses of silymarin (200 and 400 mg/Kg) administered by oral gavage against the chronic consumption of ethanol (solution: 92 mL of water/8 mL of ethanol) during a week with alkaline single cell electrophoresis (comet) assay.

Figure 2 shows the comet measurements obtained in our assay. To summarize, at the 24 hours of the schedule we found no significant DNA damage induced in the control group (only water) and the silymarin group (400 mg/kg), both groups had a mean T/N index of 1.1. On the contrary, the mice (strain CD-1) that consumed the solution of ethanol showed a slight comet increase during this same time. But at 48, 72 and 96 hours, this group showed a T/N index increase of about four times as much. During the last times (120, 144, 168 and 192) there is a decrease of DNA damage, suggesting that hepatocytes are in the process of cell regeneration. With respect to the groups treated with the combination of chemicals, a clear antigenotoxic effect was found with the two doses of silymarin; particularly with 400 mg/kg, the prevention of DNA damage was about 70% during the 48, 72, 96 and 120 hours of treatment. At the end of the experiment, these groups reached similar values to the negative control [129].

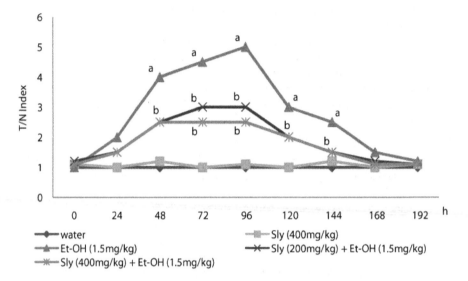

Figure 2. Antigenotoxic effect of silymarin (Sly) against the DNA damage induced by the chronic consumption of etanol (Et-OH). Results are the mean ± SD of 5 mice per group (100 nuclei per doses) [a] statistically significant difference with respect to the value of the control group and, [b] with respect to the value obtained in mice treated with Et-OH only. ANOVA and Student-Newman Keuls tests, $p \leq 0.05$.

Author details

Eduardo Madrigal-Santillán[1], Eduardo Madrigal-Bujaidar[2], Sandra Cruz-Jaime[1],
María del Carmen Valadez-Vega[1], María Teresa Sumaya-Martínez[3],
Karla Guadalupe Pérez-Ávila[1] and José Antonio Morales-González[1]

1 Instituto de Ciencias de la Salud, UAEH, México

2 Escuela Nacional de Ciencias Biológicas, IPN, México

3 Universidad Autónoma de Nayarit, Tepic, México

References

[1] Doll R. Chronic and degenerative disease: Major causes of morbidity and death. The-American Journal of ClinicalNutrition.1995; 62(6) 1301S-1305S.

[2] Nava-Chapa G., Ortiz-Espinosa RM., Reyes-Gómez D. Epidemiologia de las enfermedades crónico degenerativas. In: Morales-González JA., Fernández-Sánchez AM.,

Bautista-Ávila M., Madrigal-Santillán E. (ed.) Los antioxidantes y las enfermedades crónico degenerativas. México: UAEH; 2009. p269-310.

[3] Ugalde A., Jackson, JT. The World Bank and international health policy: a critical review. Journal of International Development 1995; 7(3) 525-41.

[4] Mejía-Median JI., Hernández-Torres I., Moreno-Aguilera F., Bazan-Castro M. Asociación de factores de riesgo con el descontrol metabólico de diabetes mellitus en pacientes de la clínica oriente del ISSSTE. Revista de Especialidades Médico Quirúrgicas 2007; 12(2) 25-30.

[5] Sies, H. Oxidative stress: Introductory remarks, In: Sies H.(ed.) Oxidative stress. London: Academic Press; 1985. p1-8.

[6] Lushchak VI. Environmentally induced oxidative stress in aquatic animals. Aquatic Toxicology 2011; 101(1) 13-30.

[7] Halliwell B. Antioxidants and human disease: a general introduction. Nutrition Reviews. 1997; 55(1Pt2) S44-S52.

[8] Yu BP. Cellular defenses against damage from reactive oxygen species. Physiological Reviews 1994; 74(1) 139-162.

[9] ChihuailafRH., Contreras PA., Wittwer FG. Pathogenesis of oxidative stress: Consequences and evaluation in animal health. Veterinaria México 2002; 33(3) 265-283.

[10] Halliwell, B. Free radicals and antioxidants-quo vadis? Trends in Pharmacological Sciences 2011; 32(3) 125–130.

[11] Halliwell B, Gutteridge JMC. Oxygen toxicity, oxygen radical, transition metals and disease. The Biochemical Journal 1984; 219(1) 1-14.

[12] Erejuwa OO., Sulaiman SA., AbWahab MS. Honey: a novel antioxidant. Molecules 2012; 17(4) 4400-4423.

[13] Halliwell B., GutteridgeJMC., Cross CE. Free radicals, antioxidants, and human disease: Where are we now?. The Journal of Laboratory and Clinical Medicine 1992; 119(6) 598-620.

[14] Chaudière J., Ferrari-Iliou R. Intracellular antioxidants: from chemical to biochemical mechanisms. Food and Chemical and Toxicology 1999; 37(9-10) 949-962.

[15] Harris ED. Regulation of antioxidant enzymes. FASEB Journal 1992; 6(9) 2675-2683.

[16] Ho YS., Magnenat JL., Gargano M., Cao J. The nature of antioxidant defense mechanism: a lesson from transgenic studies. Environmental Health Perspective 1998; 106(5) 1219-1228.

[17] Maxwell SRJ. Prospects for the use of antioxidants therapies. Drugs 1995; 49(3) 345-361.

[18] Bandyopadhyay U, Das D, Banerjee RK. Reactive oxygen species: oxidative damage and pathogenesis (review). Current Science 1999; 77(5) 658-666.

[19] Lushchak VI., Gospodaryov DV. Introductory Chapter. In: Lushchak V &Gospodaryov D (ed.) Oxidative Stress and Diseases.Rijeka: InTech; 2012. p3-10.

[20] Shibata N., Kobayashi M. The role for oxidative stress in neurodegenerative diseases. Brain and Nerve 2008; 60(2) 157-170.

[21] Kadenbach B., Ramzan R., Vogt S. Degenerative diseases, oxidative stress and cytochrome c oxidase function. Trends in Molecular Medicine 2009; 15(4) 139-147.

[22] GospodaryovDV.,Volodymyr IL. Oxidative Stress: Cause and Consequence of Diseases. In: Lushchak V. &Gospodaryov D (ed.) Oxidative Stress and Diseases.Rijeka: InTech; 2012. p13-38.

[23] Beutler E. (). Glucose-6-phosphate dehydrogenase deficiency: a historical perspective. Blood. 2008; 111(1) 16-24.

[24] Gaskin RS.,Estwick D. Peddi R. G6PDH deficiency: its role in the high prevalence of hypertension and diabetes mellitus. Ethnicity & Disease. 2001; 11(4) 749-754.

[25] Carette C., Dubois-Laforgue D., Gautier JF. Timsit J. Diabetes mellitus and glucose-6-phosphate dehydrogenase deficiency: from one crisis to another. Diabetes & Metabolism. 2011; 37(1) 79-82.

[26] Ho HY., Cheng ML., Chiu D TY. G6PDH-an old bottle with new wine. Chang Gung Medical Journal. 2005; 28(9) 606-612.

[27] Winterbourn CC., Metodiewa D. The reaction of superoxide with reduced glutathione. Archives of Biochemistry and Biophysics. 1994; 314(2) 284-290.

[28] Lubos E., Handy DE., Loscalzo J. Role of oxidative stress and nitric oxide in atherothrombosis. Frontiers in Bioscience. 2008; 13 5323-5344.

[29] Förstermann U. Nitric oxide and oxidative stress in vascular disease, PflügersArchiv: European Journal of Physiology 2010; 459(6) 923-939.

[30] Matsui R., Xu S., Maitland KA., Hayes A., Leopold J.A., Handy DE., Loscalzo J., Cohen RA. Glucose-6 phosphate dehydrogenase deficiency decreases the vascular response to angiotensin II. Circulation. 2005; 112(2) 257-263.

[31] Rimm EB., Stampfer MJ. Folate and cardiovascular disease: one size does not fitall. Lancet. 2011; 378(9791) 544-546.

[32] Leopold JA. Loscalzo J. Oxidative enzymopathies and vascular disease. Arteriosclerosis, Thrombosis, and Vascular Biology. 2005; 25(7) 1332-1340.

[33] Pavel S., Smit NP., van der Meulen H., Kolb RM., de Groot AJ., van der Velden PA., Gruis NA., Bergman W. Homozygous germline mutation of CDKN2A/p16 and glucose-6-phosphate dehydrogenase deficiency in a multiple melanoma case. Melanoma Research. 2003; 13(2) 171-178.

[34] Cocco P., Ennas M.G., Melis MA., Sollaino C., Collu S., Fadda D., Gabbas A., Massarelli G., Rais M., Todde P., Angelucci E. Glucose-6-phosphate dehydrogenase polymorphism and lymphoma risk. Tumori 2007; 93(2) 121-123.

[35] TianWN.,Braunstein LD., Pang J., Stuhlmeier KM., Xi QC., Tian X., Stanton RC. Importance of glucose-6-phosphate dehydrogenase activity for cell growth. The Journal of Biological Chemistry. 1998; 273(17) 10609-10617.

[36] Batetta B., Pulisci D., Bonatesta RR., Sanna F., Piras S., Mulas MF., Spano O., Putzolu M., Broccia G., Dessì S. G6PDH activity and gene expression in leukemic cells from G6PDH-deficient subjects. Cancer Letters. 1999; 140(1-2) 53-58.

[37] Lushchak VI. Adaptive response to oxidative stress: Bacteria, fungi, plants and animals. Comparative Biochemistry and Physiology. Toxicology & Pharmacology CBP. 2011; 153(2) 175-190.

[38] Gupte SA. Targeting the Pentose Phosphate Pathway in Syndrome X-related Cardiovascular Complications. Drug Development Research. 2010; 71(3) 161-167.

[39] Amemiya-Kudo M., Shimano H., Hasty AH., Yahagi N., Yoshikawa T., Matsuzaka T., Okazaki H., Tamura Y., Iizuka Y., Ohashi K., Osuga J.-ichi., Harada K., Gotoda T., Sato R., Kimura S., Ishibashi S., Yamada N. Transcriptional activities of nuclear SREBP-1a, -1c, and -2 to different target promoters of lipogenic and cholesterogenic genes. Journal of Lipid Research. 2002; 43(8) 1220-1235.

[40] Kletzien RF., Harris PK., Foellmi LA. Glucose-6-phosphate dehydrogenase: a "housekeeping" enzyme subject to tissue-specific regulation by hormones, nutrients, and oxidant stress. The FASEB Journal: Official Publication of the Federation of American Societies for Experimental Biology. 1994; 8(2) 174-181.

[41] Franzè A., Ferrante MI., Fusco F., Santoro A., Sanzari E., Martini G., Ursini MV. Molecular anatomy of the human glucose 6-phosphate dehydrogenase core promoter. FEBS Letters. 1998; 437(3) 313-318.

[42] Lee JW., Choi AH., Ham M., Kim JW., Choe SS., Park J., Lee GY., Yoon KH., Kim JB. G6PDH up-regulation promotes pancreatic beta-cell dysfunction. Endocrinology. 2011; 152(3) 793-803.

[43] Ralser M., Benjamin IJ. Reductive stress on life span extension in C. elegans, BMC Research Notes. 2008; 1, 19.

[44] KirkmanHN.,Gaetani GF. Mammalian catalase: a venerable enzyme with new mysteries, Trends in Biochemical Sciences. 2007; 32(1) 44-50.

[45] Ogata M., Wang DH., Ogino K. Mammalian acatalasemia: the perspectives of bioinformatics and genetic toxicology. Acta Medica Okayama. 2008; 62(6) 345-361.

[46] Góth L. Catalasedeficiency and type 2 diabetes. Diabetes Care. 2008; 31(12) e93.

[47] Lubos E., Loscalzo J., Handy DE. Homocysteine and glutathione peroxidase-1, Antioxidants & Redox Signaling. 2007; 9(11) 1923-1940.

[48] Valentine JS., Doucette PA., Zittin Potter S. Copper-zinc superoxide dismutase and amyotrophic lateral sclerosis, Annual Review of Biochemistry. 2005; 74 563-593.

[49] Vucic S., Kiernan MC. Pathophysiology of neurodegeneration in familial amyotrophic lateral sclerosis. Current Molecular Medicine. 2009; 9(3) 255-272.

[50] Liochev SI., Fridovich I. Mutant Cu, Zn superoxide dismutases and familial amyotrophic lateral sclerosis: evaluation of oxidative hypotheses. Free Radical Biology & Medicine, 2003; 34(11) 1383-1389.

[51] Tortarolo M., Grignaschi G., Calvaresi N., Zennaro E., Spaltro G., Colovic M., Fracasso C., Guiso G., Elger B., Schneider H., Seilheimer B., Caccia S., Bendotti C. Glutamate AMPA receptors change in motor neurons of SOD1 G93A transgenic mice and their inhibition by a noncompetitive antagonist ameliorates the progression of amyotrophic lateral sclerosis-like disease. Journal of Neuroscience Research 2006; 83(1) 134-146.

[52] Beckman JS., Estévez AG., Crow JP., Barbeito L. Superoxide dismutase and the death of motoneurons in ALS, Trends in Neurosciences. 2001; 24(11) S15-20.

[53] Liochev SI., Fridovich I. Copper- and zinc-containing superoxide dismutase can act as a superoxide reductase and a superoxide oxidase. The Journal of Biological Chemistry. 2000; 275(49) 38482-38485.

[54] YimMB., Chock PB, Stadtman ER. Copper, zinc superoxide dismutase catalyzes hydroxyl radical production from hydrogen peroxide. Proceedings of the National Academy of Sciences of the United States of America. 1990; 87(13) 5006-5010.

[55] Kim KS., Choi SY., Kwon HY., Won MH., Kang TC., Kang JH. Aggregation of alpha-synuclein induced by the Cu,Zn-superoxide dismutase and hydrogen peroxide system. Free Radical Biology & Medicine. 2002; 32(6) 544-550.

[56] Rakhit R., Cunningham P., Furtos-Matei A., Dahan S., Qi XF., Crow JP., Cashman NR., Kondejewski LH. Chakrabartty A. Oxidation-induced misfolding and aggregation of superoxide dismutase and its implications for amyotrophic lateral sclerosis. The Journal of Biological Chemistry. 2002; 277(49) 47551-47556.

[57] Poon H.F., Hensley K., Thongboonkerd V., Merchant ML., Lynn BC., Pierce WM., Klein J. B., Calabrese V., Butterfield DA. Redox proteomics analysis of oxidatively modified proteins in G93A-SOD1 transgenic mice-a model of familial amyotrophic lateral sclerosis. Free Radical Biology & Medicine. 2005; 39(4) 453-462.

[58] Lightfoot TJ., Skibola CF., Smith AG., Forrest MS., Adamson PJ., Morgan GJ., Bracci PM., Roman E., Smith MT., Holly EA. Polymorphisms in the oxidative stress genes superoxide dismutase, glutathione peroxidase and catalase and risk of non-Hodgkin's lymphoma. Haematologica. 2006; 91(9) 1222-1227.

[59] Forsberg L., de Faire U., Morgenstern R. Oxidative stress, human genetic variation, disease. Archives of Biochemistry and Biophysics. 2001; 389(1) 84-93.

[60] GongoraMC., Harrison DG. Sad heart from no SOD. Hypertension. 2008; 51, 28-30.

[61] Hamanishi T., Furuta H., Kato H., Doi A., Tamai M., Shimomura H., Sakagashira S., Nishi M., Sasaki H., Sanke T., Nanjo K. Functional variants in the glutathione peroxidase-1 (GPx-1) gene are associated with increased intima-media thickness of carotid arteries and risk of macrovascular diseases in japanese type 2 diabetic patients. Diabetes. 2004; 53(9) 2455-2460.

[62] Sun LM., Shang Y., Zeng, YM., Deng YY., Cheng JF. hOGG1 polymorphism in atrophic gastritis and gastric cancer after Helicobacter pylori eradication. World Journal of Gastroenterology. 2010; 16(35) 4476-4482.

[63] Bravard A., Vacher M., Moritz E., Vaslin L., Hall J., Epe B., Radicella JP. Oxidative status of human OGG1-S326C polymorphic variant determines cellular DNA repair capacity. Cancer Research. 2009; 69, 3642-3649.

[64] Nakabeppu Y., Kajitani K., Sakamoto K., Yamaguchi H., Tsuchimoto D. MTH1, an oxidized purine nucleoside triphosphatase, prevents the cytotoxicity and neurotoxicity of oxidized purine nucleotides. DNA Repair. 2006; 5(7) 761-772.

[65] Halliwell B. Oxidative stress and cancer: have we moved forward?.Biochemical Journal. 2007; 401, 1-11.

[66] Hayes JD., Flanagan JU., Jowsey IR. Glutathione transferases. Annual Review of Pharmacology and Toxicology. 2005; 45, 51-88.

[67] Konig-Greger D., Riechelmann H., Wittich U., Gronau S. Genotype and phenotype of glutathione-S-transferase in patients with head and neck carcinoma, Otolaryngology-Head and Neck Surgery: Official Journal of American Academy of Otolaryngology-Head and Neck Surgery. 2004; 130(6) 718-725.

[68] Mohr LC., Rodgers JK., Silvestri GA. Glutathione S-transferase M1 polymorphism and the risk of lung cancer. Anticancer Research. 2003; 23(3A) 2111-2124.

[69] Benhamou S., Lee WJ., Alexandrie AK., Boffetta P., Bouchardy C., Butkiewicz D., Brockmöller J., Clapper ML., Daly A., Dolzan V., Ford J., Gaspari L., Haugen A., Hirvonen A., Husgafvel-Pursiainen K., Ingelman-Sundberg M., Kalina I., Kihara M., Kremers P., Le Marchand L., London SJ., Nazar-Stewart V., Onon-Kihara M., Rannug A., Romkes M., Ryberg D., Seidegard J., Shields P., Strange RC., Stücker I., To-Figueras J., Brennan P., Taioli, E. Meta- and pooled analyses of the effects of glutathione S-transferase M1 polymorphisms and smoking on lung cancer risk, Carcinogenesis. 2002; 23(8) 1343-1350.

[70] Cote M.L., KardiaSLR.,Wenzlaff AS., Land SJ., Schwartz AG. Combinations of glutathione S-transferase genotypes and risk of early-onset lung cancer in Caucasians and African Americans: a population-based study. Carcinogenesis. 2005; 26(4) 811-819.

[71] Lushchak VI. Oxidative stress and mechanisms of protection against it in bacteria. Biochemistry. 2001; 66(5) 592-609.

[72] Wautier JL. Schmidt Ann Marie. Protein glycation: a firm link to endothelial cell dysfunction. Circulation Research. 2004; 95(3) 233-238.

[73] Davis JS., Wu X. Current state and future challenges of chemoprevention. Discovery Medicine. 2012; 13(72), 385-90.

[74] Stavric B. Role of chemopreventers in human diet. Clinical Biochemistry. 1994; 27(5) 319-32.

[75] Ferguson L. Dietary influences on mutagenesis--where is this field going?.Enviromental and Molecular Mutagenesis. 2010; 51(8-9) 909-918.

[76] Stavric B. Antimutagens and anticarcinogens in foods. Food and Chemical Toxicology. 1994; 32(1) 79-90.

[77] Tanaka T., Shnimizu M., Moriwaki H. Cancer chemoprevention by carotenoids. Molecules. 2012; 17(3) 3202-3242.

[78] Tsuda H., Ohshima Y., Nomoto H., Fujita K., Matsuda E., Iigo M., Takasuka N., Moore MA. Cancer prevention by natural compounds. Drug Metabolism and Pharmacokinetics. 2004; 19(4) 245-263.

[79] Madrigal-BujaidarE., Viveros Martha E. La prevención química del cáncer. Revista del Instituto Nacional de Cancerología México. 1996; 42(1) 37-41.

[80] Gullett NP., RuhulAmin AR, Bayraktar S, Pezzuto JM, Shin DM, Khuri FR, Aggarwal BB, Surh YJ, Kucuk O. Cancer prevention with natural compounds. Seminars in Oncology. 2010; 37(3) 258-281.

[81] Steele VE. Current mechanistic approaches to the chemoprevention of cancer. Journal of Biochemistry and Molecular Biology. 2003; 36(1) 78-81.

[82] Ferguson LR, Bronzetti G, De Flora S. Mechanistic approaches to chemoprevention of mutation and cancer. Mutation Research. 2005; 591(1-2) 3-7.

[83] de Wit M., Nel P., Osthoff G., Labuschagne, MT. The effect of variety and location on cactus pear (Opuntiaficus-indica) fruit quality. PlantFoodsfor Human Nutrition.2010; 65(2) 136-145.

[84] Jolalpa-Barrera JL., Aguilar-Zamora A., Ortiz-Barreto O., García-López L. Producción y comercialización de tuna en fresco bajo diferentes modalidades en Hidalgo. Revista Mexicana de Agronegocios. 2011; 28, 605-614.

[85] Sumaya-Martínez MT., Suárez-Diéguez T., Cruz-Cansino NS., Alanís-García E., Sampedro JG. Innovación de productos de alto valor agregado a partir de la tuna mexicana. Revista Mexicana de Agronegocios.2010; 27, 435-441.

[86] Becerra-Jiménez J., Andrade-Cetto A. Effect of OpuntiastreptacanthaLem. on alphaglucosidase activity. Journal of Ethnopharmacology. 2012; 139(2) 493-496.

[87] Hahm SW., Park J., Son YS. Opuntiahumifusa stems lower blood glucose and choles-terol levels in streptozotocin-induced diabetic rats. Nutrition Research. 2011; 31(6) 479-87.

[88] Sumaya-Martínez MT., Cruz-Jaime S., Madrigal-Santillán EO., García-Paredes JD.,Cariño-Cortés R., Cruz-Cansino N., Valadez-Vega C., Martínez-Cardenas L., Ala-nís-García E. Betalain, Acid Ascorbic, Phenolic Contents and Antioxidant Propierties of Purple, Red, Yellow and White Cactus Pears. International Journal of Molecular Science. 2011; 12(10) 6452-6468.

[89] Fernández-López J., Almela L., Obón J., Castellar R. Determination of Antioxidant Constituents in Cactus Pear Fruits. Plant Foods for Human Nutrition. 2010; 65(3) 253-259

[90] Castellar R., Obón JM., Alacid M., Fernández-López JA. Color properties and stabili-ty of betacyanins from Opuntiafruits. Journal of Agricultural and Food Chemistry. 2003, 51(9) 2772-2776.

[91] Livrea MA., Tesoriere L. Antioxidative effects of cactus pear [Opuntiaficus-indica(L) Mill] fruits from Sicily and bioavailability of betalain components in healthy humans. ActaHorticulturae.2009; 811, 197-204.

[92] Dok-Go H., Lee KH., Kim HJ., Lee EH., Lee J., Song YS., Lee YH., Jin Ch., Lee YS., Cho J. Neuroprotective effects of antioxidative flavonoids, quercetin, (+)-dihydro-quercetin and quercetin 3-methyl ether, isolated from Opuntiaficus-indicavar. Sabo-ten. Brain Research. 2003; 965(1-2) 130-136.

[93] Galati EM., Mondello MR., Giuffrida D., Dugo G., Miceli N., Pergolizzi S., Taviano MF. Chemical characterization and biological effects of Sicilian Opuntiaficus-indi-ca(L.) Mill. Fruit juice: Antioxidant and antiulcerogenic activity. Journal of Agricul-tural and Food Chemistry.2003; 51(17) 4903-4908.

[94] García-Melo LF. Degree Thesis. Evaluación de la capacidad quimioprotectora del ju-go de tuna mediante la técnica de micronúcleos. Institute of Health Sciences, Autono-mous University of Hidalgo State, México. 2009.

[95] Seeram NP. Berry fruits for cancer prevention: Current status and future prospects. Journal of Agricultural and Food Chemistry. 2008; 56(3) 630-635.

[96] Howell AB., Vorsa N., Der Marderosian A., Foo L. Inhibition of adherence of P-fim-bricatedEscherischia coli to uroepithelial-cell surfaces by proanthocyanidin extracts from cranberries. The New England Journal of Medicine. 1998; 339(15) 1085-1086.

[97] Foo LY., Lu Y., Howell AB., Vorsa N. A-type proanthocyanidintrimers from cranber-ry that inhibit adherence of uropathogenic P-fimbriatedEscherichia coli. Journal of Natural Products. 2000; 63(9) 1225-1228.

[98] Burger O., Ofek I., Tabak M., Weiss EI., Sharon N., Neeman I. A high molecular mass constituent of cranberry juice inhibits Helicobacter pylori adhesion to human gastric mucus. FEMS Immunology and Medical Microbiology. 2000; 29(4) 295-301.

[99] Bomser J., MadhaviDL., Singletary K., Smith MA. In vitro anticancer activity of fruit extracts from Vaccinium species. PlantaMedica. 1996; 62(3) 212-216.

[100] Seeram NP., Adams LS., Zhang Y., Sand D., Heber D. Blackberry, black raspberry, blueberry, cranberry, red raspberry and strawberry extracts inhibit growth and stimulate apoptosis of human cancer cells in vitro. Journal of Agricultural and Food Chemistry. 2006; 54(25) 9329-9339.

[101] Sun J., Liu RH. Cranberry phytochemical extracts induce cell cycle arrest and apoptosis in human MCF-7 breast cancer cells. Cancer Letters. 2006; 241(1) 124-134.

[102] Neto CC., Amoroso JW., Liberty AM. Anticancer activities of cranberry phytochemicals: An update. Molecular Nutrition and Food Research. 2008; 52(1) S18-S27.

[103] Coates EM., Popa G., Gill CI., McCann MJ., McDougall GJ., Stewart D., Rowland I. Colon-available raspberry polyphenols exhibit anti-cancer effects on in vitro models of colon cancer. Journal of Carcinogenesis. 2007; 6, 4.

[104] Schmidt BM., Erdman JW., Lila MA. Differential effects of blueberry proanthocyanidins on androgen sensitive and insensitive human prostate cancer cell lines. Cancer Letters. 2006; 231(2) 240-246.

[105] Boateng J., Verghese M., Shackelford L., Walker LT., Khatiwada J., Ogutu S., Williams DS., Jones J., Guyton M., Asiamah D., Henderson F., Grant L., DeBruce M., Johnson A., Washington S., Chawan CB. Selected fruits reduce azoxymethane (AOM)-induced aberrant crypt foci (ACF) in Fisher 344 male rats. Food and Chemical Toxicology. 2007; 45(5) 725-732.

[106] Barros D., Amaral OB., Izquierdo I., Geracitano L., do CarmoBassolsRaseira M., Henriques AT., Ramírez MR. Behavioral and genoprotective effects of Vaccinium berries intake in mice. Pharmacology, Biochemistry and Behavior. 2006; 84(2) 229-234.

[107] Madrigal-Santillán E., Fragoso-Antonio S., Valadez-Vega C., Solano-Solano G., Pérez CZ., Sánchez-Gutiérrez M., Izquierdo-Vega JA., Gutiérrez-Salinas J., Esquivel-Soto J., Esquivel-Chirino C., Sumaya-Martínez T., Fregoso-Aguilar T., Mendoza-Pérez J., Morales-González JA. Investigation on the protective effects of cranberry against the DNA damage induced by benzo[a]pyrene. Molecules. 2012; 17(4) 4435-4451.

[108] Fellers PJ, Nikdel S, Lee HS. Nutrient content and nutrition labeling of several processed Florida citrus juice products. Journal of the American Dietetic Association. 1990; 90(8) 1079-84.

[109] Kumar A. Dogra S. Prakash A. Protective effect of naringin, a citrus flavonoid, against colchicine-induced cognitive dysfunction and oxidative damage in rats. Journal of Medicinal Food. 2010; 13(4) 976-984.

[110] Monroe KR., Murphy SP., Kolonel LN., Pike MC. Prospective study of grapefruit intake and risk of breast cancer in postmenopausal women: the Multiethnic Cohort Study. British Journal of Cancer. 2007; 97(3) 440-5.

[111] Kim EH., Hankinson SE., Eliassen AH., Willett WC. A prospective study of grape-fruit and grapefruit juice intake and breast cancer risk.British Journal of Cancer. 2008; 98(1) 240-241.

[112] Bressler R. Grapefruit juice and drug interactions. Exploring mechanisms of this in-teraction and potential toxicity for certain drugs. Geriatrics.2006; 61(11) 12-18.

[113] Alvarez-González I., Madrigal-Bujaidar E., Dorado V., Espinosa-Aguirre JJ. Inhibito-ry effect of naringin on the micronuclei induced by ifosfamide in mouse, and evalua-tion of its modulatory effect on the Cyp3a subfamily. Mutation Research. 2001; 1(480-481) 171-178.

[114] Alvarez-González I., Madrigal-Bujaidar E., Martino-Roaro L., Espinosa-Aguirre JJ. Antigenotoxic and antioxidant effect of grapefruit juice in mice treated with daunor-ubicin. Toxicology Letters. 2004; 152(3) 203-211.

[115] Razo-Aguilera G., Baez-Reyes R., Alvarez-González I., Paniagua-Pérez R., Madrigal-Bujaidar E. Inhibitory effect of grapefruit juice on the genotoxicity induced by hydro-gen peroxide in human lymphocytes. Food and Chemical Toxicology. 2011; 49(11) 2947-2953.

[116] Hernández-Ceruelos A., Madrigal-Santillán E., Morales-González JA., Chamorro-Ce-vallos G., Cassani-Galindo M., Madrigal-Bujaidar E. Antigenotoxic Effect of Chamo-millarecutita (L.) Rauschertssential Oil in Mouse Spermatogonial Cells, and Determination of Its Antioxidant Capacity in Vitro. International Journal of Molecu-lar Sciences. 2010; 11(10) 3793-3802.

[117] Jakolev V., Issac O., Flaskamp E. Pharmacological investigation with compounds of chamazulene and matricine. PlantaMedica. 1983; 49(10) 67-73.

[118] Viola H., Wasowski C., Levi de Stein M., Wolfman C., Silveira R., Dajas F., Medina JH., Paladini AC. Apigenin, a component of Matricariarecutitaflowers is a central benzodiazepine receptor-ligand with anxiolytic effects. PlantaMedica. 1995; 61(3) 213-216.

[119] McKay DL., Blumberg JB. A review of the bioactivity and potential health benefits of chamomile tea (MatricariarecutitaL.). Phytotherary Research. 2006; 20(7) 519-530.

[120] Hernández-Ceruelos A., Madrigal-Bujaidar E., de la Cruz C. Inhibitory effect of cha-momile essential oil on the sister chromatid exchanges induced by daunorubicin and methyl methanesulfonate in mouse bone marrow. Toxicology Letters. 2002; 135(1-2) 103-110.

[121] Hamid S., Sabir A., Khan S., Aziz P. Experimental cultivation of Silybummarianu-mand chemical composition of its oil. Pakistan Journal of Scientific and Industrial Re-search. 1983; 26, 244-246

[122] Morazzoni P., Bombardelli E. Silybummarianum(Cardusarianum). Fitoterapia. 1995; 66(1) 3-42.

[123] Lee DY-W., Liu Y. Molecular structure and stereochemistry of silybinA, silybin B, isosilybin A, and isosilybin B, isolated from Silybummarianum(Milk thistle). Journal of Natural Products. 2003; 66(9) 1171-1174.

[124] Ligeret H., Brault A., Vallerand D., Haddad Y., Haddad PS. Antioxidant and mitochondrial protective effects of silibinin in cold preservation-warm reperfusion liver injury. Journal of Ethnopharmacology. 2008; 115(3) 507-514.

[125] Shaker E., Mahmoud H., Mnaa S. Silymarin, the antioxidant component and Silybummarianumextracts prevents liver damage. Food and Chemical Toxicology 2010; 48(3) 803-806.

[126] AbouZid S. Silymarin, Natural Flavonolignans from Milk Thistle. In: RaoVenketeshwer (ed.) Phytochemicals-A Global Perspective of Their Role in Nutrition and Health. Rijeka: InTech; 2012. p255-272.

[127] Deep G., Oberlies NH., Kroll DJ., Agarwal R. Isosilybin B and isosilybin A inhibit growth, induce G1 arrest and cause apoptosis in human prostate cancer LNCaP and 22Rv1 cells. Carcinogenesis. 2007; 28(7) 1533-1542.

[128] Svobodová A., Zdařilová A., Walterová D., Vostálová J. Flavonolignans from Silybummarianummoderate UVA-induced oxidative damage to HaCaT keratinocytes. Journal of DermatologicalScience. 2007; 48(3) 213-224.

[129] Zermeño-Ayala, P. DegreeThesis. Evaluación del efecto quimiopreventivo de la silimarina sobre el daño genotóxico hepático producido por el consumo subcrónico de etanol. Institute of Health Sciences, Autonomous University of Hidalgo State, México. 2011

Oxidative Stress and Antioxidant Therapy in Chronic Kidney and Cardiovascular Disease

David M. Small and Glenda C. Gobe

Additional information is available at the end of the chapter

1. Introduction

Chronic kidney disease (CKD) and cardiovascular disease (CVD) have major impacts upon the health of populations worldwide, especially in Western societies. The progression of CKD or CVD independently exerts synergistic deleterious effects on the other, for example, patients with CKD are more likely to die of CVD than to develop renal failure. This overlap between CKD and CVD, in part, relates to common etiologies such as diabetes mellitus and hypertension, but important dynamic and bidirectional interactions between the cardiovascular system and kidneys may also explain the occurrence of concurrent organ dysfunction [1]. Cardio-renal syndrome (or reno-cardiac syndrome, the prefix depending on the primary failing organ) is becoming increasingly recognised [2]. Conventional treatment targeted at either syndrome generally reduces the onset or progression of the other [3]. Even though our understanding of various factors and steps involved in the pathogenesis of CKD and CVD and their obvious links has improved, a complete picture of the mechanisms involved is still unclear. Oxidative stress has been identified as one unifying mechanism in the pathogenesis of CKD and CVD [4]. This current chapter gives a brief review of recent literature on the relationship between CKD, CVD and oxidative stress and indicates how, by applying knowledge of the molecular controls of oxidative stress, this information may help improve targeted therapy with antioxidants for these diseases.

2. Pathogenesis of chronic kidney and cardiovascular disease – The links

It is, in fact, very difficult to separate these chronic diseases, because one is a complication of the other in many situations. The development and progression of CKD are closely linked with hypertension and dyslipidemia, both causes of renal failure. Diabetic nephropathy is arguably the leading cause of renal failure. CKD, hypertension and diabetes mellitus all involve endothelial dysfunction, a change well known in the development of atherosclerosis and CVD that includes coronary artery disease, heart failure, stroke and peripheral arterial disease [5]. Vascular calcification occurs in progressive atherosclerosis and CVD, but it is also an important part of vascular injury in end-stage renal disease (ESRD), where patients need renal replacement therapy to survive. It is paradoxical that approximately 50% of individuals with ESRD die from a cardiovascular cause [6]. Thus, CKD and CVD patients have closely-linked diseases with increasing morbidity and mortality. Prevention and treatment of these diseases are major aims in health systems worldwide.

The initiating causes of CKD are highly variable, with previously-mentioned hypertension and diabetes being two of the key ones [7]. Epidemiological studies reveal other strong risk factors for CKD, such as a previous episode of acute kidney damage, exposure to nephrotoxins, obesity, smoking, and increasing age [8, 9]. However, no matter the cause, the progressive structural changes that occur in the kidney are characteristically unifying [10]. The characteristics of CKD are tubulointerstitial inflammation and fibrosis, tubular atrophy, glomerulosclerosis, renal vasculopathy, and presence of granulation tissue. Alterations in the glomerulus include mesangial cell expansion and contraction of the glomerular tuft, followed by a proliferation of connective tissue which leads to significant damage at this first point of the filtration barrier. Structural changes that occur in the kidney produce a vicious cycle of cause and effect, thereby enhancing kidney damage and giving CKD its progressive nature. Whilst early pathological changes in the kidney can occur without clinical presentations, due to the high adaptability of the kidney [10], once the adaptive threshold is reached, the progression of CKD is rapid and the development of ESRD imminent. Vascular pathology exacerbates development of CKD, and it is perhaps here that the links with CVD are closest. Hypertension induces intimal and medial hypertrophy of the intrarenal arteries, leading to hypertensive nephropathy. This is followed by outer cortical glomerulosclerosis with local tubular atrophy and interstitial fibrosis. Compensatory hypertrophy of the inner-cortical glomeruli results, leading to hyperfiltration injury and global glomerulosclerosis. Note, however, that although glomerulopathy is an important characteristic of CKD, the incidence of tubulointerstitial fibrosis has the best correlation with CKD development [11]. As such, kidney tubular cells and renal fibroblasts may be the founding cell types in the progressive development of CKD.

The main clinical manifestation of CKD is a loss of glomerular filtration rate (GFR), allowing for staging of CKD with progressively decreasing (estimated) GFR. CKD staging was facilitated by the National Kidney Foundation (NKF) Kidney Disease Outcomes Quality Initiative (KDOQI) and the Kidney Disease - Improving Global Outcomes (KDIGO), an outcome that highlighted the condition and facilitated its increased diagnosis [12]. The first two stages have normal, or slightly reduced kidney function but some indication of structural deficit in two samples at least 90 days apart. Stages 3-5 are considered the most concerning, with Stage 3 now being sub-classified into Stages 3a and b because of their diagnostic importance. It is thought that stages 2 and 3 should be targeted with prophylactic therapies, such as lipid lowering drugs or RAS modifiers [13], to minimize the progression of CKD. Table 1 summarises GFR classification and staging for CKD.

Stage	GFR*	Description
1	90mL/Min	Normal renal function but abnormal urine findings, or structural abnormalities, or a genetic trait indicating kidney disease
2	60-89mL/min	Mildly reduced renal function, and other findings (as for stage 1) indicate kidney disease
3A	45-59mL/min	Moderately reduced kidney function
3B	30-44mL/min	
4	15-29mL/min	Severely reduced kidney function
5	<15mL/min or on dialysis	Very severe, or end-stage kidney failure (sometimes called established renal failure)

* Measured using the MDRD formula (MDRD= Modification of Diet in Renal Disease). All GFR values are normalized to an average surface area (size) of 1.73m²

Table 1. Classification and description of the different stages of CKD

Similar to CKD, the initiating causes for CVD are complex. Although exposure to cardiovascular risk factors such as hypertension, dyslipidemia and diabetes mellitus contributes to CVD, obesity, lack of physical exercise, smoking, genetics, and even depression, also play a role [14]. Common themes for causality are oxidative stress and inflammation, be they local or systemic. The prevalence of CVD also has a strong positive correlation with age, with more than 80% of cases of coronary artery disease and 75% of cases of congestive heart failure observed in geriatric patients [14]. Intrinsic cardiac aging, defined as the development of structural and functional alterations during aging, may render the heart more vulnerable to various stressors, and this ultimately favours the development of CVD. In the early stages of CVD, left ventricular hypertrophy and myocardial fibrosis may be seen in many patients [15]. The processes involved in their development, particularly in association with CKD, can be attributed to hypervolaemia, systemic arterial resistance, elevated blood pressure, large vessel compliance, and activation of pathways related to the parathyroid hormone–vitamin D–phosphate axis. Left ventricular hypertrophy and myocardial fibrosis also predispose to an increase in electric excitability and ventricular arrhythmias [16].

Heart failure resulting from CVD may be staged in a system similar to CKD. In its 2001 guidelines, the American College of Cardiology (ACC) and the American Heart Association working groups introduced four stages of heart failure [17]: Stage A with patients at high risk for developing heart failure in the future but no functional or structural heart disorder; Stage B with a structural heart disorder but no symptoms at any stage; Stage C with previous or current symptoms of heart failure in the context of an underlying structural heart problem, but managed with medical treatment; and Stage D with advanced disease requiring hospital-based support, a heart transplant or palliative care. The ACC staging system is useful in that Stage A may be considered pre-heart failure where intervention with treatment may prevent progression to overt symptoms.

The links between CKD and CVD are so close that it is often difficult to tease out individual causes and mechanisms, given their chronic nature. However, children with CKD present as a particular population without pre-existing symptomatic cardiac disease. This population could also receive significant benefit from preventing and treating CKD and thereby minimising the forthcoming development of CVD which is a major cause of death in children with advanced CKD. Left ventricular hypertrophy and dysfunction, and early markers of atherosclerosis such as increased intimal-medial thickness and stiffness of the carotid artery, and coronary artery calcification, may develop in children with CKD. Early CKD, before needing dialysis, is the optimal time to identify and modify risk factors and intervene in an effort to avert risk of premature cardiac disease and death in these children [18]. These observations have sparked added interest in the mechanisms of the chronic diseases, and in ways to target these mechanisms with additional therapies, such as antioxidants.

2.1. Inflammation and chronic kidney and cardiovascular disease

The circulating nature of many inflammatory mediators such as cytokines, and inflammatory or immune cells, indicates that the immune system can act as a mediator of kidney-heart cross-talk and may be involved in the reciprocal dysfunction that is encountered commonly in the cardio-renal syndromes. Chronic inflammation may follow acute inflammation, but in many chronic diseases like CKD and CVD, it is likely that it begins as a low-grade response with no initial manifestation of an acute reaction. There are many links with visceral obesity and with increased secretion of inflammatory mediators seen in visceral fat [15]. Proinflammatory cytokines are produced by adipocytes, and also cells in the adipose stroma. The links with oxidative stress as an endogenous driver of the chronic diseases become immediately obvious when one admits the close association between oxidative stress and inflammation. The characteristics of dyslipidaemia (elevated serum triglycerides, elevated low-density lipoprotein cholesterol, and/or low high-density lipoprotein cholesterol) are also often seen in obese patients and these are all recognized as risk factors for atherosclerosis. The links between obesity, inflammation, dyslipidaemia, CKD and CVD also occur through yet another syndrome, metabolic syndrome. An improved understanding of the precise molecular mechanisms by which chronic inflammation modifies disease is required before the full implications of its presence, including links with persistent oxidative stress as a cause of chronic disease can be realized.

3. Oxidative stress and chronic kidney and cardiovascular disease

3.1. Understanding oxidative stress

Oxidative stress has been implicated in various pathological systems that are prevalent in both CKD and CVD, most importantly inflammation and fibrosis. Chronic inflammation is induced by biological (eg. infections, autoimmune disease), chemical (eg. drugs, environmental toxins), and physical factors (eg. lack of physical activity) [19]. The inflammatory cells are then a source of free radicals in the forms of reactive oxygen and nitrogen species, although reactive oxygen species (ROS) are considered the most common. The highly reactive ROS are capable of damaging various structures and functional pathways in cells. In consequence, the presence of inflammatory cells is stimulated by cell damage caused by ROS, creating a cycle of chronic damage that is difficult to break. Oxidative stress arises from alterations in the oxidation-reduction balance of cells. Normally, ROS are countered by endogenous natural defences known as antioxidants, and it is the imbalance between ROS and antioxidants which favours greater relative levels of ROS, thereby giving rise to a state of oxidative stress [20-22]. The simple oxidant "imbalance" theory has now grown to incorporate the various crucial pathways and cell metabolism that are also controlled by the interplay between oxidants and antioxidants [23-27]. The rationale for antioxidant therapies lies in restoring imbalances in the redox environment of cells.

The main ROS are superoxide ($O_2^{•-}$), the hydroxyl radical ($OH^•$) and hydrogen peroxide (H_2O_2). Mitochondria are considered the major source of ROS, however other contributing sites of ROS generation include the endoplasmic reticulum, peroxisomes and lysosomes [28-30]. Estimated levels of ROS within mitochondria are 5-10 fold higher than cystolic and nuclear compartments in cells [31] due to the presence of the electron transport chain (ETC) within the mitochondrial inner membrane. 1-3% of inspired molecular oxygen (O_2) is converted to the most common of the ROS, $O_2^{•-}$ [32, 33], a powerful precursor of H_2O_2. Although cellular H_2O_2 is stable in this form, it has the potential to interact with a variety of substrates to cause damage, especially in the presence of the ferrous iron (Fe^{2+}), which leads to cleavage and formation of the most reactive and damaging of the ROS, the $OH^•$ [34]. In healthy metabolic cells, the production of the potentially harmful H_2O_2 is countered by the catalizing actions of mitochondrial or cystolic catalase (CAT) or thiol peroxidases into water and O_2. The ETC consists of 5 multi-enzyme complexes responsible for maintaining the mitochondrial membrane potential and ATP generation. Each of these complexes presents a site of ROS generation, however complexes I and III have been identified as primary sites of $O_2^{•-}$ generation [35-38]. ROS generation from mitochondrial complexes increases with age in mice [39]. In humans, Granata and colleagues [40] have demonstrated that patients with CKD and haemodialysis patients display impaired mitochondrial respiration.

Agreement on the role of oxidative stress in the pathogenesis of chronic disease is, however, not complete. Oxidants are involved in highly conserved basic physiological processes and are effectors of their downstream pathways [41, 42]. The specific mechanisms for "oxidative stress" are difficult to define because of the rapidity of oxidant signalling [31]. For example, protein tyrosine phosphatases are major targets for oxidant signalling since they contain the amino acid residue cysteine that is highly susceptible to oxidative modification [43]. Meng and colleagues [25] demonstrated the oxidation of the SH2 domain of the platelet-derived growth factor (PDGF) receptor, which contains protein tyrosine phosphatases, in response to PDGF binding. This may indicate the induction of free radicals in response to receptor activation by a cognate ligand in a process that is similar to phosphorylation cascades of intracellular signalling.

3.2. Endogenous antioxidants – Metabolism or disease modifiers

The production of ROS is usually in balance with the availability and cellular localisation of antioxidant enzymes such as superoxide dismutase (SOD), CAT and glutathione peroxidase (Gpx). *In vivo* studies have found accumulated oxidative damage occurs from decreased levels of these enzymes rather than increased ROS production [44, 45]. However, adequate levels of both are likely to be vital for normal cell function. Mitochondria possess their own pool of antioxidants to counteract their generation of ROS. Mitochondrial manganese-SOD (Mn-SOD) converts $O_2^{\cdot-}$ to H_2O_2 which is then decomposed to harmless H_2O and O_2 by CAT and Gpx [46]. Copper/zinc-SOD (Cu/Zn-SOD) has been implicated in stabilizing $O_2^{\cdot-}$ within other cellular compartments, especially peroxisomes, and must be considered in maintenance of the redox state of the whole cell [47, 48]. Limited antioxidant actions of Cu/Zn-SOD may also occur within the inter-membrane space [49]. There is no evidence to indicate that glutathione synthesis occurs within mitochondria, however the mitochondria have their own distinct pool of glutathione required for the formation of Gpx [50].

Among the various endogenous defences against ROS, glutathione homeostasis is critical for a cellular redox environment. Glutathione-linked enzymatic defences of this family include Gpx, glutathione-S-transferase (GST), glutaredoxins (Grx), thioredoxins (Trx), and peroxiredoxins (Prx) [51]. Many of these proteins are known to interact with each other, forming redox networks that have come under investigation for their contribution to dysfunctional oxidant pathways. Mitochondrial-specific isoforms of these proteins also exist and include Grx2, Grx5, Trx2 and Prx3 [52-54], which may be more critical for cell survival compared to their cystolic counterparts [50]. Mitochondrial dysfunction, resulting in depleted ATP synthesis, has the potential to reduce the redox control of glutathione since the rate of glutathione synthesis is ATP-dependent [55]. Intracellular synthesis of glutathione from amino acid derivatives (glycine, glutamic acid and cysteine) accounts for the majority of cellular glutathione compared with extracellular glutathione uptake [56]. Antioxidant networks in which there is interplay, crosstalk and synergism to efficiently and specifically scavenge ROS, may also exist. If this is the case, these antioxidant networks could be harnessed to develop poly-therapeutic antioxidant supplements to combat oxidant-related pathologies, like CKD and CVD.

3.3. Oxidative stress and transcriptional control

The role of oxidative stress in upstream transcriptional gene regulation is becoming increasingly recognised. Not only does this provide insight into the physiological role of oxidative stress, but presents regulatory systems that are possibly prone to deregulation. Furthermore, these sites present targets for pharmacological intervention. Peroxisome proliferator-activated receptors (PPARs) are members of the nuclear hormone receptor superfamily of ligand-dependant transcription factors which have been shown to alter during CKD and CVD [57-59]. They have important roles in the transcriptional regulation of cell differentiation, lipid metabolism, glucose homeostasis, cell cycle progression, and inflammation. There are three PPAR isoforms – α, β/δ and γ. Peroxisome proliferator gamma coactivator (PGCα), in association with PPARγ activation, leads to a variety of cellular protective responses including mitochondrial biogenesis [57]. PPARγ regulation in chronic disease is increasingly recognised, with oxidative stress as the unifying initiating feature. Omega-3 polyunsaturated fatty acids (PUFA) reduce inflammation in kidney tubular epithelial cells by upregulating PPARγ [60]. PPARγ activation by pioglitazone reduced cyclo-oxygenase 2 (COX2) expression in smooth muscle cells from hypertensive rats, and upregulated endogenous antioxidants Mn- and Cu/Zn-SOD [61].

Recently, the protective responses of the nuclear factor E2-related factor 2/Kelch-like ECH-associated protein 1 (Nrf2/Keap1)/antioxidant response element (ARE) were noted [62]. Nrf2 is a nuclear transcription factor that is suppressed in the cytoplasm by the physical binding of Keap1 preventing its translocation into the nucleus. Nrf2 is activated by a loss of Keap1 binding by alterations in cellular redox status, such as increased ROS, by-products of oxidative damage, and reduced antioxidant capacity, thereby promoting its transcriptional response at the ARE [63]. The ARE is a vital component of the promoter regions of genes encoding detoxifying, antioxidant, and glutathione-regulatory enzymes such as quinone-reductase, glutathione-peroxidases, glutathione-reductase, thioredoxins and thioredoxin-reductase, peroxiredoxins, gamma-glutamyl cysteine, heme-oxygenase-1 (HO-1), CAT, SOD metallothionein and ferritin [64-67]. Important to note is that by-products of oxidative damage such a 4-hydroxynoneal and J-isoprostanes act as endogenous activators of Nrf2 [68, 69]. Thus, NRF2/Keap1 and the ARE play a crucial role in cellular defence against ROS. Recent pharmacological protocols have allowed the modulation of this pathway to enhance the capabilities of cells to combat oxidative stress and inflammation [70].

3.4. CKD and CVD are unified by oxidative stress

Chronic diseases of the kidney possess various commonalities to chronic disease of the cardiovascular system which can be linked through pathways controlled by oxidative stress, as shown in Figure 1. Vascular, cellular and biochemical factors all contribute. Increased serum uric acid levels (hyperuricaemia) can arise from increased purine metabolism, increasing age and decreased renal excretion, and have harmful systemic effects. Hyperuricaemia is associated with an increased risk for development and progression of CKD. Hyperuricemia is also a risk factor associated with coronary artery disease [71], left ventricular hypertrophy [72], atrial fibrillation [73], myocardial infarction [74] and ischemic stroke [75]. A 20.6% prevalence of hyperuricemia was found in a cross-sectional study of 18,020 CKD patients [76], and a positive correlation was found between serum uric acid and serum creatinine with impaired renal function [77]. Retention of uremic toxins promotes inflammation and oxidative stress, by priming the acute inflammatory polymorphonuclear lymphocytes, activating interleukin (IL)-1β and IL-8 [78] and stimulating the innate immune response through CD8+ cells [79]. Additionally, uric acid synthesis can promote oxidative stress directly through the activity of xanthine oxidoreductase. This enzyme is synthesized as xanthine dehydrogenase, which can be converted to xanthine oxidase by calcium-dependant proteolysis [80] or modification of cysteine residues [81]. In doing so, the enzyme loses its capacity to bind NADH by alterations in its catalytic site and, instead, transfers electrons from O_2, thereby generating O_2^- [82]. However, the role of uric acid in many conditions associated with oxidative stress is not clear and there are experimental and clinical data showing that uric acid also has a role *in vivo* as an anti-oxidant [83].

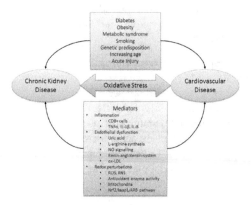

Figure 1. Chronic kidney disease and cardiovascular disease are unified by oxidative stress. Mutual risk factors influence the development and progression of CKD and CVD and can either be modifiable (diabetes, obesity, metabolic syndrome, smoking) or non-modifiable (genetic predisposition, increasing age, acute injury). Oxidative stress has been implicated in the majority of initiating factors. The progression of CKD to CVD, or vice versa, is mediated through: (1) inflammation and the release of pro-inflammatory cytokines such as tumor necrosis factor-α (TNFα), interleukin-1β (IL-1β) and IL-8 from activated lymphocytes; (2) endothelial dysfunction due to increased retention of uremic toxins, and decreased L-arginine synthesis which causes alterations in nitric oxide (NO) signalling - dyslipidaemia and associated pro-oxidative/inflammatory state lead to increased oxidised-low density lipoproteins (ox-LDL), a major component in the pathogenesis of atherosclerosis; (3) redox perturbations that ultimately underlie oxidative stress due to an imbalance between the production of reactive oxygen species (ROS)/reactive nitrogen species (RNS) and endogenous antioxidants, leading to mitochondrial dysfunction and alterations in redox sensitive pathways such as Nrf2/keap1/ARB.

The kidney is a vital source of L-arginine which is a precursor for nitric oxide (NO). A reduction in renal mass can therefore reduce the production of L-arginine and NO activity. NO is vital for regular vascular endothelial cell function, and decreased amounts have the potential to manifest into CVD [84]. Additionally, oxidized low density lipoprotein (ox-LDL), a by-product of oxidative damage in human blood, plays a pivotal role in the pathogenesis of atherosclerosis [85]. There is also a possible link between CVD and CKD that is regulated by oxidative stress through a functional mitochondrial angiotensin system [86]. Angiotensin type II receptors were co-localised with angiotensin on the inner mitochondrial membrane of human mononuclear cells and mouse renal tubular cells. This system was found to modulate mitochondrial NO production and respiration.

4. Antioxidant therapies in chronic kidney and cardiovascular disease

The current state of antioxidant therapies for CKD and CVD is one of promise, but not without controversy. *In vitro* studies commonly identify agents that are able to detoxify harmful oxidants. However, these studies are criticised for their isolated, non-holistic, nature [87, 88]. It is largely the positive pre-clinical results from *in vivo* studies, usually in rodents, which drive progress for applicability in chronic human disease, but even these show considerable discrepancies in translation into patients. Despite the well-documented dysregulated endogenous oxidant/antioxidant profile in chronic degenerative disorders such as CVD and CKD, there is still evidence that certain antioxidants have no effect [89-92]. It may first be important to identify patients having an altered oxidative stress profile, since this population provides an ideal "intention to treat" cohort. The following trials of antioxidants need then to be rigorous, identifying not only any positive patient outcomes, but also the underlying mechanism, and of course any deleterious outcome. Various approaches have been taken to reduce oxidative stress in models of CKD and accelerated CVD, ranging from reducing oxidant intake in food stuffs [93, 94] to targeted polypharmaceutical compounds. The benefit of rigorous review of outcome from antioxidant therapies in either CKD or CVD is that the primary and secondary outcomes related to both can be measured. In the following section, some antioxidants used for CKD or CVD are reviewed, as shown in Figure 2.

4.1. N-acetylcysteine – An antioxidant with promise

N-acetyl cysteine (NAC) acts as an essential precursor to many endogenous antioxidants involved in the decomposition of peroxides [95]. NAC attenuates oxidative stress from various underlying causes by replenishing intracellular glutathione stores. Glutathione is synthesized in the body by three amino acids by the catalysing of intracellular enzymes gamma-glutamylcysteine synthetase and glutathione synthetase. L-glutamic acid and glycine are two precursors of glutathione that are biologically and readily available. However, the limiting precursor to glutathione biosynthesis and the third amino acid, L-cysteine, is not readily available in a human diet. Although the primary basis for NAC supplementation is to replenish cellular cysteine levels to maintain intracellular glutathione and thus redox control, the sulfhydral-thiol group of L-cysteine is also able to exert direct antioxidant effects by scavenging free radicals, and NAC may also exert its protective effects against 2,3,5-tris(glutathion-S-yl)-hydroquinone toxicity. This was demonstrated in isolated renal tubular epithelial cells, in part by the activation of extracellular signal regulated protein kinase (ERK) 1/2 [96].

The results of NAC supplementation in kidney disease have been variable and largely dependent on the type and cause of kidney injury and also the timing of treatment. In cultured human proximal tubular epithelial cells, NAC reduced lipid peroxidation and maintained the mitochondrial membrane potential, thereby preventing apoptosis following H_2O_2 administration [97]. Although NAC had no significant effect on markers of oxidative stress and inflammation in rats following unilateral ureteral obstruction [98], it reduced kidney malondialdehyde (MDA) levels in a diabetic mouse model [99]. The treatment of CKD patients with NAC with the aim of improving renal function and preventing ESKD has been largely disappointing, with no evidence of reduction in proteinuria [100, 101]. However, NAC seems to exert the greatest antioxidant and anti-inflammatory properties when used against the greatest injury, such as in ESKD patients receiving either haemodialysis or peritoneal dialysis. In those cases, NAC reduced serum 8-isoprostane and the inflammatory cy-tokine IL-6 [102, 103]. A recent systemic review on antioxidant therapy in hemodialysis patients highlighted NAC as the most efficacious agent in decreasing oxidative stress [104].

The effect of NAC on cardiovascular pathologies is less well investigated than CKD. Crespo *et al.*, (2011) demonstrated *in vivo* that, although long-term NAC supplementation improved cardiac function, it did not delay progression to cardiomyopathy [105]. Endothelial dysfunc-tion caused by uremic toxins such as indoxyl sulphate induced ROS-dependent expression of the pro-inflammatory and pro-oxidant nuclear factor-κB (NF-κB), which was ameliorated by NAC pre-treatment [106].

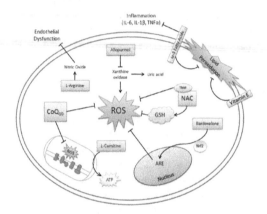

Figure 2. Cellular sites for antioxidant therapy targets in CKD and CVD. Inflammation, lipid peroxidation and reactive oxygen species (ROS) from mitochondrial, cytoplasmic and extracellular sources contribute to oxidative stress. Vitamin E incorporates into the phospholipid bilayer halting lipid peroxidation chain reactions. Omega (ω)-3 fatty acids dis-place arachadonic acid in the cell membrane and thus reduce arachadonic acid-derived ROS, but also significantly re-duce inflammation and subsequent fibrosis. The cysteine residue of N-acetyl-cysteine (NAC) is a precursor for glutathione (GSH) synthesis, and the thiol group is able to scavenge ROS directly. Bardoxolone exerts transcriptional control by promoting nuclear translocation of Nrf2, facilitating antioxidant response element (ARE) binding that upre-gulates endogenous antioxidant enzyme activity. Allopurinol inhibits xanthine oxidase-derived ROS and the damag-ing effects of hyperuricemia. Coenzyme Q_{10} (CoQ_{10}) enhances the efficacy of electron transport in the mitochondria, thereby reducing mitochondrial-derived ROS – it is also able to directly scavenge ROS. L-carnitine enhances mitochon-drial fatty acid synthesis and subsequent ATP production and thereby maintains cell health. L-arginine is a precursor for nitric oxide which restores endothelial function.

4.2. Vitamin E – An established antioxidant with controversial outcomes

Vitamin E, or α-tocopherol, is a lipid-soluble antioxidant that incorporates into the plasma membrane of cells, thereby scavenging free radicals, mainly the peroxyl radical, and halting lipid peroxidation chain reactions [107]. A benefit of α-tocopherol is its ability to restore its antioxidant capacity from its oxidized form following free radical scavenging, and incorpo-rate back into the plasma membrane. Vitamin C (ascorbic acid) is able to directly reduce α-tocopherol [108-110], and intracellular glutathione and lipoic acid can restore α-tocopherol indirectly by restoring vitamin C [111]. This is a prime example of a cellular antioxidant network prone to dysregulation. Administration of α-tocopherol to kidney proximal tubular cells in culture decreased cisplatin-induced ROS and increased cell viability [112]. The beneficial effects of α-tocopherol are not limited to its antioxidant properties, and recently attention has focused on its blood oxygenising and endogenous cell signalling functions [113]. Vitamin E foodstuffs primarily consist of α-tocotreinol, an isoform of α-tocopherol which has higher antioxidant efficacy in biological membranes. Despite this, the uptake and distribution of α-tocotreinol is far less than α-tocopherol. Therefore, the basis of vitamin E supplementation is to enhance α-tocopherol levels in cell plasma membranes to prevent lipid peroxidation and resultant oxidative stress. One drawback of α-tocopherol is that it takes several days of pre-treatment to exhibit antioxidant effects [114].

Vitamin E therapy has been extensively researched for renal and cardiovascular benefits in human disease populations. Nevertheless, confounding reports mean there is a lack of consensus as to whether vitamin E therapy induces an overall benefit. It is known that patients with CKD stage 4 display the largest decrease in serum α-tocopherol levels following a progressive decline from stage 1 indicating an increased need for α-tocopherol in the CKD population [115]. Interestingly, within the same cohort of patients, a positive correlation of serum α-tocopherol levels and GFR was found [115]. A large scale trial concluded that vitamin E supplementation to cardiovascular high-risk patients over 4.5 years induced no benefit to cardiovascular outcome [92]. The results from the Selenium and Vitamin E Cancer Prevention Trial (SELECT) are of greater concern. They suggest that vitamin E supplementation significantly increases the risk of prostate cancer for young healthy men [116]. Most studies finding beneficial outcomes of α-tocopherol supplementation have largely focused on the ESKD dialysis populations compared to healthy controls and found a reduced risk of CVD, decreased oxidative stress and increased erythrocyte antioxidants SOD, Gpx and CAT [117-119]. The use of α-tocopherol in CKD patients is not without controversy. Miller and colleagues (2005) concluded that high-dose (≥400 IU/day) vitamin E supplementation may increase all cause mortality which may be due to α-tocopherol displacing gamma-(γ)-tocopherol and delta-(δ)-tocopherol in the body [120]. However, this study was highly criticized owing to a bias in data analysis and numerous methodological flaws [121-130]. The apparent lack of clarity surrounding vitamin E supplementation and associated renal and cardiovascular outcomes appears to stem largely from differences in trial design and failure to specify the form of tocopherol used.

4.3. Coenzyme Q_{10} - Maintaining mitochondrial health

The heart and kidneys contain the highest endogenous levels of co-enzymes $(Co)Q_9$ and CoQ_{10} compared to all other organs [131, 132]. This is likely due to the respective reliance on aerobic metabolism and high density of mitochondria in the intrinsic functioning cells from these organs. It is imperative that endogenous CoQ_{10} levels are maintained to ensure mitochondrial health, and this forms the rationale for CoQ_{10} therapy. CoQ_{10} is a fundamental lipid-soluble component of all cell membranes including those enclosing subcellular compartments. The physiological roles of CoQ_{10} act mostly within the mitochondria where it has three well-characterised functions: (1) the transfer of electrons from complexes I and II to complex III along the ETC of the inner mitochondrial membrane and subsequent membrane polarisation and ATP generation [133, 134]; (2) the pro-oxidant generation of $O_2^{•-}$ and H_2O_2 [135, 136]; and (3) the anti-oxidant quenching of free radicals [137]. The continual oxidation-reduction cycle, and existence of CoQ_{10} in three different redox states, explains its actions as an important cellular redox modulator through its pro-oxidant and antioxidant actions. The fully oxidised form of CoQ_{10}, or ubiquinone, is able to accept electrons, primarily from NADH, to become fully reduced (ubiquinol - CoQ_{10}-H_2). The reduced form of CoQ_{10} is able to give up electrons, thereby scavenging free radicals. The intermediate of ubiquinone and ubiquinol is the univalently-reduced ubisemiquinone (CoQ_{10}-$H^•$) which acts as a pro-oxidant to form $O_2^{•-}$ and, subsequently, H_2O_2.

The major antioxidant role of CoQ_{10} is in preventing lipid peroxidation directly, and by interactions with α-tocopherol [138]. Ubiquinol is able to donate a hydrogen atom and thus quench peroxyl radicals, preventing lipid peroxidation chain reactions. CoQ_{10} and α-tocopherol co-operate as antioxidants through the actions of CoQ_{10}-H_2 restoring α-tocopheroxyl back to α-tocopherol [109, 139]. However, the reactivity of α-tocopherol with peroxy radicals far exceeds that of ubiquinol with peroxyl radicals, suggesting that, *in vivo*, ubiquinols do not act as antioxidants but regenerate the antioxidant properties of α-tocopherols [140]. This is in accordance with *in vivo* studies investigating the effects of CoQ_{10} supplementation which have primarily found a limited antioxidant capacity. CoQ_{10}, acting as a pro-oxidant in all biological membranes including the Golgi, endosome/lysosome systems, as well as mitochondria, has led to much criticism regarding the claimed antioxidant power of CoQ_{10} supplementation in humans [141]. Nonetheless, many *in vitro* studies demonstrate antioxidant properties of CoQ_{10} in single cells, and benefits of CoQ_{10} supplementation in humans are attributed to its ability to maintain efficient mitochondrial energy metabolism and thus prevent mitochondrial dysfunction, rather than act as a direct cellular antioxidant. CoQ_{10} supplementation *in vivo* reduced protein oxidation in skeletal muscle of rats but had no effect on mitochondrial H_2O_2 production in the kidney [142]. However, Ishikawa and colleagues (2011) demonstrated a decrease in kidney $O_2^{•-}$ levels in hemi-nephrectomised rats on a CoQ_{10} supplemented diet, and increased renal function compared with rats on a control diet [143]. Recently, CoQ_{10} supplementation improved left ventricular diastolic dysfunction and remodelling and reduced oxidative stress in a mouse model of type 2 diabetes [144]. CoQ_{10} supplementation in CVD patients also receiving statin therapy is becoming increasingly popular due to the CoQ_{10}-inhibitory actions of statins. CoQ_{10} levels decrease with age, but there are no studies measuring endogenous CoQ_{10} levels in CKD or CVD patients and this could prove vital in the identification of population where CoQ_{10} therapy may have beneficial outcomes.

4.4. Omega-3 poly-unsaturated fatty acids – Inflammation and oxidative stress

Inflammation and fibrosis are causes, as well as consequences, of oxidative stress [145, 146]. Direct targeting of inflammatory and fibrotic pathways with more specific modifying compounds presents a way to indirectly decrease oxidative stress in chronic pathologies. Long chain omega-3 PUFA, including docosahexanoic acid (DHA) and eicosapentanoic acid (EPA), have been investigated in a large range of *in vitro* and *in vivo* models and found to possess anti-inflammatory properties. Recently, omega-3 fatty acid treatment of peripheral blood mononuclear cells from pre-dialysis CKD patients reduced the inflammatory markers IL-6, IL-1β, tumor necrosis factor (TNF)-α and C-reactive protein to levels observed in healthy subjects [147]. Although the beneficial effects of EPA/DHA are attributed to their anti-inflammatory properties, they are also known to enhance endogenous antioxidant defence systems such as γ-glutamyl-cysteinyl ligase and glutathione reductase [148]. DHA and EPA incorporate into the phospholipid bilayer of cells where they displace arachidonic acid. Arachidonic acid can generate ROS through the COX2 and xanthine oxidase inflammatory pathways. DHA/EPA administration to renal epithelial cells and macrophages suppresses this pro-oxidant pathway [149]. Furthermore, chemoattractants derived from EPA are less potent that those derived from arachidonic acid [150, 151]. Recently, *in vitro* studies determined that EPA and DHA attenuated TNF-α-stimulated monocyte chemoattractant protein (MCP)-1 gene expression by interacting with ERK and NF-κB in rat mesangial cells [152]. Earlier evidence had shown that EPA and DHA inhibit NF-κB expression by stimulating PPARs in human kidney-2 cells *in vitro* [60]. *In vivo* studies have now confirmed an improvement in kidney function and structure using EPA/DHA supplementation, with reduced oxidative stress, inflammation and tubulointerstitial fibrosis through the reversal of inflammatory and oxidant pathways [153, 154]. Recently, a highly beneficial outcome of fish oil supplementation was found with heart failure patients with co-morbid diabetes [155]. Clinical studies have found fish oil treatment modulates lipid levels [156, 157], and has antithrombotic [158, 159] and anti-hypertensive effects due to its vascular and endothelial actions [160].

4.5. Allopurinol – A xanthine oxidase inhibitor

Allopurinol treatment aims is to inhibit xanthine oxidase to decrease serum uric acid and its associated toxic effects. Allopurinol and its metabolite, oxypurinol, act as competitive substrates for xanthine oxidase. They enhance urinary urate excretion and block uric acid reabsorption by urate transporters in the proximal tubule, thereby facilitating enhanced uric acid excretion [161-163]. Allopurinol treatment of diabetic mice attenuated hyperuricaemia, albuminuria, and tubulointerstitial injury [164]. Allopurinol may also have antioxidant activities in addition to its enzyme inhibitory activities, by scavenging OH$^\bullet$ as well as chlorine dioxide and HOCl [165, 166]. Although later *in vivo* studies revealed that rat serum obtained after oral administration of allopurinol did not contain allopurinol levels sufficient to scavenge free radicals [167], inhibition of xanthine oxidase-dependent production of NO$^\bullet$ and ROS provides allopurinol an indirect mechanism for decreasing oxidative stress in hyperuricaemic CKD patients. Interventional studies of use of allopurinol in renal disease have shown improved uric acid levels, GFR, cardiovascular outcomes and delayed CKD progression. A prospective randomised trial of 113 patients with GFR <60 ml/min/1.73m^3 given allopurinol 100mg/d for 2 years found an increase in GFR of 1.31 ml/min/1.73m^3 compared to the controls which decreased, and a 71% decreased risk of CVD [168]. Interestingly, Kanbay and colleagues (2007) found that allopurinol at 300mg/d over 3 months improved GFR, uric acid and C-reactive protein levels but made no change to proteinuria [169]. Allopurinol given to ESKD patients on hemodialysis reduced the risk of CVD by decreasing serum low density lipoproteins, triglycerides and uric acid [170]. Large, long-term interventional studies investigating kidney function in the CKD, and CVD, populations are needed to fully determine if allopurinol is cardio- and reno-protective via anti-oxidant mechanisms.

4.6. Bardoxolone methyl - Targeting the Nrf/Keap1/ARE pathway

A different approach has been investigated by modulating pathways that respond to oxidative stress, rather than targeting ROS by directly increasing endogenous antioxidants. The Nrf2/keap1/ARE pathway presents an exciting target to enhance the oxidant detoxifying capabilities of cells. Bardoxolone methyl [2-cyano-3,12-dioxooleana-1,9(11)-dien-28-oic acid (CDDO-Me)] is a potent activator of the Nrf2/keap1/ARE pathway and currently shows promise for halting the progressive decline of GFR in type 2 diabetic CKD patients [171, 172]. Bardoxolone methyl is a triterperoid derived from natural plant products that has undergone oleanolic acid-based modification [173]. Its mechanism of action is largely unknown, however, it induces an overall antioxidative protective effect with anti-inflammatory and cytoprotective characteristics [174, 175]. Bardoxolone methyl administered to mice ameliorated ischemia-reperfusion induced acute kidney injury by Nrf2-dependant expression of HO-1 and PPARγ [176]. Its mechanism may also reside in regulating mitochondrial biogenesis given the involvement of PPARγ. A large international study evaluating the full scale of bardoxolone methyl's effects on CKD progression is in progress, the results of which could determine if bardoxolone methyl should become a standard treatment in renal disease patients. Concurrent benefits to CVD will undoubtedly also be measured.

4.7. L-Carnitine – Improving cardiovascular health in dialysis

Carnitine is an essential cofactor required for the transformation of free fatty acids into acylcarnitine and its subsequent transport into the mitochondria for β-oxidation [177]. This underlies its importance in the production of ATP for cellular energy. Acylcarnitine is also essential for the removal of toxic fat metabolism by-products. Carnitine is obtained primarily from food stuffs, however it can be synthesised endogenously from the amino acid L-lysine and methionine [177]. L-carnitine supplementation primarily benefits ESRD patients on hemodialysis and their associated cardiovascular complications, especially anemia. This is primarily due to the well-described decrease in serum free carnitine in maintenance hemodialysis patients compared to non-dialysis CKD and healthy patients [178]. L-carnitine supplementation offsets renal anemia, lipid abnormalities and cardiac dysfunction in hemodialysis patients [179]. Left ventricular hypertrophy regressed in hemodialysis patients receiving 10mg/kg of L-carnitine immediately following hemodialysis for a 12 month period. [180]. Other measures of cardiac morbidity such as reduced left ventricular ejection fraction and increased left ventricular mass also significantly improved following low dose L-carnitine supplementation [181]. Benefits to the peripheral vasculature have also been demonstrated by L-carnitine through a mechanism thought to involve an associated decrease in homocysteine levels [182]. Interestingly, oxidative stress is a major characteristic of hemodialysis patients [183].

As well as the physiological role of L-carnitine in mitochondrial fatty acid synthesis, oxidant reducing capabilities have also been demonstrated and may underlie the health benefits of L-carnitine therapy in CKD and CVD. L-carnitine infusions significantly improved blood urea nitrogen (BUN) and creatinine levels in a 5/6 nephrectomy model of CKD with a concomitant increase in plasma SOD, Gpx, CAT and GSH, and decrease in the oxidative stress marker malondialdehyde [184]. Ye et al., (2010) suggest that L-carnitine attenuates renal tubular cell oxidant injury and subsequent apoptosis by reducing mitochondrial-derived ROS [97]. They suggest that this anti-apoptotic mechanism may also explain the demonstrated reduction in morbidity from cardiomyopathies in L-carnitine supplemented hemodialysis patients.

4.8. L-Arginine - Maintaining endothelial function

The premise of L-arginine supplementation is to maintain NO signalling and thereby maintain vascular endothelial cell function. L-arginine is a physiological precursor to NO and its availability and transport determine the rate of NO biosynthesis. CKD patients most often present with atherosclerosis, thromboembolitic complications, and endothelial dysfunction, primarily due to altered endothelium-dependant relaxation factors [185]. It is believed that the impaired NO synthesis, common in CKD individuals, contributes significantly to their disease pathogenesis [186]. L-arginine synthesis occurs in the liver and kidney, with the kidney functioning to maintain homeostatic plasma levels since the liver processes NO from the diet [187]. The addition of L-aspartic acid or L-glutamic acid with L-citrulline and arginirosuccinic acid synthase as the rate determining enzyme forms L-arginine [188]. The proximal tubular cells account for the majority of kidney NO synthesis [189, 190], thus kidney damage and atrophy, a primary corollary of CKD, results in decreased synthesis of L-arginine. The majority of research demonstrates decreased levels of NO production in CKD and CVD patients [191-193]. However, some research suggests NO activity increases [194, 195]. These disparate findings highlight the need to measure L-arginine levels in patients before commencing L-arginine supplementation. Rajapaske et al. (2012) demonstrated impaired kidney L-arginine transport and a contributing factor to hypertension in rats, irrespective of an underlying renal disease [196]. During a state of oxidative stress, L-arginine supplementation was shown to decrease MDA, myeloperoxidase and xanthine oxidase and increase gluta-thionine in both heart and kidney tissue from rats [197]. As such, L-arginine supplementation represents an approach to restoring a dysregulation of NO signalling and subsequent endothelial dysfunction in both chronic kidney and heart diseases.

4.9. Combination antioxidants

Compounds commonly used to alleviate oxidative stress exhibit different antioxidant actions, and so there exists the potential for different antioxidants to work together to improve whole cell and organ function through a targeted polypharmaceutical approach to decrease oxidative stress. However, most clinical studies investigating the effects of combination antioxidants have demonstrated confounding results. Mosca *et al.*, (2002) demonstrated that daily intake of NAC 100mg, L-carntine 100mg, selenomethionine 0.05mg, α-tocopherol 10mg, CoQ_{10} 100mg and α-lipoic acid 100mg successfully increased plasma CAT, Gpx and total antioxidant capacity whilst decreasing lipid peroxides and ROS generation by lymphocyte mitochondria [198]. However, this trial only included healthy participants and cannot be extrapolated to the CKD and CVD populations.

In a murine model of diabetic nephropathy, a major cause of CKD with associated CVD, the beneficial effects of NAC, L-ascorbic acid (vitamin C) and α-tocopherol were demonstrated [199]. Daily supplementation for 8 weeks decreased lipid peroxidation, BUN, serum creatinine and blood glucose, mainly due to a reduction in the inflammatory response induced by hyperglycemia. In comparison, a prospective trial investigating oral supplementation of mixed tocopherols and α-lipoic acid in stage 3 and 4 CKD patients has revealed disappointing results. Over 2 months, supplementation did not reduce biomarkers of oxidative stress (F_2-isoprostanes and protein thiol concentration) or inflammation (CRP and IL-6). The short period of time (2 months) of the intervention may explain this result and longer trials need to be carried out. The inclusion of vitamin E in these interventions has polarized discussion on the outcomes, because of its negligible benefits when cardiovascular outcomes were measured [91, 92, 200] and also because of contraindications, discussed previously. Despite this, long-term treatment in with the antioxidants vitamin C, vitamin E, CoQ_{10} and selenium has been shown to reduce multiple cardiovascular risk factors [201]. Recently, multiple antioxidants in combination with L-arginine have shown promise in animal models of CKD and associated CVD. Korish (2010) has demonstrated in a 5/6 nephrectomy CKD model that L-arginine improved the effects of L-carnitine, catechin and vitamins E and C on blood pressure, dyslipidemia, inflammation and kidney function [84].

5. Conclusion

CKD is a progressive disease with increasing incidence, having very little success in current conventional therapies once CKD reaches stage 4. Stages 2 and 3 are best to target to slow or stop further development of the disease. There is an almost inseparable connection between CKD and CVD, with many patients with CKD dying of the cardiovascular complications before renal failure reaches its fullest extent. Oxidative stress and inflammation are closely interrelated with development of CKD and CVD, and involve a spiralling cycle that leads to progressive patient deterioration. Given the complex nature of oxidative stress and its molecular pathways, antioxidants may need to be given as a polypharmacotherapy to target each aberrant pathway, with the aim of reducing the burden of these chronic diseases. It is vital for the progression of antioxidant therapy research in CKD and CVD that measures of oxidative stress are compared with pathophysiological outcome in the diseases, especially in connection with antioxidant therapies that may be delivered with or without more conventional CKD therapies.

Author details

David M. Small and Glenda C. Gobe*

*Address all correspondence to: g.gobe@uq.edu.au

Centre for Kidney Disease Research, School of Medicine, The University of Queensland, Brisbane, Australia

References

[1] Rosner MH, Ronco C, Okusa MD. The role of inflammation in the cardio-renal syndrome: a focus on cytokines and inflammatory mediators. Semin Nephrol. 2012 Jan; 32(1):70-8.

[2] Ronco C, McCullough P, Anker SD, Anand I, Aspromonte N, Bagshaw SM, et al. Cardio-renal syndromes: report from the consensus conference of the acute dialysis quality initiative. Eur Heart J. 2010 Mar;31(6):703-11.

[3] Leung FP, Yung LM, Laher I, Yao X, Chen ZY, Huang Y. Exercise, vascular wall and cardiovascular diseases: an update (Part 1). Sports Med. 2008;38(12):1009-24.

[4] Bongartz LG, Cramer MJ, Doevendans PA, Joles JA, Braam B. The severe cardiorenal syndrome: 'Guyton revisited'. Eur Heart J. 2005 Jan;26(1):11-7.

[5] Sallam N, Fisher A, Golbidi S, Laher I. Weight and inflammation are the major determinants of vascular dysfunction in the aortae of db/db mice. Naunyn Schmiedebergs Arch Pharmacol. 2011 May;383(5):483-92.

[6] Schiffrin EL, Lipman ML, Mann JF. Chronic kidney disease: effects on the cardiovascular system. Circulation. 2007 Jul 3;116(1):85-97.

[7] McDonald SP, Chang S, Excell L, editors. ANZDATA Registry Report. Adelaide 2007.

[8] Tanner RM, Brown TM, Muntner P. Epidemiology of obesity, the metabolic syndrome, and chronic kidney disease. Curr Hypertens Rep. 2012 Apr;14(2):152-9.

[9] Graf J, Ryan C, Green F. An overview of chronic kidney disease in Australia, 2009. Canberra: Australian Inst Health Welfare 2009.

[10] Tesch GH. Review: Serum and urine biomarkers of kidney disease: A pathophysiological perspective. Nephrol (Carlton). 2010 Sep;15(6):609-16.

[11] Rodriguez-Iturbe B, Johnson RJ, Herrera-Acosta J. Tubulointerstitial damage and progression of renal failure. Kidney Int Suppl. 2005 Dec(99):S82-6.

[12] Fassett RG, Venuthurupalli SK, Gobe GC, Coombes JS, Cooper MA, Hoy WE. Bio-
      markers in chronic kidney disease: a review. Kidney Int. 2011 Oct;80(8):806-21.

[13] Choudhury D, Luna-Salazar C. Preventive health care in chronic kidney disease and
      end-stage renal disease. Nat Clin Pract Nephrol. 2008 Apr;4(4):194-206.

[14] Dutta D, Calvani R, Bernabei R, Leeuwenburgh C, Marzetti E. Contribution of im-
      paired mitochondrial autophagy to cardiac aging: mechanisms and therapeutic op-
      portunities. Circ Res. 2012 Apr 13;110(8):1125-38.

[15] Manabe I. Chronic inflammation links cardiovascular, metabolic and renal diseases.
      Circ J. 2011;75(12):2739-48.

[16] Glassock RJ, Pecoits-Filho R, Barberato SH. Left ventricular mass in chronic kidney
      disease and ESRD. Clin J Am Soc Nephrol: CJASN. 2009 Dec;4 Suppl 1:S79-91.

[17] Hunt SA, Abraham WT, Chin MH, Feldman AM, Francis GS, Ganiats TG, et al.
      ACC/AHA 2005 Guideline Update for the Diagnosis and Management of Chronic
      Heart Failure in the Adult: a report of the American College of Cardiology/American
      Heart Association Task Force on Practice Guidelines (Writing Committee to Update
      the 2001 Guidelines for the Evaluation and Management of Heart Failure): developed
      in collaboration with the American College of Chest Physicians and the International
      Society for Heart and Lung Transplantation: endorsed by the Heart Rhythm Society.
      Circulation. 2005 Sep 20;112(12):e154-235.

[18] Mitsnefes MM. Cardiovascular disease in children with chronic kidney disease. J Am
      Soc Nephrol: JASN. 2012 Apr;23(4):578-85.

[19] Whaley-Connell A, Pavey BS, Chaudhary K, Saab G, Sowers JR. Renin-angiotensin-
      aldosterone system intervention in the cardiometabolic syndrome and cardio-renal
      protection. Ther Adv Cardiovasc Dis. 2007 Oct;1(1):27-35.

[20] Gomes P, Simao S, Silva E, Pinto V, Amaral JS, Afonso J, et al. Aging increases oxida-
      tive stress and renal expression of oxidant and antioxidant enzymes that are associat-
      ed with an increased trend in systolic blood pressure. Oxid Med Cell Longev. 2009
      Jul-Aug;2(3):138-45.

[21] Pias EK, Aw TY. Apoptosis in mitotic competent undifferentiated cells is induced by
      cellular redox imbalance independent of reactive oxygen species production. FASEB
      J. 2002 Jun;16(8):781-90.

[22] Zhuang S, Yan Y, Daubert RA, Han J, Schnellmann RG. ERK promotes hydrogen per-
      oxide-induced apoptosis through caspase-3 activation and inhibition of Akt in renal
      epithelial cells. Am J Physiol Renal Physiol. 2007 Jan;292(1):F440-7.

[23] Blanchetot C, Tertoolen LG, den Hertog J. Regulation of receptor protein-tyrosine
      phosphatase alpha by oxidative stress. EMBO J. 2002 Feb 15;21(4):493-503.

[24] Jones DP. Redefining oxidative stress. Antioxid Redox Signal. 2006 Sep-Oct;8(9-10):
      1865-79.

[25] Meng TC, Fukada T, Tonks NK. Reversible oxidation and inactivation of protein tyrosine phosphatases in vivo. Mol Cell. 2002 Feb;9(2):387-99.

[26] Rao RK, Clayton LW. Regulation of protein phosphatase 2A by hydrogen peroxide and glutathionylation. Biochem Biophys Res Commun. 2002 Apr 26;293(1):610-6.

[27] Tavakoli S, Asmis R. Reactive Oxygen Species and Thiol Redox Signaling in the Macrophage Biology of Atherosclerosis. Antioxid Redox Signal. 2012 Jun 11.

[28] Madesh M, Hajnoczky G. VDAC-dependent permeabilization of the outer mitochondrial membrane by superoxide induces rapid and massive cytochrome c release. J Cell Biol. 2001 Dec 10;155(6):1003-15.

[29] Soubannier V, McBride HM. Positioning mitochondrial plasticity within cellular signaling cascades. Biochim Biophys Acta. 2009 Jan;1793(1):154-70.

[30] Vay L, Hernandez-SanMiguel E, Lobaton CD, Moreno A, Montero M, Alvarez J. Mitochondrial free [Ca2+] levels and the permeability transition. Cell Calcium. 2009 Mar;45(3):243-50.

[31] Cadenas E, Davies KJ. Mitochondrial free radical generation, oxidative stress, and aging. Free Radic Biol Med. 2000 Aug;29(3-4):222-30.

[32] Boveris A, Chance B. The mitochondrial generation of hydrogen peroxide. General properties and effect of hyperbaric oxygen. Biochem J. 1973 Jul;134(3):707-16.

[33] Nohl H, Hegner D. Do mitochondria produce oxygen radicals in vivo? Eur J Biochem. 1978 Jan 16;82(2):563-7.

[34] Lipinski B. Is it oxidative stress or free radical stress and why does it matter? Oxid Antioxid Med Sci. 2012 March;1(1):5-9.

[35] Cadenas E, Boveris A, Ragan CI, Stoppani AO. Production of superoxide radicals and hydrogen peroxide by NADH-ubiquinone reductase and ubiquinol-cytochrome c reductase from beef-heart mitochondria. Arch Biochem Biophys. 1977 Apr 30;180(2):248-57.

[36] Turrens JF, Boveris A. Generation of superoxide anion by the NADH dehydrogenase of bovine heart mitochondria. Biochem J. 1980 Nov 1;191(2):421-7.

[37] Turrens JF, Alexandre A, Lehninger AL. Ubisemiquinone is the electron donor for superoxide formation by complex III of heart mitochondria. Arch Biochem Biophys. 1985 Mar;237(2):408-14.

[38] Turrens JF. Mitochondrial formation of reactive oxygen species. J Physiol. 2003 Oct 15;552(Pt 2):335-44.

[39] Choksi KB, Nuss JE, Boylston WH, Rabek JP, Papaconstantinou J. Age-related increases in oxidatively damaged proteins of mouse kidney mitochondrial electron transport chain complexes. Free Radic Biol Med. 2007 Nov 15;43(10):1423-38.

[40] Granata S, Zaza G, Simone S, Villani G, Latorre D, Pontrelli P, et al. Mitochondrial dysregulation and oxidative stress in patients with chronic kidney disease. BMC Genomics. 2009;10:388.

[41] Nemoto S, Takeda K, Yu ZX, Ferrans VJ, Finkel T. Role for mitochondrial oxidants as regulators of cellular metabolism. Mol Cell Biol. 2000 Oct;20(19):7311-8.

[42] Werner E, Werb Z. Integrins engage mitochondrial function for signal transduction by a mechanism dependent on Rho GTPases. J Cell Biol. 2002 Jul 22;158(2):357-68.

[43] Cooper CE, Patel RP, Brookes PS, Darley-Usmar VM. Nanotransducers in cellular redox signaling: modification of thiols by reactive oxygen and nitrogen species. Trends Biochem Sci. 2002 Oct;27(10):489-92.

[44] Kokoszka JE, Coskun P, Esposito LA, Wallace DC. Increased mitochondrial oxidative stress in the Sod2 (+/-) mouse results in the age-related decline of mitochondrial function culminating in increased apoptosis. Proc Natl Acad Sci U S A. 2001 Feb 27;98(5): 2278-83.

[45] Meng Q, Wong YT, Chen J, Ruan R. Age-related changes in mitochondrial function and antioxidative enzyme activity in fischer 344 rats. Mech Ageing Dev. 2007 Mar; 128(3):286-92.

[46] Raha S, McEachern GE, Myint AT, Robinson BH. Superoxides from mitochondrial complex III: the role of manganese superoxide dismutase. Free Radic Biol Med. 2000 Jul 15;29(2):170-80.

[47] Angermuller S, Islinger M, Volkl A. Peroxisomes and reactive oxygen species, a lasting challenge. Histochem Cell Biol. 2009 Apr;131(4):459-63.

[48] Islinger M, Li KW, Seitz J, Volkl A, Luers GH. Hitchhiking of Cu/Zn superoxide dismutase to peroxisomes-evidence for a natural piggyback import mechanism in mammals. Traffic. 2009 Nov;10(11):1711-21.

[49] Sturtz LA, Diekert K, Jensen LT, Lill R, Culotta VC. A fraction of yeast Cu,Zn-superoxide dismutase and its metallochaperone, CCS, localize to the intermembrane space of mitochondria. A physiological role for SOD1 in guarding against mitochondrial oxidative damage. J Biol Chem. 2001 Oct 12;276(41):38084-9.

[50] Soderdahl T, Enoksson M, Lundberg M, Holmgren A, Ottersen OP, Orrenius S, et al. Visualization of the compartmentalization of glutathione and protein-glutathione mixed disulfides in cultured cells. FASEB J. 2003 Jan;17(1):124-6.

[51] Godoy JR, Oesteritz S, Hanschmann EM, Ockenga W, Ackermann W, Lillig CH. Segment-specific overexpression of redoxins after renal ischemia and reperfusion: protective roles of glutaredoxin 2, peroxiredoxin 3, and peroxiredoxin 6. Free Radic Biol Med. 2011 Jul 15;51(2):552-61.

[52] Lillig CH, Holmgren A. Thioredoxin and related molecules--from biology to health and disease. Antioxid Redox Signal. 2007 Jan;9(1):25-47.

[53] Lonn ME, Hudemann C, Berndt C, Cherkasov V, Capani F, Holmgren A, et al. Expression pattern of human glutaredoxin 2 isoforms: identification and characterization of two testis/cancer cell-specific isoforms. Antioxid Redox Signal. 2008 Mar; 10(3):547-57.

[54] Hanschmann EM, Lonn ME, Schutte LD, Funke M, Godoy JR, Eitner S, et al. Both thioredoxin 2 and glutaredoxin 2 contribute to the reduction of the mitochondrial 2-Cys peroxiredoxin Prx3. J Biol Chem. 2010 Dec 24;285(52):40699-705.

[55] Lash LH, Putt DA, Matherly LH. Protection of NRK-52E cells, a rat renal proximal tubular cell line, from chemical-induced apoptosis by overexpression of a mitochondrial glutathione transporter. J Pharmacol Exp Ther. 2002 Nov;303(2):476-86.

[56] Visarius TM, Putt DA, Schare JM, Pegouske DM, Lash LH. Pathways of glutathione metabolism and transport in isolated proximal tubular cells from rat kidney. Biochem Pharmacol. 1996 Jul 26;52(2):259-72.

[57] Funk JA, Odejinmi S, Schnellmann RG. SRT1720 induces mitochondrial biogenesis and rescues mitochondrial function after oxidant injury in renal proximal tubule cells. J Pharmacol Exp Therapeut. 2010 May;333(2):593-601.

[58] Lepenies J, Hewison M, Stewart PM, Quinkler M. Renal PPARgamma mRNA expression increases with impairment of renal function in patients with chronic kidney disease. Nephrology. 2010 Oct;15(7):683-91.

[59] Sakamoto A, Hongo M, Saito K, Nagai R, Ishizaka N. Reduction of renal lipid content and proteinuria by a PPAR-gamma agonist in a rat model of angiotensin II-induced hypertension. Eur J Pharmacol. 2012 May 5;682(1-3):131-6.

[60] Li H, Ruan XZ, Powis SH, Fernando R, Mon WY, Wheeler DC, et al. EPA and DHA reduce LPS-induced inflammation responses in HK-2 cells: evidence for a PPARgamma-dependent mechanism. Kidney Int. 2005 Mar;67(3):867-74.

[61] Martin A, Perez-Giron JV, Hernanz R, Palacios R, Briones AM, Fortuno A, et al. Peroxisome proliferator-activated receptor-gamma activation reduces cyclooxygenase-2 expression in vascular smooth muscle cells from hypertensive rats by interfering with oxidative stress. J Hypertens. 2012 Feb;30(2):315-26.

[62] Hybertson BM, Gao B, Bose SK, McCord JM. Oxidative stress in health and disease: the therapeutic potential of Nrf2 activation. Mol Aspects Med. 2011 Aug;32(4-6): 234-46.

[63] Wilmes A, Crean D, Aydin S, Pfaller W, Jennings P, Leonard MO. Identification and dissection of the Nrf2 mediated oxidative stress pathway in human renal proximal tubule toxicity. Toxicol In Vitro. 2011 Apr;25(3):613-22.

[64] Nelson SK, Bose SK, Grunwald GK, Myhill P, McCord JM. The induction of human superoxide dismutase and catalase in vivo: a fundamentally new approach to antioxidant therapy. Free Radic Biol Med. 2006 Jan 15;40(2):341-7.

[65] Prestera T, Talalay P, Alam J, Ahn YI, Lee PJ, Choi AM. Parallel induction of heme oxygenase-1 and chemoprotective phase 2 enzymes by electrophiles and antioxidants: regulation by upstream antioxidant-responsive elements (ARE). Mol Med. 1995 Nov;1(7):827-37.

[66] Li Y, Jaiswal AK. Regulation of human NAD(P)H:quinone oxidoreductase gene. Role of AP1 binding site contained within human antioxidant response element. J Biol Chem. 1992 Jul 25;267(21):15097-104.

[67] Okuda A, Imagawa M, Maeda Y, Sakai M, Muramatsu M. Structural and functional analysis of an enhancer GPEI having a phorbol 12-O-tetradecanoate 13-acetate responsive element-like sequence found in the rat glutathione transferase P gene. J Biol Chem. 1989 Oct 5;264(28):16919-26.

[68] Itoh K, Mochizuki M, Ishii Y, Ishii T, Shibata T, Kawamoto Y, et al. Transcription factor Nrf2 regulates inflammation by mediating the effect of 15-deoxy-Delta(12,14)-prostaglandin j(2). Mol Cell Biol. 2004 Jan;24(1):36-45.

[69] Levonen AL, Landar A, Ramachandran A, Ceaser EK, Dickinson DA, Zanoni G, et al. Cellular mechanisms of redox cell signalling: role of cysteine modification in controlling antioxidant defences in response to electrophilic lipid oxidation products. Biochem J. 2004 Mar 1;378(Pt 2):373-82.

[70] Rojas-Rivera J, Ortiz A, Egido J. Antioxidants in kidney diseases: the impact of bardoxolone methyl. Int J Nephrol. 2012;2012:321714.

[71] Brand FN, McGee DL, Kannel WB, Stokes J, 3rd, Castelli WP. Hyperuricemia as a risk factor of coronary heart disease: The Framingham Study. Am J Epidemiol. 1985 Jan;121(1):11-8.

[72] Mitsuhashi H, Tamura K, Yamauchi J, Ozawa M, Yanagi M, Dejima T, et al. Effect of losartan on ambulatory short-term blood pressure variability and cardiovascular remodeling in hypertensive patients on hemodialysis. Atherosclerosis. 2009 Nov; 207(1):186-90.

[73] Letsas KP, Korantzopoulos P, Filippatos GS, Mihas CC, Markou V, Gavrielatos G, et al. Uric acid elevation in atrial fibrillation. Hellenic J Cardiol. 2010 May-Jun;51(3): 209-13.

[74] Car S, Trkulja V. Higher serum uric acid on admission is associated with higher short-term mortality and poorer long-term survival after myocardial infarction: retrospective prognostic study. Croat Med J. 2009 Dec;50(6):559-66.

[75] Chen JH, Chuang SY, Chen HJ, Yeh WT, Pan WH. Serum uric acid level as an independent risk factor for all-cause, cardiovascular, and ischemic stroke mortality: a Chinese cohort study. Arthritis Rheum. 2009 Feb 15;61(2):225-32.

[76] Shan Y, Zhang Q, Liu Z, Hu X, Liu D. Prevalence and risk factors associated with chronic kidney disease in adults over 40 years: a population study from Central China. Nephrology (Carlton). 2010 Apr;15(3):354-61.

[77] Chen YC, Su CT, Wang ST, Lee HD, Lin SY. A preliminary investigation of the association between serum uric acid and impaired renal function. Chang Gung Med J. 2009 Jan-Feb;32(1):66-71.

[78] Martinon F, Petrilli V, Mayor A, Tardivel A, Tschopp J. Gout-associated uric acid crystals activate the NALP3 inflammasome. Nature. 2006 Mar 9;440(7081):237-41.

[79] Sakamaki I, Inai K, Tsutani Y, Ueda T, Tsutani H. Binding of monosodium urate crystals with idiotype protein efficiently promote dendritic cells to induce cytotoxic T cells. Cancer Sci. 2008 Nov;99(11):2268-73.

[80] Amaya Y, Yamazaki K, Sato M, Noda K, Nishino T. Proteolytic conversion of xanthine dehydrogenase from the NAD-dependent type to the O2-dependent type. Amino acid sequence of rat liver xanthine dehydrogenase and identification of the cleavage sites of the enzyme protein during irreversible conversion by trypsin. J Biol Chem. 1990 Aug 25;265(24):14170-5.

[81] Nishino T, Okamoto K, Kawaguchi Y, Hori H, Matsumura T, Eger BT, et al. Mechanism of the conversion of xanthine dehydrogenase to xanthine oxidase: identification of the two cysteine disulfide bonds and crystal structure of a non-convertible rat liver xanthine dehydrogenase mutant. J Biol Chem. 2005 Jul 1;280(26):24888-94.

[82] Maia L, Duarte RO, Ponces-Freire A, Moura JJ, Mira L. NADH oxidase activity of rat and human liver xanthine oxidoreductase: potential role in superoxide production. J Biol Inorg Chem. 2007 Aug;12(6):777-87.

[83] Miller NJ, RiceEvans CA. Spectrophotometric determination of antioxidant activity. Redox Report. 1996 Jun;2(3):161-71.

[84] Korish AA. Multiple antioxidants and L-arginine modulate inflammation and dyslipidemia in chronic renal failure rats. Ren Fail. 2010 Jan;32(2):203-13.

[85] Ehara H, Yamamoto-Honda R, Kitazato H, Takahashi Y, Kawazu S, Akanuma Y, et al. ApoE isoforms, treatment of diabetes and the risk of coronary heart disease. World J Diabetes. 2012 Mar 15;3(3):54-9.

[86] Abadir PM, Foster DB, Crow M, Cooke CA, Rucker JJ, Jain A, et al. Identification and characterization of a functional mitochondrial angiotensin system. Proc Nat Acad Sci USA. 2011 Sep 6;108(36):14849-54.

[87] Halliwell B. The wanderings of a free radical. Free Radic Biol Med. 2009 Mar 1;46(5): 531-42.

[88] Halliwell B, Whiteman M. Measuring reactive species and oxidative damage in vivo and in cell culture: how should you do it and what do the results mean? Br J Pharmacol. 2004 May;142(2):231-55.

[89] Golbidi S, Ebadi SA, Laher I. Antioxidants in the treatment of diabetes. Curr Diabetes Rev. 2011 Mar;7(2):106-25.

[90] Ramos LF, Kane J, McMonagle E, Le P, Wu P, Shintani A, et al. Effects of combination tocopherols and alpha lipoic acid therapy on oxidative stress and inflammatory biomarkers in chronic kidney disease. J Ren Nutr. 2011 May;21(3):211-8.

[91] Yusuf S, Dagenais G, Pogue J, Bosch J, Sleight P. Vitamin E supplementation and cardiovascular events in high-risk patients. The Heart Outcomes Prevention Evaluation Study Investigators. N Engl J Med. 2000 Jan 20;342(3):154-60.

[92] Mann JF, Lonn EM, Yi Q, Gerstein HC, Hoogwerf BJ, Pogue J, et al. Effects of vitamin E on cardiovascular outcomes in people with mild-to-moderate renal insufficiency: results of the HOPE study. Kidney Int. 2004 Apr;65(4):1375-80.

[93] Harcourt BE, Sourris KC, Coughlan MT, Walker KZ, Dougherty SL, Andrikopoulos S, et al. Targeted reduction of advanced glycation improves renal function in obesity. Kidney Int. 2011 Jul;80(2):190-8.

[94] Vlassara H, Torreggiani M, Post JB, Zheng F, Uribarri J, Striker GE. Role of oxidants/inflammation in declining renal function in chronic kidney disease and normal aging. Kidney Int Suppl. 2009 Dec(114):S3-11.

[95] Zafarullah M, Li WQ, Sylvester J, Ahmad M. Molecular mechanisms of N-acetylcysteine actions. Cell Mol Life Sci. 2003 Jan;60(1):6-20.

[96] Zhang F, Lau SS, Monks TJ. The cytoprotective effect of N-acetyl-L-cysteine against ROS-induced cytotoxicity is independent of its ability to enhance glutathione synthesis. Toxicol Sci. 2011 Mar;120(1):87-97.

[97] Ye J, Li J, Yu Y, Wei Q, Deng W, Yu L. L-carnitine attenuates oxidant injury in HK-2 cells via ROS-mitochondria pathway. Regul Pept. 2010 Apr 9;161(1-3):58-66.

[98] Pat B, Yang T, Kong C, Watters D, Johnson DW, Gobe G. Activation of ERK in renal fibrosis after unilateral ureteral obstruction: modulation by antioxidants. Kidney Int. 2005 Mar;67(3):931-43.

[99] Ribeiro G, Roehrs M, Bairros A, Moro A, Charao M, Araujo F, et al. N-acetylcysteine on oxidative damage in diabetic rats. Drug Chem Toxicol. 2011 Aug 16.

[100] Moist L, Sontrop JM, Gallo K, Mainra R, Cutler M, Freeman D, et al. Effect of N-acetylcysteine on serum creatinine and kidney function: results of a randomized controlled trial. Am J Kidney Dis. 2010 Oct;56(4):643-50.

[101] Renke M, Tylicki L, Rutkowski P, Larczynski W, Aleksandrowicz E, Lysiak-Szydlowska W, et al. The effect of N-acetylcysteine on proteinuria and markers of tubular injury in non-diabetic patients with chronic kidney disease. A placebo-controlled, randomized, open, cross-over study. Kidney Blood Press Res. 2008;31(6):404-10.

[102] Hsu SP, Chiang CK, Yang SY, Chien CT. N-acetylcysteine for the management of anemia and oxidative stress in hemodialysis patients. Nephron Clin Pract. 2010;116(3):c207-16.

[103] Nascimento MM, Suliman ME, Silva M, Chinaglia T, Marchioro J, Hayashi SY, et al. Effect of oral N-acetylcysteine treatment on plasma inflammatory and oxidative stress markers in peritoneal dialysis patients: a placebo-controlled study. Perit Dial Int. 2010 May-Jun;30(3):336-42.

[104] Coombes JS, Fassett RG. Antioxidant therapy in hemodialysis patients: a systematic review. Kidney Int. 2012 Feb;81(3):233-46.

[105] Crespo MJ, Cruz N, Altieri PI, Escobales N. Chronic treatment with N-acetylcysteine improves cardiac function but does not prevent progression of cardiomyopathy in Syrian cardiomyopathic hamsters. J Cardiovasc Pharmacol Ther. 2011 Jun;16(2): 197-204.

[106] Tumur Z, Shimizu H, Enomoto A, Miyazaki H, Niwa T. Indoxyl sulfate upregulates expression of ICAM-1 and MCP-1 by oxidative stress-induced NF-kappaB activation. Am J Nephrol. 2010;31(5):435-41.

[107] Serbinova E, Kagan V, Han D, Packer L. Free radical recycling and intramembrane mobility in the antioxidant properties of alpha-tocopherol and alpha-tocotrienol. Free Radic Biol Med. 1991;10(5):263-75.

[108] Fujisawa S, Ishihara M, Atsumi T, Kadoma Y. A quantitative approach to the free radical interaction between alpha-tocopherol or ascorbate and flavonoids. In Vivo. 2006 Jul-Aug;20(4):445-52.

[109] Kagan VE, Serbinova EA, Packer L. Recycling and antioxidant activity of tocopherol homologs of differing hydrocarbon chain lengths in liver microsomes. Arch Biochem Biophys. 1990 Nov 1;282(2):221-5.

[110] Kagan VE, Serbinova EA, Forte T, Scita G, Packer L. Recycling of vitamin E in human low density lipoproteins. J Lipid Res. 1992 Mar;33(3):385-97.

[111] Guo Q, Packer L. Ascorbate-dependent recycling of the vitamin E homologue Trolox by dihydrolipoate and glutathione in murine skin homogenates. Free Radic Biol Med. 2000 Aug;29(3-4):368-74.

[112] Schaaf GJ, Maas RF, de Groene EM, Fink-Gremmels J. Management of oxidative stress by heme oxygenase-1 in cisplatin-induced toxicity in renal tubular cells. Free Radic Res. 2002 Aug;36(8):835-43.

[113] Sen CK, Khanna S, Roy S, Packer L. Molecular basis of vitamin E action. Tocotrienol potently inhibits glutamate-induced pp60(c-Src) kinase activation and death of HT4 neuronal cells. J Biol Chem. 2000 Apr 28;275(17):13049-55.

[114] Machlin LJ, Gabriel E. Kinetics of tissue alpha-tocopherol uptake and depletion following administration of high levels of vitamin E. Ann N Y Acad Sci. 1982;393:48-60.

[115] Karamouzis I, Sarafidis PA, Karamouzis M, Iliadis S, Haidich AB, Sioulis A, et al. Increase in oxidative stress but not in antioxidant capacity with advancing stages of chronic kidney disease. Am J Nephrol. 2008;28(3):397-404.

[116] Klein EA, Thompson IM, Jr., Tangen CM, Crowley JJ, Lucia MS, Goodman PJ, et al. Vitamin E and the risk of prostate cancer: the Selenium and Vitamin E Cancer Prevention Trial (SELECT). JAMA. 2011 Oct 12;306(14):1549-56.

[117] Boaz M, Smetana S, Weinstein T, Matas Z, Gafter U, Iaina A, et al. Secondary prevention with antioxidants of cardiovascular disease in endstage renal disease (SPACE): randomised placebo-controlled trial. Lancet. 2000 Oct 7;356(9237):1213-8.

[118] Giray B, Kan E, Bali M, Hincal F, Basaran N. The effect of vitamin E supplementation on antioxidant enzyme activities and lipid peroxidation levels in hemodialysis patients. Clin Chim Acta. 2003 Dec;338(1-2):91-8.

[119] Islam KN, O'Byrne D, Devaraj S, Palmer B, Grundy SM, Jialal I. Alpha-tocopherol supplementation decreases the oxidative susceptibility of LDL in renal failure patients on dialysis therapy. Atherosclerosis. 2000 May;150(1):217-24.

[120] Huang HY, Appel LJ. Supplementation of diets with alpha-tocopherol reduces serum concentrations of gamma- and delta-tocopherol in humans. J Nutr. 2003 Oct;133(10): 3137-40.

[121] Baggott JE. High-dosage vitamin E supplementation and all-cause mortality. Ann Intern Med. 2005 Jul 19;143(2):155-6; author reply 6-8.

[122] Blatt DH, Pryor WA. High-dosage vitamin E supplementation and all-cause mortality. Ann Intern Med. 2005 Jul 19;143(2):150-1; author reply 6-8.

[123] Carter T. High-dosage vitamin E supplementation and all-cause mortality. Ann Intern Med. 2005 Jul 19;143(2):155; author reply 6-8.

[124] DeZee KJ, Shimeall W, Douglas K, Jackson JL. High-dosage vitamin E supplementation and all-cause mortality. Ann Intern Med. 2005 Jul 19;143(2):153-4; author reply 6-8.

[125] Hemila H. High-dosage vitamin E supplementation and all-cause mortality. Ann Intern Med. 2005 Jul 19;143(2):151-2; author reply 6-8.

[126] Krishnan K, Campbell S, Stone WL. High-dosage vitamin E supplementation and all-cause mortality. Ann Intern Med. 2005 Jul 19;143(2):151; author reply 6-8.

[127] Lim WS, Liscic R, Xiong C, Morris JC. High-dosage vitamin E supplementation and all-cause mortality. Ann Intern Med. 2005 Jul 19;143(2):152; author reply 6-8.

[128] Marras C, Lang AE, Oakes D, McDermott MP, Kieburtz K, Shoulson I, et al. High-dosage vitamin E supplementation and all-cause mortality. Ann Intern Med. 2005 Jul 19;143(2):152-3; author reply 6-8.

[129] Meydani SN, Lau J, Dallal GE, Meydani M. High-dosage vitamin E supplementation and all-cause mortality. Ann Intern Med. 2005 Jul 19;143(2):153; author reply 6-8.

[130] Possolo AM. High-dosage vitamin E supplementation and all-cause mortality. Ann Intern Med. 2005 Jul 19;143(2):154; author reply 6-8.

[131] Lass A, Forster MJ, Sohal RS. Effects of coenzyme Q10 and alpha-tocopherol administration on their tissue levels in the mouse: elevation of mitochondrial alpha-tocopherol by coenzyme Q10. Free Radic Biol Med. 1999 Jun;26(11-12):1375-82.

[132] Lass A, Sohal RS. Effect of coenzyme Q(10) and alpha-tocopherol content of mitochondria on the production of superoxide anion radicals. FASEB J. 2000 Jan;14(1): 87-94.

[133] Merker MP, Audi SH, Lindemer BJ, Krenz GS, Bongard RD. Role of mitochondrial electron transport complex I in coenzyme Q1 reduction by intact pulmonary arterial endothelial cells and the effect of hyperoxia. Am J Physiol Lung Cell Mol Physiol. 2007 Sep;293(3):L809-19.

[134] Ohnishi T, Ohnishi ST, Shinzawa-Ito K, Yoshikawa S. Functional role of coenzyme Q in the energy coupling of NADH-CoQ oxidoreductase (Complex I): stabilization of the semiquinone state with the application of inside-positive membrane potential to proteoliposomes. Biofactors. 2008;32(1-4):13-22.

[135] James AM, Smith RA, Murphy MP. Antioxidant and prooxidant properties of mitochondrial Coenzyme Q. Arch Biochem Biophys. 2004 Mar 1;423(1):47-56.

[136] Linnane AW, Kios M, Vitetta L. Coenzyme Q(10) - its role as a prooxidant in the formation of superoxide anion/hydrogen peroxide and the regulation of the metabolome. Mitochondrion. 2007 Jun;7 Suppl:S51-61.

[137] Nohl H, Gille L, Kozlov AV. Critical aspects of the antioxidant function of coenzyme Q in biomembranes. Biofactors. 1999;9(2-4):155-61.

[138] Frei B, Kim MC, Ames BN. Ubiquinol-10 is an effective lipid-soluble antioxidant at physiological concentrations. Proc Natl Acad Sci U S A. 1990 Jun;87(12):4879-83.

[139] Stoyanovsky DA, Osipov AN, Quinn PJ, Kagan VE. Ubiquinone-dependent recycling of vitamin E radicals by superoxide. Arch Biochem Biophys. 1995 Nov 10;323(2): 343-51.

[140] Lass A, Sohal RS. Electron transport-linked ubiquinone-dependent recycling of alpha-tocopherol inhibits autooxidation of mitochondrial membranes. Arch Biochem Biophys. 1998 Apr 15;352(2):229-36.

[141] Linnane AW, Kios M, Vitetta L. Healthy aging: regulation of the metabolome by cellular redox modulation and prooxidant signaling systems: the essential roles of superoxide anion and hydrogen peroxide. Biogerontology. 2007 Oct;8(5):445-67.

[142] Kwong LK, Kamzalov S, Rebrin I, Bayne AC, Jana CK, Morris P, et al. Effects of coenzyme Q(10) administration on its tissue concentrations, mitochondrial oxidant generation, and oxidative stress in the rat. Free Radic Biol Med. 2002 Sep 1;33(5):627-38.

[143] Ishikawa A, Kawarazaki H, Ando K, Fujita M, Fujita T, Homma Y. Renal preservation effect of ubiquinol, the reduced form of coenzyme Q10. Clin Exp Nephrol. 2011 Feb;15(1):30-3.

[144] Huynh K, Kiriazis H, Du XJ, Love JE, Jandeleit-Dahm KA, Forbes JM, et al. Coenzyme Q10 attenuates diastolic dysfunction, cardiomyocyte hypertrophy and cardiac fibrosis in the db/db mouse model of type 2 diabetes. Diabetologia. 2012 May;55(5): 1544-53.

[145] Dendooven A, Ishola DA, Jr., Nguyen TQ, Van der Giezen DM, Kok RJ, Goldschmeding R, et al. Oxidative stress in obstructive nephropathy. Int J Exp Pathol. 2011 Jun;92(3):202-10.

[146] Irita J, Okura T, Jotoku M, Nagao T, Enomoto D, Kurata M, et al. Osteopontin deficiency protects against aldosterone-induced inflammation, oxidative stress, and interstitial fibrosis in the kidney. Am J Physiol Renal Physiol. 2011 Jul 6.

[147] Shing CM, Adams MJ, Fassett RG, Coombes JS. Nutritional compounds influence tissue factor expression and inflammation of chronic kidney disease patients in vitro. Nutrition. 2011 Sep;27(9):967-72.

[148] Arab K, Rossary A, Flourie F, Tourneur Y, Steghens JP. Docosahexaenoic acid enhances the antioxidant response of human fibroblasts by upregulating gamma-glutamylcysteinyl ligase and glutathione reductase. Br J Nutr. 2006 Jan;95(1):18-26.

[149] Kim YJ, Chung HY. Antioxidative and anti-inflammatory actions of docosahexaenoic acid and eicosapentaenoic acid in renal epithelial cells and macrophages. J Med Food. 2007 Jun;10(2):225-31.

[150] Mayer K, Meyer S, Reinholz-Muhly M, Maus U, Merfels M, Lohmeyer J, et al. Short-time infusion of fish oil-based lipid emulsions, approved for parenteral nutrition, reduces monocyte proinflammatory cytokine generation and adhesive interaction with endothelium in humans. J Immunol. 2003 Nov 1;171(9):4837-43.

[151] Sperling RI, Benincaso AI, Knoell CT, Larkin JK, Austen KF, Robinson DR. Dietary omega-3 polyunsaturated fatty acids inhibit phosphoinositide formation and chemotaxis in neutrophils. J Clin Invest. 1993 Feb;91(2):651-60.

[152] Diaz Encarnacion MM, Warner GM, Cheng J, Gray CE, Nath KA, Grande JP. n-3 Fatty acids block TNF-alpha-stimulated MCP-1 expression in rat mesangial cells. Am J Physiol Renal Physiol. 2011 May;300(5):F1142-51.

[153] An WS, Kim HJ, Cho KH, Vaziri ND. Omega-3 fatty acid supplementation attenuates oxidative stress, inflammation, and tubulointerstitial fibrosis in the remnant kidney. Am J Physiol Renal Physiol. 2009 Oct;297(4):F895-903.

[154] Peake JM, Gobe GC, Fassett RG, Coombes JS. The effects of dietary fish oil on inflammation, fibrosis and oxidative stress associated with obstructive renal injury in rats. Mol Nutr Food Res. 2011 Mar;55(3):400-10.

[155] Kazemian P, Kazemi-Bajestani SM, Alherbish A, Steed J, Oudit GY. The use of omega-3 poly-unsaturated fatty acids in heart failure: a preferential role in patients with diabetes. Cardiovasc Drugs Ther. 2012 May 30.

[156] Bouzidi N, Mekki K, Boukaddoum A, Dida N, Kaddous A, Bouchenak M. Effects of omega-3 polyunsaturated fatty-acid supplementation on redox status in chronic renal failure patients with dyslipidemia. J Ren Nutr. 2010 Sep;20(5):321-8.

[157] Harris WS. n-3 fatty acids and serum lipoproteins: human studies. Am J Clin Nutr. 1997 May;65(5 Suppl):1645S-54S.

[158] Cohen MG, Rossi JS, Garbarino J, Bowling R, Motsinger-Reif AA, Schuler C, et al. Insights into the inhibition of platelet activation by omega-3 polyunsaturated fatty acids: Beyond aspirin and clopidogrel. Thromb Res. 2011 May 26.

[159] Woodman RJ, Mori TA, Burke V, Puddey IB, Watts GF, Beilin LJ. Effects of purified eicosapentaenoic and docosahexaenoic acids on glycemic control, blood pressure, and serum lipids in type 2 diabetic patients with treated hypertension. Am J Clin Nutr. 2002 Nov;76(5):1007-15.

[160] Matsumoto T, Nakayama N, Ishida K, Kobayashi T, Kamata K. Eicosapentaenoic acid improves imbalance between vasodilator and vasoconstrictor actions of endothelium-derived factors in mesenteric arteries from rats at chronic stage of type 2 diabetes. J Pharmacol Exp Ther. 2009 Apr;329(1):324-34.

[161] El-Sheikh AA, van den Heuvel JJ, Koenderink JB, Russel FG. Effect of hypouricaemic and hyperuricaemic drugs on the renal urate efflux transporter, multidrug resistance protein 4. Br J Pharmacol. 2008 Dec;155(7):1066-75.

[162] Riegersperger M, Covic A, Goldsmith D. Allopurinol, uric acid, and oxidative stress in cardiorenal disease. Int Urol Nephrol. 2011 Jun;43(2):441-9.

[163] Sanders SA, Eisenthal R, Harrison R. NADH oxidase activity of human xanthine oxidoreductase--generation of superoxide anion. Eur J Biochem. 1997 May 1;245(3): 541-8.

[164] Kosugi T, Nakayama T, Heinig M, Zhang L, Yuzawa Y, Sanchez-Lozada LG, et al. Effect of lowering uric acid on renal disease in the type 2 diabetic db/db mice. Am J Physiol Renal Physiol. 2009 Aug;297(2):F481-8.

[165] Das DK, Engelman RM, Clement R, Otani H, Prasad MR, Rao PS. Role of xanthine oxidase inhibitor as free radical scavenger: a novel mechanism of action of allopurinol and oxypurinol in myocardial salvage. Biochem Biophys Res Commun. 1987 Oct 14;148(1):314-9.

[166] Moorhouse PC, Grootveld M, Halliwell B, Quinlan JG, Gutteridge JM. Allopurinol and oxypurinol are hydroxyl radical scavengers. FEBS Lett. 1987 Mar 9;213(1):23-8.

[167] Klein AS, Joh JW, Rangan U, Wang D, Bulkley GB. Allopurinol: discrimination of antioxidant from enzyme inhibitory activities. Free Radic Biol Med. 1996;21(5):713-7.

[168] Goicoechea M, de Vinuesa SG, Verdalles U, Ruiz-Caro C, Ampuero J, Rincon A, et al. Effect of allopurinol in chronic kidney disease progression and cardiovascular risk. Clin J Am Soc Nephrol. 2010 Aug;5(8):1388-93.

[169] Kanbay M, Ozkara A, Selcoki Y, Isik B, Turgut F, Bavbek N, et al. Effect of treatment of hyperuricemia with allopurinol on blood pressure, creatinine clearence, and proteinuria in patients with normal renal functions. Int Urol Nephrol. 2007;39(4):1227-33.

[170] Shelmadine B, Bowden RG, Wilson RL, Beavers D, Hartman J. The effects of lowering uric acid levels using allopurinol on markers of metabolic syndrome in end-stage renal disease patients: a pilot study. Anadolu Kardiyol Derg. 2009 Oct;9(5):385-9.

[171] Pergola PE, Krauth M, Huff JW, Ferguson DA, Ruiz S, Meyer CJ, et al. Effect of bardoxolone methyl on kidney function in patients with T2D and Stage 3b-4 CKD. Am J Nephrol. 2011;33(5):469-76.

[172] Pergola PE, Raskin P, Toto RD, Meyer CJ, Huff JW, Grossman EB, et al. Bardoxolone methyl and kidney function in CKD with type 2 diabetes. N Engl J Med. 2011 Jul 28;365(4):327-36.

[173] Honda T, Yoshizawa H, Sundararajan C, David E, Lajoie MJ, Favaloro FG, Jr., et al. Tricyclic compounds containing nonenolizable cyano enones. A novel class of highly potent anti-inflammatory and cytoprotective agents. J Med Chem. 2011 Mar 24;54(6): 1762-78.

[174] Eskiocak U, Kim SB, Roig AI, Kitten E, Batten K, Cornelius C, et al. CDDO-Me protects against space radiation-induced transformation of human colon epithelial cells. Radiat Res. 2010 Jul;174(1):27-36.

[175] Nagaraj S, Youn JI, Weber H, Iclozan C, Lu L, Cotter MJ, et al. Anti-inflammatory triterpenoid blocks immune suppressive function of MDSCs and improves immune response in cancer. Clin Cancer Res. 2010 Mar 15;16(6):1812-23.

[176] Wu QQ, Wang Y, Senitko M, Meyer C, Wigley WC, Ferguson DA, et al. Bardoxolone methyl (BARD) ameliorates ischemic AKI and increases expression of protective genes Nrf2, PPARgamma, and HO-1. Am J Physiol Renal Physiol. 2011 May; 300(5):F1180-92.

[177] Kelly GS. L-Carnitine: therapeutic applications of a conditionally-essential amino acid. Altern Med Rev. 1998 Oct;3(5):345-60.

[178] Fouque D, Holt S, Guebre-Egziabher F, Nakamura K, Vianey-Saban C, Hadj-Aissa A, et al. Relationship between serum carnitine, acylcarnitines, and renal function in patients with chronic renal disease. J Ren Nutr. 2006 Apr;16(2):125-31.

[179] Eknoyan G, Latos DL, Lindberg J. Practice recommendations for the use of L-carnitine in dialysis-related carnitine disorder. National Kidney Foundation Carnitine Consensus Conference. Am J Kidney Dis. 2003 Apr;41(4):868-76.

[180] Sakurabayashi T, Miyazaki S, Yuasa Y, Sakai S, Suzuki M, Takahashi S, et al. L-carnitine supplementation decreases the left ventricular mass in patients undergoing hemodialysis. Circ J. 2008 Jun;72(6):926-31.

[181] Matsumoto Y, Sato M, Ohashi H, Araki H, Tadokoro M, Osumi Y, et al. Effects of L-carnitine supplementation on cardiac morbidity in hemodialyzed patients. Am J Nephrol. 2000 May-Jun;20(3):201-7.

[182] Signorelli SS, Fatuzzo P, Rapisarda F, Neri S, Ferrante M, Oliveri CG, et al. Propionyl-L-carnitine therapy: effects on endothelin-1 and homocysteine levels in patients with peripheral arterial disease and end-stage renal disease. Kidney Blood Pressure Res. 2006;29(2):100-7.

[183] Zhou Q, Wu S, Jiang J, Tian J, Chen J, Yu X, et al. Accumulation of circulating advanced oxidation protein products is an independent risk factor for ischemic heart disease in maintenance hemodialysis patients. Nephrology. 2012 Jun 28.

[184] Sener G, Paskaloglu K, Satiroglu H, Alican I, Kacmaz A, Sakarcan A. L-carnitine ameliorates oxidative damage due to chronic renal failure in rats. J Cardiovasc Pharmacol. 2004 May;43(5):698-705.

[185] Pecoits-Filho R, Lindholm B, Stenvinkel P. The malnutrition, inflammation, and atherosclerosis (MIA) syndrome -- the heart of the matter. Nephrol Dial Transplant. 2002;17 Suppl 11:28-31.

[186] Brunini TM, da SCD, Siqueira MA, Moss MB, Santos SF, Mendes-Ribeiro AC. Uremia, atherothrombosis and malnutrition: the role of L-arginine-nitric oxide pathway. Cardiovasc Hematol Disord Drug Targets. 2006 Jun;6(2):133-40.

[187] Reyes AA, Karl IE, Klahr S. Role of arginine in health and in renal disease. Am J Physiol. 1994 Sep;267(3 Pt 2):F331-46.

[188] Morris SM, Jr. Enzymes of arginine metabolism. J Nutr. 2004 Oct;134(10 Suppl): 2743S-7S; discussion 65S-67S.

[189] Stuehr DJ. Enzymes of the L-arginine to nitric oxide pathway. J Nutr. 2004 Oct;134(10 Suppl):2748S-51S; discussion 65S-67S.

[190] Morel F, Hus-Citharel A, Levillain O. Biochemical heterogeneity of arginine metabolism along kidney proximal tubules. Kidney Int. 1996 Jun;49(6):1608-10.

[191] Mendes RAC, Brunini TM, Ellory JC, Mann GE. Abnormalities in L-arginine transport and nitric oxide biosynthesis in chronic renal and heart failure. Cardiovasc Res. 2001 Mar;49(4):697-712.

[192] Brunini TM, Roberts NB, Yaqoob MM, Ellory JC, Mann GE, Mendes RAC. Activation of L-arginine transport in undialysed chronic renal failure and continuous ambulatory peritoneal dialysis patients. Clin Exp Pharmacol Physiol. 2006 Jan-Feb;33(1-2): 114-8.

[193] Mendes RAC, Hanssen H, Kiessling K, Roberts NB, Mann GE, Ellory JC. Transport of L-arginine and the nitric oxide inhibitor NG-monomethyl-L-arginine in human erythrocytes in chronic renal failure. Clin Sci (Lond). 1997 Jul;93(1):57-64.

[194] Noris M, Benigni A, Boccardo P, Aiello S, Gaspari F, Todeschini M, et al. Enhanced nitric oxide synthesis in uremia: implications for platelet dysfunction and dialysis hypotension. Kidney Int. 1993 Aug;44(2):445-50.

[195] Aiello S, Noris M, Remuzzi G. Nitric oxide synthesis and L-arginine in uremia. Miner Electrolyte Metab. 1997;23(3-6):151-6.

[196] Rajapakse NW, Kuruppu S, Hanchapola I, Venardos K, Mattson DL, Smith AI, et al. Evidence that renal arginine transport is impaired in spontaneously hypertensive rats. Am J Physiol Renal Physiol. 2012 Jun;302(12):F1554-62.

[197] Huang CC, Tsai SC, Lin WT. Potential ergogenic effects of L-arginine against oxidative and inflammatory stress induced by acute exercise in aging rats. Exp Gerontol. 2008 Jun;43(6):571-7.

[198] Mosca L, Marcellini S, Perluigi M, Mastroiacovo P, Moretti S, Famularo G, et al. Modulation of apoptosis and improved redox metabolism with the use of a new antioxidant formula. Biochem Pharmacol. 2002 Apr 1;63(7):1305-14.

[199] Park NY, Park SK, Lim Y. Long-term dietary antioxidant cocktail supplementation effectively reduces renal inflammation in diabetic mice. Br J Nutr. 2011 Nov;106(10): 1514-21.

[200] Dietary supplementation with n-3 polyunsaturated fatty acids and vitamin E after myocardial infarction: results of the GISSI-Prevenzione trial. Gruppo Italiano per lo Studio della Sopravvivenza nell'Infarto miocardico. Lancet. 1999 Aug 7;354(9177): 447-55.

[201] Shargorodsky M, Debby O, Matas Z, Zimlichman R. Effect of long-term treatment with antioxidants (vitamin C, vitamin E, coenzyme Q10 and selenium) on arterial compliance, humoral factors and inflammatory markers in patients with multiple cardiovascular risk factors. Nutr Metab (Lond). 2010;7:55.

Chapter 9

Oxidative Stress in Diabetes Mellitus and the Role Of Vitamins with Antioxidant Actions

Maria-Luisa Lazo-de-la-Vega-Monroy and
Cristina Fernández-Mejía

Additional information is available at the end of the chapter

1. Introduction

Diabetes mellitus is a group of metabolic diseases characterized by hyperglycemia resulting from defects of insulin action, insulin secretion or both [1]. Diabetes has taken place as one of the most important diseases worldwide, reaching epidemic proportions. Global estimates predict that the proportion of adult population with diabetes will increase 69% for the year 2030 [2].

Hyperglycemia in the course of diabetes usually leads to the development of microvascular complications, and diabetic patients are more prone to accelerated atherosclerotic macrovascular disease. These complications account for premature mortality and most of the social and economical burden in the long term of diabetes [3].

Increasing evidence suggests that oxidative stress plays a role in the pathogenesis of diabetes mellitus and its complications [4]. Hyperglycemia increases oxidative stress, which contributes to the impairment of the main processes that fail during diabetes, insulin action and insulin secretion. In addition, antioxidant mechanisms are diminished in diabetic patients, which may further augment oxidative stress [5, 6]. Several studies have addressed the possible participation of dietary antioxidants, such as vitamins, in ameliorating the diabetic state and retarding the development of diabetes complications [7, 8].

The aim of this chapter is to revise the current knowledge of the role of oxidative stress in the pathogenesis of diabetes mellitus and its complications, and to discuss the existing evidence of the effects of vitamins as antioxidant therapy for this disease.

2. Oxidative stress

At the beginning of life, the organisms obtained their energy (ATP) by anoxygenic photosynthesis, for which oxygen was toxic. Most of the metabolic pathways were developed during this anaerobic stage of life, in which oxygen came later. Cyanobacteria started producing oxygen from photosynthesis, which raised the atmospheric oxygen, and favored those organisms which have evolved into eukaryotic cells with mitochondria, able to use oxygen for a more efficient energy production [9].

Whenever a cell's internal environment is perturbed by infections, disease, toxins or nutritional imbalance, mitochondria diverts electron flow away from itself, forming reactive oxygen species (ROS) and reactive nitrogen species (RNS), thus lowering oxygen consumption. This "oxidative shielding" acts as a defense mechanism for either decreasing cellular uptake of toxic pathogens or chemicals from the environment, or to kill the cell by apoptosis and thus avoid the spreading to neighboring cells [9]. Therefore, ROS formation is a physiological response to stress.

The term "oxidative stress" has been used to define a state in which ROS and RNS reach excessive levels, either by excess production or insufficient removal. Being highly reactive molecules, the pathological consequence of ROS and RNS excess is damage to proteins, lipids and DNA [10]. Consistent with the primary role of ROS and RNS formation, this oxidative stress damage may lead to physiological dysfunction, cell death, pathologies such as diabetes and cancer, and aging of the organism [11].

2.1. ROS and RNS production

ROS and RNS are highly reactive molecules, which can be free radicals such as superoxide ($^\bullet O_2^-$), hydroxyl ($^\bullet OH$), peroxyl ($^\bullet RO_2$), hydroperoxyl ($^\bullet HRO_2^-$), nitric oxide ($^\bullet NO$) and nitrogen dioxide ($^\bullet NO_2^-$), or nonradicals such as hydrogen peroxide (H_2O_2), hydrochlorous acid (HOCl), peroxynitrite ($ONOO^-$), nitrous oxide (HNO_2), and alkyl peroxynitrates (RONOO). Most of the studies regarding diabetes and its complications have addressed the role of superoxide ($^\bullet O_2^-$), nitric oxide ($^\bullet NO$), and peroxynitrite ($ONOO^-$) in this disease. There are basically two pathways for $^\bullet O_2^-$ production: NADPH oxidases and mitochondrial function, while $^\bullet NO$ and $ONOO^-$ are produced by the Nitric Oxide Synthase pathway [10].

2.1.1. NADPH oxidases

Oxidases are enzymes which catalyze redox reactions involving molecular oxygen (O_2). Superoxide is generated by oxidases via one-electron reduction of oxygen and the oxidation of their substrates. Several oxidases exist in the body, such as xantine oxidase, glucose oxidase, monoamine oxidase, cytochrome P450 oxidase, and NADPH oxidases.

NADP in the cell exists in its reduced (NADPH) and oxidized (NADP+) forms. NADPH supplies reducing power in reactions for biosynthesis, and it also serves as electron donor substrate for the NADPH oxidase. This enzyme is a membrane-bound electron transport complex which pumps electrons from NADPH in the cytosol across biological membranes and into intracellular and extracellular compartments, such as nucleus, endoplasmic reticulum, endosome, phagosome, mitochondria and extracellular space. It is the only enzyme whose primary function is generating superoxide and/or hydrogen peroxide, mainly for preventing the transfer of pathogens and for cellular bactericidal function[12, 13].

2.1.2. Mitochondrial electron transport chain

Mitochondrion is the site of eukaryotic oxidative metabolism. It contains the enzymes needed for converting pyruvate into Acetyl-CoA, the citric acid cycle (also known as the Krebs cycle) and for fatty acid oxidation. Additionally, it performs the electron transport and oxidative phosphorylation. Substrate (amino acid, fatty acid and carbohydrate) oxidation in the citric acid cycle release electrons, which are transferred to the coenzymes NAD+ and FAD to form NADH and FADH2. These electrons then pass into the mitochondrial electron-transport chain, a system of linked electron carrier proteins comprised by Complexes I, II, III and IV. Complex I, III, and IV drive the exit of protons from the mitochondrial matrix, producing a proton gradient across the inner mitochondrial membrane. The free energy stored in this electrochemical gradient drives the condensation of ADP with inorganic phosphate in order to form ATP by oxidative phosphorylation. Along this electron transport, molecular oxygen is the final electron acceptor, which will be then reduced to H_2O [14, 15]. However, between 0.4 and 4% of all oxygen consumed will be converted into superoxide anion [16]. There is also a normal threshold for protonic potential above which electron transfer is inhibited at complex III, causing the electrons to go back to complex II where there are transferred to molecular oxygen prematurely and not to complex IV as it naturally occurs. Therefore, the endproduct of this transfer is superoxide [17].

Mitochondria play an important role in the maintenance of cellular redox status, acting as a redox sink and limiting NADPH oxidase activity. However, when the proton potential threshold is surpassed, mitochondria is also a significant source of ROS, which may further stimulate NADPH oxidases, creating a vicious cycle of ROS production [18]. When mitochondria cannot further extract oxygen, cell and tissue oxygen levels rise, decreasing the tissue extraction of oxygen from the blood. This results in tissue vascularity reduction, which may be associated with peripheral vascular disease and, in time, chronic tissue hypoxia and ischemia [9].

2.1.3. NO and RNS production

Nitric oxide •NO is produced by the enzyme nitric oxide synthase (NOS), of which there are three isoforms: neural (nNOS or NOS-I) expressed in neurons, inducible (iNOS or NOS-II) expressed in smooth muscle of bold vessels, hepatocites, macrophagues and neuroendocirne tissue, and endothelial (eNOS or NOS-III) expressed constitutively in endothelial cells. iNOS and eNOS can be stimulated by the redox state in the cell, cytokines, hormones and nutrients [19, 20]. NOS catalyze the oxidation of the terminal guanidine nitrogen of the L–arginine, in presence of oxygen and NADPH, to yield L-citruline and •NO [21].

Once produced and released, •NO can diffuse freely through membranes or act on different cellular targets. •NO participates as mediator of several physiological effects such as vasore-laxation, macrophague activation, gene expression and apoptosis. Usually, •NO is consid-ered as a vasculoprotective molecule. However, one of its multiple effects is also protein nitrosilation at the thiol groups and RNS generation such as peroxynitrite (ONOO-), as •NO easily reacts with $•O_2^-$. Therefore, the amount of $•O_2^-$ determines whether •NO acts as a pro-tective or harmful molecule [10, 22].

2.2. Antioxidant defenses in the organism

As a small part the oxygen consumed for aerobic processes will be converted into superox-ide anion [16], which will have to be scavenged or converted into less reactive (and harmful) molecules. The main enzymes that regulate this process are Superoxide dismutase (SOD), Glutathione Peroxidase (GSH-Px) and Catalase (Figure 1). When ROS overproduction or chronic hyperglycemia occurs, the activity of these enzymes is insufficient, leading to more ROS and RNS formation and activation oxidative stress pathways.

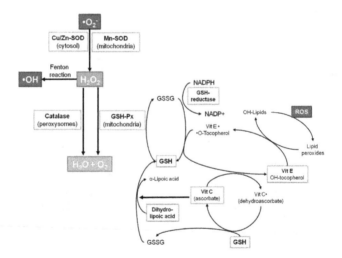

Figure 1. Antioxidant defenses in the organism.

SOD is considered a first-line defense against ROS. This enzyme is present in nearly all cells, and converts $•O_2^-$ into H_2O_2. Mitochondrial and bacterial SOD contain Mn, while cytosolic SOD is a dimer containing Cu and Zn. As the H_2O_2 may still react with other ROS, it needs to be degraded by either one of the other two antioxidant enzymes, GSH-Px or catalase [10, 12].

GSH peroxidase is located in the mitochondria. It catalyzes degradation of H_2O_2 by reduc-tion, where two gluthathione (GSH) molecules are oxidized to glutathione disulfide (GSSG).

Regeneration of GSH by GSH-reductase, requires NADPH, which is oxidized to NADP+. Catalase, on the other hand, is localized primarily in peroxisomes, and so it detoxifies the H_2O_2 that diffuses from the mitochondria to the cytosol, converting it into water and molecular oxygen [10, 12].

There are also nonenzymatic antioxidant mechanisms, which mostly help regenerate GSSG back into GSH. Antioxidant vitamins such as A, C, E and alpha-lipoic acid are among these mechanisms. Although all these antioxidant defenses work together to eliminate H_2O_2 (and thus superoxide) from the cell, in the presence of reduced transition metals (Cu, Fe), H_2O_2 can be transformed into ·OH, which is a highly reactive ROS, by the Fenton reaction [10, 23].

2.3. Metabolic and signaling pathways involved in oxidative stress in diabetes

There are several molecular pathways involved in ROS formation and ROS induced damage. Here we will review the ones that have been related to oxidative stress in diabetes. Not surprisingly, most of them are related to glucose and/or lipid metabolism.

2.3.1. Glucose oxidation and GAPDH

In order to generate energy, glucose needs to be first oxidized inside the cells by glycolysis. In this process, once glucose enters the cells, it is phosphorylated to form glucose-6-phosphate, a reaction mediated by hexocinases. Glucose-6-P is then converted to Fructose-6-P by phosphoglucoisomerase, which can undergo two fates: the pentose phosphate pathway, where reduction of NADP+ to NADPH occurs, or to continue glycolysis to yield Gliceraldehyde-3-P. Glyceraldehyde-3-P dehydrogenase (GAPDH) phosphorylates this product and glycolysis is further completed until its end product pyruvate, which enters the Krebs cycle and mitochondrial metabolism (Figure 2).

It has been proposed that hyperglycemia-induced mitochondrial superoxide production activates damaging pathways by inhibiting glyderaldehyde-3-phosphate dehydrogenase (GAPDH) [4, 24], an enzyme that normally translocates in and out of the nucleus [25, 26]. ROS inhibit glyderaldehyde-3-phosphate dehydrogenase through a mechanism involving the activation of enzyme poly-ADP-ribose polymerase-1 (PARP-1). This enzyme is involved in DNA repair and apoptotic pathways. ROS cause strand breaks in nuclear DNA which activates PARP-1. PARP-1 activation results in inhibition of glyderaldehyde-3-phosphate dehydrogenase by poly-ADP-ribosylation [27]. This results in increased levels of all the glycolytic intermediates upstream of GAPDH. Accumulation of glyceraldehyde 3-phosphate activates two major pathways involved in hyperglycemia-complications: a) It activates the AGE pathway deriving glyceraldehyde phosphate and dihydroxyacetone phosphate to the nonenzymatic synthesis of methylglyoxal. b) Increased glyceraldehyde 3-phosphate favors diacylglycerol production which activates PKC pathway. Further upstream, levels of the glycolytic metabolite fructose 6-phosphate increase, which then increases flux through the hexosamine pathway, where fructose 6-phosphate is converted by the enzyme glutamine-fructose-6-phosphate amidotransferase (GFAT) to UDP–N-Acetylglucosamine. Finally, inhibition of GAPDH favors the accumulation of the first glycolytic metabolite, glucose. This increases its flux through the polyol pathway, consuming NADPH in the process [24].

2.3.2. The polyol pathway

The family of aldo-keto reductase enzymes catalyzes the reduction of a wide variety of carbonyl compounds to their respective alcohols. These reactions utilize nicotinic acid adenine dinucleotide phosphate (NADPH). Aldo-keto reductase has a low affinity (high Km) for glucose, and at the normal glucose concentrations, metabolism of glucose by this pathway is a very small percentage of total glucose metabolism. However, in a hyperglycemic environment, increased intracellular glucose results in its increased enzymatic conversion to the polyalcohol sorbitol, with concomitant decreases in NADPH [4] (Figure 2). Since NADPH is a cofactor required to regenerate reduced glutathione, an antioxidant mechanism, and this compound is an important scavenger of reactive oxygen species (ROS), this could induce or exacerbate intracellular oxidative stress [24]. Moreover, sorbitol is oxidated to fructose by sorbitol dehydrogenase, which can lead to PKC activation via the increased NADH/NAD+ ratio [4]. Although this mechanism does not produce ROS in a direct way, it takes part in the redox imbalance causing oxidative stress.

2.3.3. Hexosamine pathway

When glucose levels are within normal range, a relatively low amount of fructose-6-P is drived away from glycolysis. If intracellular glucose rises, excess fructose-6-phosphate is diverted from glycolysis to provide substrate for the rate-limiting enzyme of this pathway, GFAT. This enzyme converts fructose 6-phosphate to glucosamine 6-phosphate, which is then converted to UDP-NAcetylglucosamine, which is essential for making the glycosyl chains of proteins and lipids. Specific O-Glucosamine-N-Acetyl transferases use this metabolite for post-translational modification of specific serine and threonine residues on cytoplasmic and nuclear proteins [24, 28].

2.3.4. Diacylglycerol formation and PKC activation

The Protein Kinase C (PKC) family comprises at least eleven isoforms of serine/threonine kinases, which participate in signaling pathways activated by phosphatidyl serine, Calcium and Diacylglycerol (DAG). DAG levels are elevated chronically in the hyperglycemic or diabetic environment due to an increase in the glycolytic intermediate dihydroxyacetone phosphate (figure 2). This intermediate is reduced to glycerol-3-phosphate, which, conjugated with fatty acids, increases de novo synthesis of DAG [29]. Evidence suggests that the enhanced activity of PKC isoforms could arise from inhibition of the glycolytic enzyme glyceraldehide-3-phosphate dehydrogenase by increased ROS intracellular levels [4, 24]. Other studies suggest that enhanced activity of PKC isoforms could also result from the interaction between AGEs and their extracellular receptors [30]. PKC isoforms constitute a wide range of cellular signals, including activation of NADPH oxidase and NF-κB, resulting in excessive ROS production. They also increase vascular permeability, stabilize vascular endothelial growth factor (VEGF) mRNA expression and increase leukocyte-endothelium interaction [11].

2.3.5. Glyceraldehyde autoxidation

Accumulation of glyceraldehyde 3-phosphate, besides activating the AGE formation and the PKC pathway, it can oxidate itself. This autoxidation generates H_2O_2, which further contributes to oxidative stress [31].

2.3.6. Advanced glycation end-products (AGEs)

Intracellular hyperglycaemia is the primary initiating event in the formation of both intracellular and extracellular AGEs [32]. AGEs can arise from intracellular auto-oxidation of glucose to glyoxal, decomposition of the Amadori product (glucose-derived 1-amino-1-deoxyfructose lysine adducts) to 3-deoxyglucosone (perhaps accelerated by an amadoriase), and nonenzymatic phosphate elimination from glyceraldehyde phosphate and dihydroxyacetone phosphate to form methylglyoxal. These reactive intracellular dicarbonyl glyoxal, methylglyoxal and 3-deoxyglucosone react with amino groups of intracellular and extracellular proteins to form AGEs [4]. Intracellular production of AGE precursors can damage cells by three general mechanisms: 1) Intracellular proteins modified by AGEs have altered function, 2) Extracellular matrix components modified by AGE precursors interact abnormally with other matrix components and with matrix receptors (integrins) that are expressed on the surface of cells, and 3) Plasma proteins modified by AGE precursors bind to AGE receptors (such as RAGE and AGE-R1,2 and 3) on cells such as macrophages, vascular endothelial cells and vascular smooth muscle cells. AGE receptors binding induces the production of ROS, which in turn activates PKC. It also activates NF-κB and NADPH oxidase, and disturbs MAPK signaling [31].

2.3.7. Stress-sensitive signaling pathways

In addition to direct damage of biomolecules in the cells, oxidative stress is also involved in activation of several stress-sensitive signaling pathways, which can result in inflammation, cytokine release, and even apoptosis. Among these pathways we find the transcription factor NF-κB, which together with PARP acts as a transcriptional coactivator of inflammation molecules such as iNOS, intracellular adhesion molecule-1 (ICAM-I), and histocompatibility complex class II [33]. p38 MAPK pathway and c-Jun Nterminal kinase (JNK) (also known as stress-activated protein kinase (SAPK) participate in cellular responses to stress due to osmotic shock, cytokines and UV light, playing a role in cellular proliferation, apoptosis, and inflammatory responses [33]. Jak/STAT is another important signaling pathway, which initiates and mediates cellular responses to cytokines such as interferons and interleukins [33].

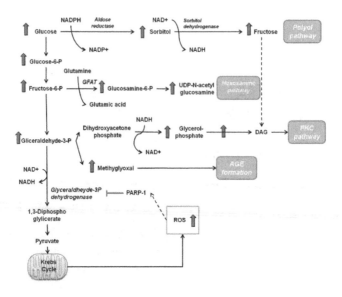

Figure 2. Oxidative stress-related pathways derived from glucose metabolism.

2.4. ROS induced damages

Being highly reactive species, ROS may modify and damage nucleic acids, proteins, lipids and carbohydrates, finally leading to cell damage. Among the motifs that can react with ROS we have the metal ligand from metalloproteases and Fe from oxihemoglobin. $^{\bullet}O_2^{\bullet}$ can also modify and inhibit catalases, while $^{\bullet}OH$ can bind to the histidine residue from SOS causing its inhibition. ROS react mostly with insaturated and sulfur containing molecules, thus, proteins with high contents of tryptophan, tyrosine, phenylalanine, histidine, methionine and cysteine can suffer ROS modifications. Finally, ROS may also break peptidic bonds after oxidation of proline residues by $^{\bullet}O_2^{\bullet}$ or $^{\bullet}OH$ [31].

ROS and RNS may also modify fatty acids, lipoproteins, and phospholipids, a process termed lipid peroxidation, where $^{\bullet}OH$ and $^{\bullet}O_2^{\bullet}$ form hydroperoxide lipids. Hydroperoxyde products cause severe damage to plasma membranes, or they can diffuse to other cells in the organisms and cause vascular permeability and inflammation by binding to (oxidized low-density lipoprotein) LOX receptors, and apoptosis [31].

H_2O_2 in cells can function as a signaling molecule leading to cellular proliferation or can result in cell death. At low concentrations, H_2O_2 serves as a second messenger to activate NF-κB and various kinases (p38 MAPK, ERK, PI3K, Akt, JAK2, STAT). H_2O_2 at slightly higher concentrations can induce the release of cytochrome c and apoptosis-inducing factor (AIF) from mitochondria into the cytosol where they trigger the activation of caspase, leading to cell death by apoptosis [12].

3. Diabetes mellitus

Diabetes mellitus is a group of metabolic diseases characterized by hyperglycemia, caused by a defect on insulin production, insulin action or both [1]. There are two main types of diabetes: type 1 and type 2 diabetes.

Type 1 diabetes is due to an autoimmune destruction of the insulin producing pancreatic beta-cells, which usually leads to absolute insulin deficiency. Patients with type 1 diabetes require insulin for survival. This type of diabetes accounts for 5-10% of the total cases of diabetes worldwide. Type 2 diabetes represents approximately 90% of the total diabetes cases, and it is characterized by impairment in insulin action and/or abnormal insulin secretion [1].

The origins of type 2 diabetes are multifactorial. Obesity, age, ethnic origin and familiar history of diabetes are among the factors that contribute to its development. Even though a strong genetic component has been recognized, genotype only establishes the conditions for the individual to be more or less prone to environmental effects and lifestyle factors [34].

Type 2 diabetes develops when insulin secretion or insulin action fails. The impairment of insulin actions is known as insulin resistance, presented as a suppression or retard in metabolic responses of the muscle, liver and adipose tissue to insulin action. This failure is located at the signaling pathways held after insulin binding to its specific receptor [35]. Chronic insulin resistance leads to hyperglycemia.

When the beta cells cannot secrete enough insulin in response to the metabolic demand caused by insulin resistance, frank diabetes type 2 occurs. This failure in the beta cell may be due to an acquired secretory dysfunction and/or a decrease in beta-cell mass [36]. All type 2 diabetic patients have some defect in the ability of beta cells to produce or secrete insulin [37].

3.1. Insulin action and insulin resistance

Once secreted to the portal circulation, insulin is transported to peripheral tissues, on which it will exert mainly anabolic actions [38]. Insulin starts its action by binding to insulin receptor, a transmembrane protein belonging to protein tyrosine kinase activity receptors superfamily, which can autophosphorylate. This initiates a series of events involving protein and membrane lipid phosphorylation, coupling proteins and cytoskeleton activity [39] [40]. The three main signaling pathways activated in response to insulin receptor phosphorylation are 1) PI3K 2)MAPK, and 3) Cb1. These pathways act in a concerted way to translate the signal of insulin receptor into biological actions in target organs, such as glucose transport by transporting GLUT4 vesicles to the membrane, protein, lipid and glycogen synthesis, mitosis and gene expression [40] (Figure 3).

As protein phosphorylation activates these signaling pathways, dephosphorylation inhibits them. Different phosphatases such as protein-tyrosine phosphatase 1B (PTP1B), Phosphatase and tensin homolog (PTEN), SH2-containing tyrosine- protein phosphatase (SHO2), and suppressor of cytokine signaling 3 (SOCS-3) dephosphorylate and shut down insulin signaling [35]. Any alteration in the insulin pathway, being inefficient phosphorylation or increment in phosphatase acticity, causes impairment in insulin action. This is the molecular mechanism leading to insulin resistance.

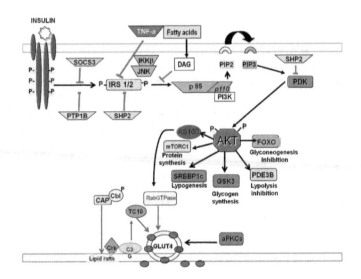

Figure 3. Molecular mechanisms of insulin signaling.

3.2. Insulin secretion

Beta-cells in the endocrine pancreas are responsible for secreting insulin in response to rises in blood nutrient levels during the postprandial state. Glucose is the most important nutrient for insulin secretion. The process by which glucose promotes insulin secretion requires glucose sensing and metabolism by the beta-cell, a process called glucose-stimulated insulin secretion (Figure 4). In the first phase of insulin secretion, glucose enters the cell by glucose transporters (GLUT2 in rodents, GLUT1 in humans). Glucose is then phosphorylated to form glucose-6-phosphate by glucokinase [41]. The generation of ATP by glycolysis, the Krebs cycle and the respiratory chain closes the ATP-sensitive K+ channel (KATP) [42], allowing sodium (Na+) entry without balance. These two events depolarize the membrane and open voltage-dependent T-type calcium (Ca2+) and sodium (Na+) channels. Na+ and Ca2+ entry further depolarizes the membrane and voltage-dependent calcium channels open. This activation increases intracellular Ca2+ ([Ca2+]i) [43], which leads to fusion of insulin-containing secretory granules with the plasma membrane and the first phase insulin secretion [44, 45].

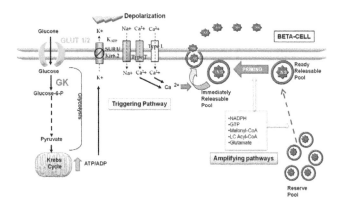

Figure 4. Mechanisms of biphasic glucose-stimulated insulin secretion.

Besides increasing ATP/ADP ratio, glucose metabolism in the beta cell can generate a series of metabolic coupling signals that can initiate and sustain a second insulin secretion phase. Some of these coupling factors participate in mitocondrial metabolism and anaplerosis, constituting cycles involving NADPH, pyruvate, malate, citrate, isocitrate, Acyl-CoA and glutamate [46]. Diverse signaling pathways can also contribute to glucose-induced insulin secretion such as CaMKII [47-49], PKA [50, 51], PKC [51, 52] y PKG [53, 54]. Most secretagogues and potentiators of insulin secretion, such as nutrients, hormones and neurotransmitters, use these pathways to modulate insulin secretion.

4. Oxidative stress in diabetes mellitus

Hyperglycemia and free fatty acid intake are among the causes for oxidative stress conditions [23]. Hence, it may not be surprising that diabetic subjects tend to have more oxidative cell and organism environments than healthy subjects, i.e. an increase in ROS generation [5, 55, 56]. Moreover, diabetic patients present a decrease in antioxidant defenses. The antioxidant enzyme levels are affected by diabetes, which further increase oxidative stress [5, 6].

Oxidative stress has been proposed as a major participant in the patophysiology of diabetic complications [27]. Nevertheless, regarding diabetes onset and development, oxidative stress has also shown to affect the two major mechanisms failing during diabetes: insulin resistance and insulin secretion.

4.1. Oxidative stress processes in insulin resistance

ROS and RNS affect the insulin signaling cascade [5]. As with other ROS effects, low doses play a physiological role in insulin signaling. After insulin stimulation of its receptor in adipocytes, H_2O_2 is produced via NADPH oxidase, which by inhibits PTP1B catalytic activity, thus increasing tyrosine phosphorylation [57].

However, oxidative stress caused by hyperglycemia in diabetes may impair insulin signaling, leading to insulin resistance. Although no mechanisms have been completely established, several responses to ROS excess in the insulin signaling have been proposed.

Disturbs in cellular redistribution of insulin signaling components may alter the insulin cascade, a process mediated by NF-κB [58]. A decrease in GLUT4 gene transcription and increase in GLUT1 (insulin independent glucose transporter) has also been observed, as well as increases in phosphorylation of IRS protein in an insulin receptor-independent fashion (perhaps by the stress kinases). Altogether, hyperglycemia and insulin resistance may also lead to altered mitochondrial function, and insulin action impairment by cytokines in response to metabolic stress [59, 60]. An increase in the hexosamine pathway has also been linked to insulin resistance. Moreover, it has been proposed that this pathway acts as a cellular sensor for the glucose excess. From that point of view, insulin resistance may be a protective mechanism from the glucose excess entrance [28].

4.2. Oxidative stress processes in insulin secretion

Pancreatic beta-cells are especially sensitive to ROS and RNS, because their natural enzymatic antioxidant defenses are lower compared to other tissues such as liver. Moreover, they lack the ability to adapt their low enzyme activity levels in response to stress such as high glucose or high oxygen [61]. Glucose enters to the beta-cell in an insulin independent fashion, because besides providing energy, glucose sensing in the beta-cell is crucial for insulin secretion. It has been suggested that hyperglycemia can generate chronic oxidative stress by the glucose oxidation pathway [62], leading to an excess in mitochondrial superoxide production, which further activates uncoupling protein-2 (UCP-2). This protein lowers ATP/ADP relationship through proton leak in the beta-cell, which reduces insulin secretion [63].

ROS also increase the stress signaling pathways in the beta cells, such as NF-κB activity, which potentially leading to beta-cell apoptosis [64], and the JNK pathway which has been related to suppression of insulin gene expression, possibly by reduction of PDX-1 DNA binding activity, a major regulator of insulin expression [65]. It has also been shown that the activation of the hexosamine pathway in beta-cells leads to suppression of PDX-1 binding to the insulin and other genes involved in insulin secretion, perhaps contributing to the beta-cell dysfunction present in diabetes mellitus [66].

As in other cell types, NO in beta-cells has physiologic roles. NO may regulate glucokinase activity by s-nitrosilation [67] in the beta-cell, and possibly increase insulin secretion. However, NO excess and concomitant NRS may cause apoptosis through caspase-3 activation and decrease in ATP levels [68].

Besides ROS hyperproduction, excess mitochondrial metabolism resulting form hyperglycemia in the beta-cell may also alter mitochondrial shape, volume and behavior, uncoupling K-ATP channels from mitochondrial activity and thus altering glucose-induced insulin secretion [69].

5. Diabetic complications

Hyperglycemia, is the responsible of the development of diabetes complications as well. Hyperglycemia damage is produced in cells in which glucose uptake is independent of insulin, which, similarly to what happens in beta-cells, explains that the cause of the complications resides inside the cells [4]. Prolonged exposure to high glucose levels, genetic determinants of susceptibility and accelerating factors such as hypertension and dyslipidemia participate in the development of diabetic complications. Moreover, the development and progression of damage is proportional to hyperglycemia, which makes the lowering of glucose levels the most important goal for preventing complications and treating diabetes.

The main tissues affected by diabetes complications at the microvasculature levels are retina, renal glomerulus, and peripheral nerves. Diabetes is also associated with accelerated atherosclerotic disease affecting arteries that supply the heart, brain, and lower extremities. In addition, diabetic cardiomyopathy is a major diabetic complication [24].

5.1. Oxidative stress in diabetic complications

Oxidative stress plays a pivotal role in the development of diabetes complications, both at the microvascular and macrovascular levels. Results derived from two decades of diabetes complications investigation point towards mitochondrial superoxide overproduction as the main cause of metabolic abnormalities of diabetes. Thus, all of the above reviewed pathways are involved in microvasculature and macrovasculature hyperglycemic damage [24].

5.2. Microvascular complications

Diabetic retinopathy: Diabetic retinopathy appears in most patients after 10 to 15 years after diabetes onset. Background retinopathy presents small hemorrhages in the middle layers of the retina, appearing as "dots". Lipid deposition occurs at the margins of the hemorraghe, and microaneurisms (small vascular dilatations) and edema may appear. Proliferative retinopathy occurs when new blood vessels on the surface of the retina cause vitreous hemorrhage, and eventually, blindness. As the cells of the retina contain high amounts of aldoketoreductase, they have high susceptibility to increase the polyol pathway in the presence of excess glucose, with concomitant decreases in NADPH [4]. Sorbitol produced in this process increases osmotic stress, which has been linked to microaneurysm formation, thickening of the basement membranes and loss of pericytes. It is also thought that retina cells are damaged by glycoproteins, particularly form AGEs. Additionally, ROS by themselves may damage the cells. Importantly, VEGF, growth hormone and TGF-beta increases during diabetes may be the cause of proliferation of blood vessels [70].

Diabetic nephropaty: this complication causes glomerular basement membrane thickness, microaneurism formation, and mesangial nodule formation, all which are reflected in proteinuria and, in the end, renal insufficiency. The mechanisms for injury also involve the increased polyol pathway and AGE formation. AGE binding to its receptors has been proven to play a role as well in renal damage, fibrosis and inflammation associated with diabetic nephropaty. This actions of AGE also potentiate oxidative stress, while synergizing with rennin-angiotensin system activation, which leads to a vicious cycle causing kidney failure. As mentioned, diabetic patients, and particularly those with nephropaty, have lowered anti-oxidant defenses. Moreover, AGE receptors are significantly increased [71].

Diabetic neuropathy: Diabetic neuropathy is defined as the presence of symptoms and/or signs of peripheral nerve dysfunction in diabetic patients after exclusion of other causes. Peripheral neuropathy in diabetes may manifest in several different forms, including sensory, focal/multifocal, and autonomic neuropathies. [72]. Mechanisms of nerve injury are less known but likely related also to the polyol pathway, AGE formation and ROS themselves [70]. Oxidized proteins and lipoproteins also interact with receptors in the membrane of neurons, initiating inflammatory signaling mechanisms which further produce ROS, damaging cellular components and leading to neuronal injury [73].

5.3. Macrovascular complications

The central pathological mechanism in macrovascular complications is atherosclerotic disease. Atherosclerosis occurs as a result of chronic inflammation and injury to the arterial wall in the peripheral or coronary vascular system. This damages cause accumulation of oxidized lipids from LDL particles in the endothelial wall of arteries, whose rupture leads to acute vascular infarction. Additionally, platelet adhesion and hypercoagulability also occurs in type 2 diabetes, increasing the risk of vascular occlusion [70]. It has been proposed that increased superoxide production is the central and major mediator of endothelial tissue damage, causing direct inactivation of two antiatherosclerotic enzymes, endothelial nitric oxide synthase and prostacyclin synthase and that the activation of oxidative stress pathways is involved in the pathogenesis of complications [24].

Endothelial cells also contain high amounts of aldo-keto reductase, and are thus prone to increased polyol pathway activation. Moreover, a large body of evidence supports hypothesis that hyperglycemia or diabetes leads to vascular diacylglycerol accumulation and subsequent PKC activation, causing a variety of cardiovascular defects [29]. PKC activation has been associated with vascular alterations such as increases in permeability, contractility, extracellular matrix synthesis, cell growth and apoptosis, angiogenesis, leukocyte adhesion, and cytokine activation and inhibition [29]. Hyperglycemia-induced activation of PKC has also been implicated in the overexpression of the fibrinolytic inhibitor, plasminogen activator inhibitor-1 (PAI-1) [74]. In smooth muscle PKC hyperactivity is associated with decreased NO production [75] and has been shown to inhibit insulin-stimulated expression of eNOs in endothelial cells.

In arterial endothelial cells O-glucosamine-acylation participates in vascular complications interfering with the action of Akt/PKB, a critical insulin signaling protein, on eNOS [76]. GFAT activity is associated with increased transcription of transforming growth factor (TGF) alpha and beta and PAI-1, factors involved in the proliferation of vascular smooth-muscle and endothelial cells. This effect appears to be mediated by O-glucosamine-acylation of the transcription factor, Sp1 [77]. Increased TGF-beta and PAI-1 are associated with capillary and vascular occlusion by mechanisms associated with collagen and fibronectin expression causing capillary occlusion, in the case of TGF-beta, and decreased fibrinolysis in the case of PAI-1. O-GlcNAcylation impairs cardiomyocyte calcium cycling decreasing sarcoplasmic reticulum calcium ATPase 2a (Serca 2a) [78-80].

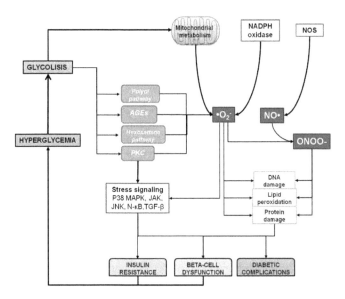

Figure 5. Oxidative stress pathways in diabetes mellitus.

6. Antioxidant vitamins and diabetes mellitus

As mentioned above, vitamins C, E, and A constitute the non enzymatic defense against oxidative stress, by regenerating endogenous antioxidants (Figure 1). Vitamin C has a role in scavenging ROS and RNS by becoming oxidized itself. The oxidized products of vitamin C, ascorbic radical and dehydroascorbic radical are regenerated by glutathione, NADH or NADPH. In addition, vitamin C can reduce the oxidized forms of vitamin E and glutathione [81]. Vitamin E is a fat-soluble vitamin which may interact with lipid hydroperoxides and scavenge them. It also participates, together with vitamin C, in gluthatione regeneration by interaction with lipoic acid [23]. Vitamin A has a plethora of cellular actions. Besides modulating gene expression, cell growth and differentiation, this vitamin may also act as antioxidant, although the mechanisms of action in this role are not fully deciphered. The antioxidant potential of carotenoids (vitamin A) depends on their distinct membrane-lipid interactions, while some carotenoids can decrease lipid peroxidation, others can stimulate it [82].

Since oxidative stress is present during the progression of diabetes and its complications, amelioration of oxidative status, mainly by increasing antioxidant non-enzymatic defenses, has been largely proposed and studied. Several clinical observational trials have particularly studied the correlation between vitamin E status in plasma and/or diet, and markers of oxidation, inflammation, type 2 diabetes incidence, and diabetic complications. Although inverse association has been found for vitamin E in some studies [7, 83, 84], the association found in other study disappeared after adjustment for cardiovascular risk factors such as obesity, smoking, and hypertension [85], or have observed no beneficial effect at all [7, 8]. Such contrasting results have also been reported for studies looking association of vitamin A and C consumption and amelioration of diabetes status and/or complications [7, 8, 81, 86].

On the other hand, in interventional trials with vitamin supplementation, the effects of vitamins E, C and A, alone or in diverse combinations, have yielded barely any promising result. There appears to be no beneficial effect of vitamin supplementation on diabetes or macrovascular complications [7, 8, 81]. Some of these studies have even evidenced associations between vitamin supplementation and an increased incidence of stroke [7]. Likewise, supplementation with antioxidant vitamins can even block beneficial ROS production during exercise, inhibiting the health-promoting effects of exercise in humans [87].

Paradoxically, in spite of the solid evidence of increased oxidative stress in diabetes, and the well established actions of vitamins as antioxidants, the association studies between antioxidant vitamin status and its beneficial effects in diabetes has no consistent results at all. What is more, interventional studies have failed in demonstrating a favorable effect of vitamin supplementation, discouraging its use as antioxidant therapy for diabetes.

Several reasons have been suggested for these contradictory observations. First, as vitamins may be easily oxidized, a vitamin may have antioxidant or oxidant properties, depending on the presence of other vitamins and the oxidative state in the cells i.e., if the oxidized form of a vitamin is not correctly reversed into the reduced form. Additionally, some vitamins may also activate oxidative stress pathways and further increase the oxidative stress, such as the activation of PKC by retinoids [88].

Vitamin doses may also be part of the problem, as the effect of vitamins depends on dietary concentrations and/or supplement intake. The wide variety of doses reached with diet and supplements, and the lack of an established "pharmacological" dose of vitamins, makes it difficult to ascertain the true net effect of vitamin status or supplementation needed to generate beneficial effects. As well, the required dose for antioxidant effects versus the required for the vitamin's role in the body may differ, which, together with vitamin's bioavailability and its interaction with other vitamins, are caveats for assessing and finding vitamins' effects, if any [7, 88].

Finally, the antioxidant effects of vitamins may not be sufficient to scavenge the great amount of ROS present in diabetes. Certainly, glucose levels have been correlated to the presence and severity of the complications. However once hyperglycemia has established, the incidence of complications after tight glycemic control remains the same. This effect has been termed glycemic memory, and is the cause for accumulative damage rendering diabetic complications. Considering that hyperglycemia is the main cause of oxidative stress in diabetes, in a similar way, the chronic undesirable effects that occur by ROS production may generate a vicious cycle difficult to break, in which ROS damage exacerbates the diabetic state, increasing glucose levels, which will further induce more oxidative imbalance [24].

7. Conclusions

Diabetes mellitus has reached epidemic proportions in the last decade, becoming one of the most important diseases worldwide. Several studies indicate oxidative stress is present in the dysfunction of insulin action and secretion that occur during diabetes, as well as in the development of diabetic complications. Nevertheless, oxidative stress is not the primary cause of diabetes, but rather a consequence of nutrient excess, given that oxidative stress is a natural response to stress, in this case, to glucose and/or lipid overload.

Vitamins such as E, C and A with antioxidant properties constitute the physiological non-enzymatic defense against oxidative stress. However, the evidence in favor of the use of vitamin supplementation as antioxidant therapy remains uncertain. Although some beneficial effects have been proven in observational studies, the results of interventional trials are still ineffective. Perhaps more studies on the physiopathology of oxidative stress and the role of vitamins in it, as well as standardizing vitamin dosage and assessing their undesirable effects are needed in order to determine a clear participation of vitamin supplementation in amelioration of the oxidative balance. More studies addressing the possibility of targeting directly at the enzymes and mechanisms involved in ROS production and not by antioxidants are needed as well.

Given that it is mostly dietary vitamin intake which has shown an association with ameliorating the diabetic state, and that oxidative stress is a response to excess of nutrients, it seems that attending the cause of excessive ROS production represents the best therapeutic option. Thus, adequate dietary interventions that reduce hyperglycemia, and increases in oxygen consumption (i.e. improve mitochondrial function) by exercise remain the primary choices for diabetes treatment and prevention of its complications.

Acknowledgements

This work was supported by grants from CONACyT and from the Dirección General de Asuntos del Personal Académico, UNAM.

Author details

Maria-Luisa Lazo-de-la-Vega-Monroy* and Cristina Fernández-Mejía

Unidad de Genética de la Nutrición, Departamento de Medicina Genómica y Toxicología Ambiental, Instituto de Investigaciones Biomédicas, Universidad Nacional Autónoma de México/ Instituto Nacional de Pediatría, México

References

[1] ADA. (2009). Diagnosis and classification of diabetes mellitus. *Diabetes care*, 32(1), 62-7.

[2] Shaw, J. E., Sicree, R. A., & Zimmet, P. Z. (2010). Global estimates of the prevalence of diabetes for 2010 and 2030. *Diabetes Res Clin Pract*, 87(1), 4-14.

[3] King, H., Aubert, R. E., & Herman, W. H. (1998). Global burden of diabetes, 1995-2025: prevalence, numerical estimates, and projections. *Diabetes Care*, 21(9), 1414-1431.

[4] Brownlee, M. (2001). Biochemistry and molecular cell biology of diabetic complications. *Nature*, 414(6865), 813-820.

[5] Rains, J. L., & Jain, S. K. (2011). Oxidative stress, insulin signaling, and diabetes. *Free Radic Biol Med*, 50(5), 567-575.

[6] Maritim, A. C., Sanders, R. A., & Watkins, J. B. (2003). Diabetes, oxidative stress, and antioxidants: A review. *Journal of Biochemical and Molecular Toxicology*, 17(1), 24.

[7] Sheikh-Ali, M., Chehade, J. M., & Mooradian, A. D. (2011). The antioxidant paradox in diabetes mellitus. *Am J Ther*, 18(3), 266-278.

[8] Cuerda, C., Luengo, L. M., Valero, Vidal. A., Burgos, R., Calvo, F. L., et al. (2011). Antioxidants and diabetes mellitus: review of the evidence]. *Nutr Hosp*, 26(1), 68-78.

[9] Naviaux, R. K. (2012). Oxidative Shielding or Oxidative Stress? *J Pharmacol Exp Ther.*

[10] Johansen, J. S., Harris, A. K., Rychly, D. J., & Ergul, A. (2005). Oxidative stress and the use of antioxidants in diabetes: linking basic science to clinical practice. *Cardiovasc Diabetol*, 4(1), 5.

[11] Ceriello, A. (2006). Oxidative stress and diabetes-associated complications. *Endocr Pract*, 12(1), 60-62.

[12] Fisher, A. B., Zhang, Q. Â. Â., Geoffrey, J. L., & Steven, D.S. (2006). Nadph and nadph oxidase. *Encyclopedia of Respiratory Medicine.*, Oxford, Academic Press, 77.

[13] Drummond, G. R., Selemidis, S., Griendling, K. K., & Sobey, C. G. (2011). Combating oxidative stress in vascular disease: NADPH oxidases as therapeutic targets. *Nat Rev Drug Discov*, 10(6), 453-471.

[14] Maechler, P., & Wollheim, C. B. (2001). Mitochondrial function in normal and diabetic beta-cells. *Nature*, 414(6865), 807-812.

[15] Giacco, F., & Brownlee, M. (2010). Oxidative stress and diabetic complications. *Circ Res*, 107(9), 1058-1070.

[16] Boveris, A. (1984). Determination of the production of superoxide radicals and hydrogen peroxide in mitochondria. *Methods Enzymol*, 105, 429-435.

[17] Korshunov, S. S., Skulachev, V. P., & Starkov, A. A. (1997). High protonic potential actuates a mechanism of production of reactive oxygen species in mitochondria. *FEBS Lett*, 416(1), 15-18.

[18] Dikalov, S. Cross talk between mitochondria and NADPH oxidases. *Free Radical Biology and Medicine*, 51(7), 1289.

[19] Schmidt, H. H., Lohmann, S. M., & Walter, U. (1993). The nitric oxide and cGMP signal transduction system: regulation and mechanism of action. *Biochim Biophys Acta*, 1178(2), 153-175.

[20] Hobbs, A. J., & Ignarro, L. J. (1996). Nitric oxide-cyclic GMP signal transduction system. *Methods Enzymol*, 269, 134-148.

[21] Murad, F. (1999). Cellular signaling with nitric oxide and cyclic GMP. *Braz J Med Biol Res*, 32(11), 1317-1327.

[22] Mc Donald, L. J., & Murad, F. (1995). Nitric oxide and cGMP signaling. *Adv Pharmacol*, 34, 263-275.

[23] Evans, J. L., Goldfine, I. D., Maddux, B. A., & Grodsky, G. M. (2002). Oxidative stress and stress-activated signaling pathways: a unifying hypothesis of type 2 diabetes. *Endocr Rev*, 23(5), 599-622.

[24] Giacco, F., & Brownlee, M. Oxidative stress and diabetic complications. *Circ Res*, 107(9), 1058-1070.

[25] Mazzola, J. L., & Sirover, M. A. (2003). Subcellular localization of human glyceraldehyde-3-phosphate dehydrogenase is independent of its glycolytic function. *Biochim Biophys Acta*, 1622(1), 50-56.

[26] Tristan, C., Shahani, N., Sedlak, T. W., & Sawa, A. The diverse functions of GAPDH: views from different subcellular compartments. *Cell Signal*, 23(2), 317-323.

[27] Brownlee, M. (2005). The pathobiology of diabetic complications: a unifying mechanism. *Diabetes*, 54(6), 1615-1625.

[28] Buse, M. G. (2006). Hexosamines, insulin resistance, and the complications of diabetes: current status. *Am J Physiol Endocrinol Metab*, 290(1), 1-8.

[29] Geraldes, P., & King, G. L. Activation of protein kinase C isoforms and its impact on diabetic complications. *Circ Res*, 106(8), 1319-1331.

[30] Scivittaro, V., Ganz, M. B., & Weiss, M. F. (2000). AGEs induce oxidative stress and activate protein kinase C-beta(II) in neonatal mesangial cells. *Am J Physiol Renal Physiol*, 278(4), 676-683.

[31] Camacho-Ruiz, A., & Esteban-Méndex, M. Diabetes y radicales libres. *In: Morales-González JA, Madrigal-Santillán EO, Nava-Chapa G, Durante-Montiel I, Jongitud-Falcón A, Esquivel-Soto J. 2010Editors. Diabetes. 2nd ed. Pachuca, Hidalgo, México.: Universidad Autónoma del Estado de Hidalgo.*

[32] Degenhardt, T. P., Thorpe, S. R., & Baynes, J. W. (1998). Chemical modification of proteins by methylglyoxal. *Cell Mol Biol (Noisy-le-grand)*, 44(7), 1139-1145.

[33] Gomperts, B. D., Kramer, I. M., & Tatham, P. E. R. (2003). Signal Transduction. London, UK, Elsevier Academic Press.

[34] Permutt, M. A., Wasson, J., & Cox, N. (2005). Genetic epidemiology of diabetes. J Clin Invest , 115(6), 1431-9.

[35] Zick, Y. (2004). Uncoupling insulin signalling by serine/threonine phosphorylation: a molecular basis for insulin resistance. *Biochem Soc Trans*, 32(5), 812-816.

[36] Weir, G. C., Laybutt, D. R., Kaneto, H., Bonner-Weir, S., & Sharma, A. (2001). Beta-cell adaptation and decompensation during the progression of diabetes. *Diabetes*, 50(1), 154-159.

[37] Leahy, J. L., Hirsch, I. B., Peterson, K. A., & Schneider, D. (2010). Targeting beta-cell function early in the course of therapy for type 2 diabetes mellitus. *J Clin Endocrinol Metab*, 95(9), 4206-4216.

[38] Lazo-de-la-Vega-Monroy, M. L., & Fernandez-Mejia, C. Bases moleculares de la diabetes tipo 2. *In: Morales-González JA, Madrigal-Santillán EO, Nava-Chapa G, Durante-Montiel I, Jongitud-Falcón A, Esquivel-Soto J. 2010Editors. Diabetes. 2nd ed. Pachuca, Hidalgo, México:Universidad Autónoma del Estado de Hidalgo.*

[39] Bjornholm, M., & Zierath, J. R. (2005). Insulin signal transduction in human skeletal muscle: identifying the defects in Type II diabetes. *Biochem Soc Trans*, 33(2), 354-357.

[40] Withers, D. J., Gutierrez, J. S., Towery, H., Burks, D. J., Ren, J. M., Previs, S., et al. (1998). Disruption of IRS-2 causes type 2 diabetes in mice. *Nature*, 391(6670), 900-904.

[41] Matschinsky, F. M. (1996). Banting Lecture 1995. A lesson in metabolic regulation inspired by the glucokinase glucose sensor paradigm. *Diabetes*, 45(2), 223-241.

[42] Aguilar-Bryan, L., Clement, J. P., Gonzalez, G., Kunjilwar, K., Babenko, A., & Bryan, J. (1998). Toward understanding the assembly and structure of KATP channels. *Physiol Rev*, 78(1), 227-245.

[43] Hiriart, M., & Aguilar-Bryan, L. (2008). Channel regulation of glucose sensing in the pancreatic beta-cell. *Am J Physiol Endocrinol Metab*, 295(6), 1298-1306.

[44] Rorsman, P., & Renstrom, E. (2003). Insulin granule dynamics in pancreatic beta cells. *Diabetologia*, 46(8), 1029-1045.

[45] Straub, S. G., & Sharp, G. W. (2002). Glucose-stimulated signaling pathways in biphasic insulin secretion. *Diabetes Metab Res Rev*, 18(6), 451-463.

[46] Jitrapakdee, S., Wutthisathapornchai, A., Wallace, J. C., & Mac Donald., M.J. (2010). Regulation of insulin secretion: role of mitochondrial signalling. *Diabetologia*, 53(6), 1019-1032.

[47] Krueger, K. A., Bhatt, H., Landt, M., & Easom, R. A. (1997). Calcium-stimulated phosphorylation of MAP-2 in pancreatic betaTC3-cells is mediated by Ca2+/calmodulin-dependent kinase II. *J Biol Chem*, 272(43), 27464-27469.

[48] Nielander, H. B., Onofri, F., Valtorta, F., Schiavo, G., Montecucco, C., Greengard, P., et al. (1995). Phosphorylation of VAMP/synaptobrevin in synaptic vesicles by endogenous protein kinases. *J Neurochem*, 65(4), 1712-1720.

[49] Easom, R. A. (1999). CaM kinase II: a protein kinase with extraordinary talents germane to insulin exocytosis. *Diabetes*, 48(4), 675-684.

[50] Sharp, G. W. (1979). The adenylate cyclase-cyclic AMP system in islets of Langerhans and its role in the control of insulin release. *Diabetologia*, 16(5), 287-296.

[51] Jones, P. M., & Persaud, S. J. (1998). Protein kinases, protein phosphorylation, and the regulation of insulin secretion from pancreatic beta-cells. *Endocr Rev*, 19(4), 429-461.

[52] Doyle, M. E., & Egan, J. M. (2003). Pharmacological agents that directly modulate insulin secretion. *Pharmacol Rev*, 55(1), 105-131.

[53] Laychock, S.G., Modica, M.E., & Cavanaugh, C.T. (1991). L-arginine stimulates cyclic guanosine 3′,5′-monophosphate formation in rat islets of Langerhans and RINm5F insulinoma cells: evidence for L-arginine:nitric oxide synthase.(6), *Endocrinology*, 129(6), 3043-3052.

[54] Russell, M. A., & Morgan, N. (2010). Expression and functional roles of guanylate cyclase isoforms in BRIN-BD11 beta-cells. *Islets*, 2(6), 23-31.

[55] Guzik, T. J., Mussa, S., Gastaldi, D., Sadowski, J., Ratnatunga, C., Pillai, R., et al. (2002). Mechanisms of increased vascular superoxide production in human diabetes mellitus: role of NAD(P)H oxidase and endothelial nitric oxide synthase. *Circulation*, 105(14), 1656-1662.

[56] Ceriello, A., Mercuri, F., Quagliaro, L., Assaloni, R., Motz, E., Tonutti, L., et al. (2001). Detection of nitrotyrosine in the diabetic plasma: evidence of oxidative stress. *Diabetologia*, 44(7), 834-838.

[57] Mahadev, K., Motoshima, H., Wu, X., Ruddy, J. M., Arnold, R. S., Cheng, G., et al. (2004). The NAD(P)H oxidase homolog Nox4 modulates insulin-stimulated genera-

tion of H2O2 and plays an integral role in insulin signal transduction. *Mol Cell Biol*, 24(5), 1844-1854.

[58] Ogihara, T., Asano, T., Katagiri, H., Sakoda, H., Anai, M., Shojima, N., et al. (2004). Oxidative stress induces insulin resistance by activating the nuclear factor-kappa B pathway and disrupting normal subcellular distribution of phosphatidylinositol 3-kinase. *Diabetologia*, 47(5), 794-805.

[59] Eriksson, J. W. (2007). Metabolic stress in insulin's target cells leads to ROS accumulation- a hypothetical common pathway causing insulin resistance. *FEBS Lett*, 581(19), 3734-3742.

[60] Bloch-Damti, A., & Bashan, N. (2005). Proposed mechanisms for the induction of insulin resistance by oxidative stress. *Antioxid Redox Signal*, 7(11-12), 1553-1567.

[61] Tiedge, M., Lortz, S., Drinkgern, J., & Lenzen, S. (1997). Relation between antioxidant enzyme gene expression and antioxidative defense status of insulin-producing cells. *Diabetes*, 46(11), 1733-1742.

[62] Robertson, R. P., Harmon, J., Tran, P. O., Tanaka, Y., & Takahashi, H. (2003). Glucose toxicity in beta-cells: type 2 diabetes, good radicals gone bad, and the glutathione connection. *Diabetes*, 52(3), 581-587.

[63] Brownlee, M. (2003). A radical explanation for glucose-induced beta cell dysfunction. *J Clin Invest*, 112(12), 1788-90.

[64] Rhodes, C. J. (2005). Type 2 diabetes-a matter of beta-cell life and death? *Science*, 307(5708), 380-384.

[65] Kaneto, H., Matsuoka, T. A., Nakatani, Y., Kawamori, D., Matsuhisa, M., & Yamasaki, Y. (2005). Oxidative stress and the JNK pathway in diabetes. *Curr Diabetes Rev*, 1(1), 65-72.

[66] Kaneto, H., Xu, G., Song, K. H., Suzuma, K., Bonner-Weir, S., Sharma, A., et al. (2001). Activation of the hexosamine pathway leads to deterioration of pancreatic beta-cell function through the induction of oxidative stress. *J Biol Chem*, 276(33), 31099-31104.

[67] Rizzo, M. A., & Piston, D. W. (2003). Regulation of beta cell glucokinase by S-nitrosylation and association with nitric oxide synthase. *J Cell Biol*, 161(2), 243-248.

[68] Tejedo, J., Bernabe, J. C., Ramirez, R., Sobrino, F., & Bedoya, F. J. (1999). NO induces a cGMP-independent release of cytochrome c from mitochondria which precedes caspase 3 activation in insulin producing RINm5F cells. *FEBS Lett*, 459(2), 238-243.

[69] Drews, G., Krippeit-Drews, P., & Dufer, M. (2010). Oxidative stress and beta-cell dysfunction. *Pflugers Arch*, 460(4), 703-718.

[70] Fowler, M. J. (2008). Microvascular and Macrovascular Complications of Diabetes. *Clinical Diabetes*, 26(2), 77-782.

[71] Thomas, M. C. (2011). Advanced glycation end products. *Contrib Nephrol*, 170, 66-74.

[72] ADA. (2012). Standards of Medical Care in Diabetes. *Diabetes Care*, 170(1), 11-63.

[73] Vincent, A. M., Callaghan, B. C., Smith, A. L., & Feldman, E.L. Diabetic neuropathy: cellular mechanisms as therapeutic targets. *Nat Rev Neurol*, 7(10), 573.

[74] Feener, E. P., Xia, P., Inoguchi, T., Shiba, T., Kunisaki, M., & King, G. L. (1996). Role of protein kinase C in glucose- and angiotensin II-induced plasminogen activator inhibitor expression. *Contrib Nephrol*, 118, 180-187.

[75] Ganz, M. B., & Seftel, A. (2000). Glucose-induced changes in protein kinase C and nitric oxide are prevented by vitamin E. *Am J Physiol Endocrinol Metab*, 278(1), 146-152.

[76] Akimoto, Y., Kreppel, L. K., Hirano, H., & Hart, G. W. (2001). Hyperglycemia and the O-GlcNAc transferase in rat aortic smooth muscle cells: elevated expression and altered patterns of O-GlcNAcylation. *Arch Biochem Biophys*, 389(2), 166-175.

[77] Du, X. L., Edelstein, D., Rossetti, L., Fantus, I. G., Goldberg, H., Ziyadeh, F., et al. (2000). Hyperglycemia-induced mitochondrial superoxide overproduction activates the hexosamine pathway and induces plasminogen activator inhibitor-1 expression by increasing Sp1 glycosylation. *Proc Natl Acad Sci USA*, 97(22), 12222-12226.

[78] Clark, R. J., Mc Donough, P. M., Swanson, E., Trost, S. U., Suzuki, M., Fukuda, M., et al. (2003). Diabetes and the accompanying hyperglycemia impairs cardiomyocyte calcium cycling through increased nuclear O-GlcNAcylation. *J Biol Chem*, 278(45), 44230-44237.

[79] Pang, Y., Bounelis, P., Chatham, J. C., & Marchase, R. B. (2004). Hexosamine pathway is responsible for inhibition by diabetes of phenylephrine-induced inotropy. *Diabetes*, 53(4), 1074-1081.

[80] Liu, J., Pang, Y., Chang, T., Bounelis, P., Chatham, J. C., & Marchase, R. B. (2006). Increased hexosamine biosynthesis and protein O-GlcNAc levels associated with myocardial protection against calcium paradox and ischemia. *J Mol Cell Cardiol*, 40(2), 303-312.

[81] Garcia-Bailo, B., El -Sohemy, A., Haddad, P. S., Arora, P., Benzaied, F., Karmali, M., et al. (2011). Vitamins D, C, and E in the prevention of type 2 diabetes mellitus: modulation of inflammation and oxidative stress. *Biologics*, 5, 7-19.

[82] Mc Nulty, H., Jacob, R. F., & Mason, R. P. (2008). Biologic activity of carotenoids related to distinct membrane physicochemical interactions. *Am J Cardiol*, 101(10A), 20-29.

[83] Salonen, J. T., Nyyssonen, K., Tuomainen, T. P., Maenpaa, P. H., Korpela, H., Kaplan, G. A., et al. (1995). Increased risk of non-insulin dependent diabetes mellitus at low plasma vitamin E concentrations: a four year follow up study in men. *BMJ*, 311(7013), 1124-1127.

[84] Mayer-Davis, E. J., Costacou, T., King, I., Zaccaro, D. J., & Bell, R. A. (2002). Plasma and dietary vitamin E in relation to incidence of type 2 diabetes: The Insulin Resistance and Atherosclerosis Study (IRAS). *Diabetes Care*, 25(12), 2172-2177.

[85] Reunanen, A., Knekt, P., Aaran, R. K., & Aromaa, A. (1998). Serum antioxidants and risk of non-insulin dependent diabetes mellitus. *Eur J Clin Nutr*, 52(2), 89-93.

[86] Sinclair, A. J., Taylor, P. B., Lunec, J., Girling, A. J., & Barnett, A. H. (1994). Low plasma ascorbate levels in patients with type 2 diabetes mellitus consuming adequate dietary vitamin C. *Diabet Med*, 11(9), 893-898.

[87] Ristow, M., Zarse, K., Oberbach, A., Kloting, N., Birringer, M., Kiehntopf, M., et al. (2009). Antioxidants prevent health-promoting effects of physical exercise in humans. *Proc Natl Acad Sci USA*, 106(21), 8665-8670.

[88] Chertow, B. (2004). Advances in diabetes for the millennium: vitamins and oxidant stress in diabetes and its complications. *MedGenMed*, 6(3), 4.

Chapter 10

Oxidative Stress in Periodontal Disease and Oral Cancer

Mario Nava-Villalba, German González-Pérez,
Maribel Liñan-Fernández and
Torres-Carmona Marco

Additional information is available at the end of the chapter

1. Introduction

The oral cavity is a region interconnected with other systems of the body; it should not be viewed as an isolated area. Diseases that it lays down can have systemic scope and significantly affect the quality of life of individuals who suffer them. Periodontal disease is one of the oral health problems that most often affect the global population, lack of treatment leads to loss of tooth organs and consequently alters the digestion and nutrition, without considering other relevant aspects as phonation, aesthetics and social or emotional impact. The importance of periodontal disease has raised possible bidirectional relationships with systemic diseases such as diabetes, metabolic syndrome and cardiovascular disease. We address herein the role of oxidative stress in the etiopathogeny of periodontal disease. In the same context, another disease that has become relevant in our days is the oral cancer. Epidemiological data show that the incidence of this neoplasm has been increasing in several countries. The impact of oral cancer on patients, who suffer it, is devastating. The role of oxidative stress in the development of this disease and some alternatives for its treatment, are topics addressed in this brief review. These two oral diseases are a sample of the plethora of effects that oxidative stress may have at local and systemic level.

2. Periodontal disease

Periodontitis is the second world health problem since it affects between 10 to 15% of the world population [1]. Although the various states in this disease depend on the degree of destruction and inflammation present, the American Dental Association classifies according to a system development based on the severity of the loss of periodontal insertion. The information obtained in clinical and radiographic examination classifies the patient in four typical cases that are:

• Type I: Gingivitis

• Type II: Mild Periodontitis

• Type III: Moderate Periodontitis

• Type IV: Advanced Periodontitis

There are other classifications of the inflammatory process [2]:

• Ulceronecrotic acute gingivitis

• Acute gingivitis

• Chronic gingivitis

• Marginal periodontitis

• Superficial marginal periodontitis

• Deep marginal periodontitis

Periodontal disease is an inflammatory process involving a set of changes that directly affect tissues that hold the teeth. The etiology plays a role which is essential within the bacterial infection. In fact, within the 300 to 400 species of bacteria located in the oral cavity consider that some of them are exclusive to the periodontal tissues. However in recent years it has been determined that the evolution and spread of the disease will play a decisive role in the host response to bacterial attack. This is reflected in the model of the critical path in the pathogenesis of this disease. Through this one can understand that there are diseases and systemic conditions that have risk factors for periodontal disease, because they are going to modify the host response and favor the development of damage [3].

When it is lost in the inclusion of periodontal fibers, usually after puberty, the cases that are reported before this stage are only 5%. Previously it has reported that there was a ratio of two to one in the frequency of periodontal disease, women being the most affected in this order. Currently known, the presence by gender of this involvement is very similar.

In adults with more than 1 mm of affected dental faces periodontal insertion loss increases with age. An epidemiological report in United States mentions that approximately 80-92% of the population between the ages of 35 and 64 years performed, lost more than 1 mm insertion in 20 to 47% of teeth. From 18 to 22% of the population of 35 to 64 years were more 2 mm deep in the probing of the periodontal bags in 11 to 13% of tooth surfaces. Periodontitis occurs when tissue destruction due to the direct effect of bacterial toxins and removal products, in addition, the effects caused indirectly by the harmful organic defense mechanisms. Microorganisms as *p. gingivalis*, *a. actinomycetemcomitans* and *Capnocytophaga*sp. produce collagenase (substances similar to trypsin) and phospholipase, among others. Extracellularly there are acid phosphatase and alkaline, lipopolysaccharides, aminopeptidase, epithelium toxin, inhibitor of fibroblasts and a toxin that induces a bone resorption.

Bacteria causes tissue destruction with its deletion, this is a feature of marginal periodontitis products. Destruction of tissues within a radius of 1.5 to 2.5 mm around the plaque has been observed (the so-called *influence radioplate*) in periodontitis. The hydrolysis of the connective tissue associated with the inflammation is due to the reactive oxygen species and the elastase/lysosomic-like enzymes. Collagenase and gelatinase are segregated to the microenvironment. Prostaglandin E, Interleukin 1-/ J and the lipopolysaccharide activates osteoclasts and induce a resorption of alveolar bone. Cellular and humoral components of the immune system, mainly involved in the periodontal immune response are leukocytes, immunoglobulins, complement system and lysozyme. If the immune defenses are working properly, the periodontium is protected from the harmful effect of pathogenic substances secreted by the microorganisms. The immunocompetent host is able to defend itself against microbial attacks that occur every day. Thus prevents *infections,* i.e. the multiplication of microorganisms within the periodontium. We can say that the periodontal inflammation is a local reaction to a tissue injury whose purpose is the destruction of the causal factor, dilution or its encapsulation.

The human immune system can be classified according to their function within the periodontium, follows:

• Secretory system

• Neutrophils, antibodies and complement system

• Leukocytes and macrophages

• Immune regulation system.

The system formed by neutrophils, antibodies and complement is crucial to the immune defense against periodontal infections. When functional defects of neutrophils occur, it increases the frequency of serious marginal periodontitis [4].

3. Oxidative stress

A phenomenon that occurs within the periodontal disease is called oxidative stress. In order to understand the phenomenon of oxidative stress it is important to know what the free radicals (FR) are, where they come from and how to act. A FR is considered that molecule presented an electron unpaired or odd in the orbital external, in its atomic structure giving it a spatial configuration that generates a high instability. In the molecule of oxygen (O_2) know the following FR or also called oxygen reactive species: anion superoxide ($O_2^{\cdot-}$), hydrogen peroxide (H_2O_2), hydroxyl radical ($OH^{\cdot}$) and singlet oxygen ($^1O_2$). The H_2O_2 is not strictly a FR but by its ability to generate the $OH^{\cdot}$ in the presence of metals such as iron, it incorporates it as such. A fundamental characteristic of the reactions of free radicals is that act of chain reactions, where a radical reaction generates another consecutively.

Oxygen (O_2) that this in the air is fundamental to life, many reactions in which participates the O_2 generates reactive oxygen species (ROS), of which some have the chemical character of being free radical (FR), whose biochemical entities in its atomic structure presented an odd or unpaired electron in the outer orbital, giving it a spatial configuration that generates a high instability with an enormous capacity for combined with the diversity of molecules members of the cell structure: carbohydrates, lipids, proteins, nucleic acids, and derivatives of each of them, causing important functional alterations. In this sense, the body has an anti-oxidant system to counteract the generation of ROS, which maintains a homeostatic balance. However there are pro-oxidant factors that favor the generation of FR, causing an imbalance in favor of the latter, generating so-called oxidative stress (OS) [5].

The tetravalent reduction of oxygen to produce water through the electron transport chain in mitochondria is relatively safe. However, the univalent reduction of oxygen generates ROS. The human organism also has antioxidant system to counteract the generation of ROS, which maintains a homeostatic balance. However, there are pro-oxidant factors that favor the generation of FR, causing an imbalance in favor of the latter, generating OS. The antioxi-dant enzyme superoxide dismutase (SOD), Glutathione peroxidase (GP), glutathione reduc-tase (GR) and catalase (CAT), as well as proteins carriers of metals (ceruplasmina, transferrin, lactoferrin, etc.), and another micronutrients as vitamins A, C and E, bilirubin, uric acid and selenium, constitute the most important elements of the antioxidant system. Also, between the most important pro-oxidant factors we can highlight the process of aging, ionizing radiation, ultraviolet rays, environmental pollution, cigarette smoke, excess of exer-cise, intake of alcoholic beverages and inadequate diet [6].

The role of Coenzyme Q_{10} is the mitochondrial energy coupling. It is an essential part of the cel-lular machinery used to produce ATP that provides the energy for muscle contraction and oth-er vital cellular functions. Most of the ATP production occurs in the inner membrane of the mitochondria, where the Coenzyme Q_{10} is located. The most important function is serving as a suppressor of primary free radicals, located in the membranes in the vicinity of unsaturated lipid chains. There are less established functions that include the oxidation/reduction of the control of the origin and transmission of signals in cells that induce the expression of gender, the control of membrane channels, the structure and solubility in lipids [7].

Free radicals cause damage to periodontal tissues by a variety of different mechanisms in-cluding:

• DNA damage

• Lipid peroxidation

• Protein damage

• The oxidation of important enzymes (anti proteases)

• Stimulation and release of pro-inflammatory cytokines

ROS covers other reactive species that are not true radicals, but are however capable of react in intra and extracellular environment: peroxide of hydrogen, hypochlorous acid, oxygen, ozone. The living organism has adapted to an existence under a continuous output of radical free flow. Between the different antioxidant defense mechanism adaptation mechanism is of great importance. Antioxidants are "those substances that when they are present in lower concentrations compared to the substrate of an oxidizable, significantly delay or inhibit the oxidation of the substrate". The various possible mechanisms that antioxidants can offer protection against damage from free radicals are:

- The prevention of the formation of radical free.

- Interception of the radical free to eliminate reactive metabolites and their conversion to less reactive molecules.

- Facilitate the repair of the damage caused by free radicals.

- Create a favourable environment for the effective functioning of other antioxidants.

Antioxidant defense system is very dynamic and responsive to any disturbance that occurs in the body redox balance. Antioxidants can be regulated and neutralize the formation of radical free that can occur due to oxidative stress, such as the factor transcription factors Activator protein 1 and nuclear-kb are redox sensitive. Redox potential is a measure of the affinity of a substance for electrons [8].

The presence of inflammatory infiltrate is a constant feature in periodontal disease. It is known that these cells release lots of free radicals; it is suspected that these metabolites are involved in the pathogenesis of the disease. The presence of a dense inflammatory infiltrate in periodontal disease leads to the suspicion that the relationship of periodontal leukocyte-tissue has a double aspect. The role of these cells in the containment of the gingival bacteria and their products must be analyzed according to a balance with the destruction of tissue due to the release of the products of its action (FR and proteases). In this way, a defensive mechanism, under the interaction of various factors, can be harmful to periodontal tissues, and they are therefore involved in the pathogenesis of inflammatory periodontal disease.

There is numerous evidence pointing to the involvement of FR in periodontal disease. It has been reported in patients with rapidly progressive periodontitis, that the polymorphonuclear neutrophils (PMN) are functionally activated, produce high levels of O_2 and have a high response the luminol-dependent (QL) chemiluminescent. There is an increase of the PMN oxidative response peripherals in patients with localized and generalized juvenile periodontitis, as well as in adult patients with periodontitis (AP). This increase is related to clinical periodontal status and is reversed by therapy.

It has also compared the generation O_2 by the activated PMN in the gingival crevicular fluid (GCF) of patients with AP. The PMN activation with phorbolmyristateacetate causes a marked increase in the release of O_2 in patients with AP, while the antioxidant activity of the gum is similar to the controls. The effect of the PMN in crevicular fluid of patients is dependent on variations in the rate of formation of O_2 relative to the intrinsic antioxidant capacity of the gingival tissue.

In gingival epithelial cells in culture studies have shown the PMN may cause lysis of these through the action of the free myeloperoxidase(MPO), a leukocyte enzyme generating radi-

cals. Its activity has been increased in the crevicular fluid of sites with gingivitis and periodontitis with respect to healthy sites.

There is a close relationship between free radical production by leukocytes and activation of
proteases. Altogether these actions could have profound effects on the function and integrity of the gingival epithelium.

The above evidence leads to consider that in the inflammatory periodontal disease, the general etiological factors causing the breakup of physiological systems of inhibition of lipid
peroxidation, creates a low level of antioxidant protection of periodontal tissues. In these circumstances, the local factors lead to the migration of neutrophils to the gingiva and gingival
fluid. The activation of these leukocytes in phagocytosis, causes the release of ROS, which
leads to the outbreak of the lipid peroxidation of the soft tissues of the periodontium and
activation of protease. This lipid peroxidation is the mechanism that triggers the development of morphofunctionalchangesin periodontium and their vessels, which results in destruction of collagen and bone resorption.

Due to numberless evidences that suggest a participation of the ROS in the pathogenesis of
the periodontal disease, it has been raised that the factors that promote a rupture of the antioxidant physiological system, contribute to the development of oxidative mechanisms that
initiate the periodontitis. The main cause of lipid peroxidation in the periodontal disease
seems to lie in the liberation of ROS by leukocytes in phagocytosis. These concepts emphasize the utility of antioxidants in the prophylaxis and treatment of periodontal disease and
therefore justify the search of new antioxidant preparations for this purpose. For example
the *p. gingivalis* is a major cause of periodontitis, and their presence is a risk factor for systemic inflammatory syndromes, such as atherosclerosis and cardiac dysfunction. The capacity of the virulence factors such as proteases and LPS to induce inflammation has been
studied intensely. In some cases, however, the inflammation occurs regardless of these factors, suggesting the existence of other stimulating immune. It was found that the cell death
induced by *p. gingivalis*in the tissues is through the production of ROS [9].

4. Oral Cancer

The oral cancer occupies 2-5 % of all whole body cancers. This percentage places this neoplasia within the ten most common cancers [10]. Although its magnitude is relatively low, its
impact on affected patients and their costs in health systems is high.There is a considerable
variation in the incidence and mortality rates around the world. The incidence is greater in
south of India, Australia, North of America, many European countries, Brazil, certain countries of Africa and of central Asia [11]. 90% of oral cancer is of epithelial origin and the
rest 10% are distributing in adenocarcinomas, sarcomas, lymphoproliferative disorders,
metastasis, melanomas and malignant odontogenic tumors. The intraoral main site of oral
squamous cell cancer (OSCC) is the posterior lateral border of tongue (Figure 1) and floor of
mouth (Figure 2). If the lips are considered within the oral territory, then this site has the

highest frequency (Figure 3).Since oral squamous cell carcinoma (OSCC) is the main malignant neoplasia we focus in it.

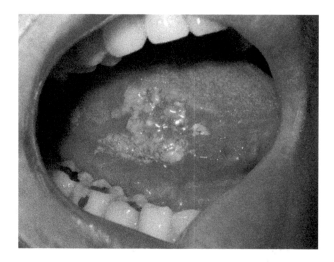

Figure 1. Squamous cell cancer of the posterior lateral border of the tongue in a 28-year-old woman.She smoked a cigarette per day for 15 years.

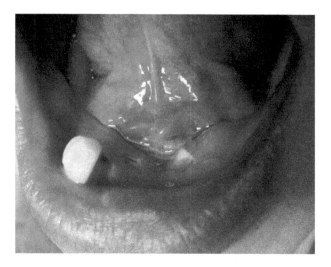

Figure 2. Squamous cell cancer of floor of mouth in a 58-year-old woman.She had a history of poorly controlled diabetes type 2 from 42 years. She also has used ill-fitting dentures since age 50. Note the linear lesion with presence of necrosis in the centre of the fissure.

There are premalignant lesions recognized like: leukoplakia, erythroplakia, oralsubmucosal fibrosis, palatal lesions of reverse cigar smoking, oral liken planus, discoid lupus erythematosus, and hereditary disorders likecongenital dyskeratosis and epidermolysisbullosa, but beyond of a clinical standpoint, diverse carcinogenic molecular mechanisms have been postulated.The main target is the DNA, since mutations that occur in it generates a wide range of deleterious effects in the cell. In a very general overview, the balance between tumor suppressor genes and those genes that induce cell cycle is altered.Allowing cells to escape cell cycle control and developing an unpredictable biological behavior. Subsequently, the cells express molecules that allow them to acquire an invasive phenotype, a phenomenon known as epithelial-mesenchymal transition. Why malignant cells colonize distant sites? Is not yet fully understood, but it is the feature that makes it lethal.

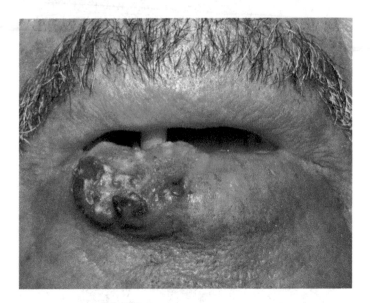

Figure 3. Squamous cell cancer of the lip in a 74-year-old man. He was a farmer and consumed alcohol chronically.

Free radicals are products of the oxidation-reduction systems of the cell and its participation in cellular metabolic functions is essential for cell survival. A classic example is the electron transport chain in mitochondria. However, in whatpathologicalconditions, free radicals can become deleterious? In fact, what are the results of its harmful effects? The involvement of free radicals in cancer development has been studied for 3 decades, and there is sufficient evidence that implicates theirs in the multistage theory of carcinogenesis. They are proposed to cause diverse DNA alterations like: punctual mutations, DNA base oxidations, strand breaks, mutation of tumor suppressor genes and can induce overexpression of proto-oncogenes [12].It should be added that oxidative protein damage participates in facilitating the development of cancer.

Several works explore the levels of oxidative stress in patients with oral cancer [13-15] most of them quantified the products of lipid-peroxidation(mainly malonilaldehyde) and contrast them with the activity of antioxidant enzymes or exogenous antioxidants levels in blood or even saliva. The results agree that there is an imbalance between the high amount of free radicals and insufficient antioxidant system activity.Added to this, some researchers have observed that high levels of lipid-peroxidation combined with low levels of thiols and anti-oxidant status, correlate with poor survival rate in patients with oral cancer [16].

The OSCC is a multifactorial disease, however, a factor strongly associated, is smoking. 90% of individuals with oral cancer are smokers. It is considered that the smoke from cigarettes have 4000 chemicals, 40 of which have carcinogenic potential. It has been shown that ciga-rette smoke contains pro-oxidants that are capable of initiating the process of lipid-peroxida-tion and deplete levels of antioxidants from the diet [17,18].

In contrast, there is epidemiological evidence that demonstrates the protective effect of diet on some populations [19-21].For example in Greece, which has the lowest rates of oral can-cer among European countries,its population is exposed to latent risk factors such as alcohol intake and smoking; micronutrients consume such as riboflavin, magnesium and iron corre-lated inversely with oral cancer [19].

Consequently, several authors have proposed the ingestion of diverse exogenous antioxi-dants; supporting in those epidemiological studies, where the diet offers protection for the development of cancer, and taking into account that the endogenous antioxidant systems have been overwhelmed by oxidative stress.

For example, vitamin C is one of the most extensively evaluated antioxidants in oral cancer alternative co-therapies. Low or even undetectable levels of vitamin C correlate with the presence of oral cancer [17, 22]; in contrast, is one of the micronutrients that have a consis-tent inverse correlation in different studies [23].Vitamin C acts as a scavenger of free radicals and impedes the detrimental chain reactions triggered by the free radicals.The l-glutamine is another antioxidant that has shown a beneficial modulating effect in patients with oral can-cer in stages III and IV. The l-glutamine is administered in the diet as a complementary ther-apy; the proposal is that restores glutathione cascade system [15].In addition, other antioxidants such as carotene, vitamin E, thiamine, vitamin B_6, folic acid, niacin and potassi-um have shown a convincing protective effect [24]. Even more,when them are administered together during the cycles of radiotherapy [25].

Author details

Mario Nava-Villalba, German González-Pérez[2], Maribel Liñan-Fernández[3] and Torres-Carmona Marco[4]

*Address all correspondence to: marionava23@gmail.com

1 Dentistry Department, School of Medicine.Autonomous University of Querétaro.andDentistry Department, Health Science Division,University of the Valley of México, Campus QuerétaroQuerétaro, México

2 Dentistry Department, School of Medicine.AutonomousUniversity of Querétaro, Querétaro, México

3 Dentistry Department, School of Medicine.AutonomousUniversity of Querétaro, Querétaro, México

4 Dentistry Department, School of Medicine.Autonomous University of Querétaro.andGenetics Department,Comprehensive Rehabilitation Center of Querétaro.Querétaro, México

References

[1] (2010). Mendes S.A, A.M Vargas Duarte Ferreira faith, Nogueira GH.Periodontitis in individuals with diabetes treated in the public health system of Belo Horizonte, Brazil., *Revista Brasileira de Epidemiología*, 13(1), 118-25.

[2] Hee-Kyung, L., Sang-Hee, C., Kyu, C. W., Anwar, T. M., Keun-Bae, S., & Seong-Hwa, J. (2009). The effect of intensive oral hygiene care on gingivitis and periodontal destruction in Type 2 diabetic patients., *Yonsei Medical Journal*, 50(4), 529-36.

[3] (2002). Garcia T.B, Saldaña B.A, Soto F.C.Oxidative stress in systemic effects of inflammatory periodontal disease., *Revista Cubana de Investigación Biomédica*, 21(3), 194-196.

[4] (2010). Newman T.C. Clinical Periodontology. McGraw-Hill, 10thEdition,.

[5] Valdez P.A, Mendoza N.V.Relationship of oxidative stress with periodontal disease in older adults with type 2 diabetes mellitus. Revista ADM (2006). , 63(5), 189-94.

[6] (2004). García M.A.Influence of oxidative stress in periodontal disease., *Revista Cubana de Ciencias Médicas*, http://www.cpicmha.sld.cu/hab/10 2_04/hab07204.htm.

[7] Shobha, P., Sunitha, J., & Mayank, H. (2010). Role of coenzyme Q_{10} as an antioxidant and bioenergizer in periodontal diseases., *Indian journal Pharmacology*, 42(6), 334-37.

[8] Gowri, P., Biju, T., & Suchetha, K. (2008). The challenge of antioxidants to free radicals in periodontitis., *Journan Indian Society Periodontology*, 12(3), 79-83.

[9] Kenichi, I., Hiroshi, H., Katsutoshi, I., Tatsuo, A., Mikio, S., Koji, N., et al. (2010). Porphyromonasgingivalis Peptidoglycans induce excessive activation of the innate immune system in silkworm slrvae., *Journal Biological Chemistry*, 285(43), 33338-33347.

[10] Stewart B.W, Kleihues P, editors. World Cancer Report. Lyon:, *WHO International Agency for Research on Cancer; 2003.*

[11] (2009). Petersen P.E. Oral cancer prevention and control- The approach of the World Health Organization., *Oral Oncology*.

[12] Halliwell, B. (2007). Oxidative stress and cancer: have we moved forward?, *Biochemical Journal*, 401(1), 1-11.

[13] Sultan-Beevi, S. S., Hassanal-Rasheed, A. M., & Geetha, A. (2004). Evaluation of oxidative stress and nitric oxide levels in patients with oral cavity cancer., *Japan Journal of Clinical Oncology*, 34(7), 379-385.

[14] Khanna, R., Thapa, P. B., Khanna, H. D., Khanna, S., Khanna, A. K., & Shukla, H. S. (2005). Lipid peroxidation and antioxidant enzyme status in oral carcinoma patients., *Kathmandu University Medical Journal*, 3(4), 334-339.

[15] Das, S., Mahapatra, S. K., Gautam, N., Das, A., & Roy, S. (2007). Oxidative stress in lymphocytes, neutrophils, and serum of oral cavity cancer patients: modulatory array of l-glutamine., *Supportive Care in Cancer*, 15(12), 1399-1405.

[16] (2007). Patel B.P, Rawal U.M, Dave T.K, Rawal R.M, Shukla S.N, Shah P.M, Patel P.S. Lipid peroxidation, total antioxidant status, and total thiol levels predict overall survival in patients with oral squamous cell carcinoma., *Integrative Cancer Therapies*, 6(4), 365-372.

[17] Hedge, N. D., Kumari, S., Hedge, M. N., Bekal, M., & Rajaram, P. (2012). Status of serum vitamin C level and peroxidation in smokers and non-smokers with oral cancer., *Research Journal of Pharmaceutical, Biological and Chemical Sciences*, 3(1), 170-175.

[18] Raghavendra, U. D., Souza, V. D., & Souza, B. (2012). Erythrocyte malonilaldheyde and antioxidant status in oral squamous cell carcinoma patients and tobacco chewers/smokers., *Biomedical Research*, 21(4), 441-444.

[19] Petridou, E., et al. (2002). The role of diet and specific micronutrients in the etiology of oral carcinoma., *Cancer*, 94(11), 2981-2988.

[20] Winn, D. M., Ziegler, R. G., Pickle, L. W., Gridley, G., Blot, W. J., & Hoover, R. N. (1984). Diet in the etiology of oral and pharyngeal cancer among women from the southern United States., *Cancer Research*, 44(3), 1216-1222.

[21] Franceschi S, Favero A, Conti E, Talamini R, Volpe R, Negri E, Barzan L, La Vecchia C. Food groups, oils and butter, and cancer of the oral cavity and pharynx. *British Journal of Cancer* 1999; 80 (3/4) 614-620.

[22] Aravindh, L., Jagathesh, P., Shanmugam, S., Sarkar, S., Kumar, P. M., & Ramasubramanian, S. (2012). Estimation of plasma antioxidants beta carotene, vitamin C and vitamin E levels in patients with OSMF and oral Cancer- Indian population., *International Journal of Biological and Medical Research*, 3(2), 1655-1657.

[23] World Cancer Research Fund & American Institute for CancerResearch. (1997). Food, nutrition and the prevention of cancer.Mouth and pharynx., *Cancers, nutrition, and food. Menasha,WI: World Cancer Research Fund & American Institutefor Cancer Research*.

[24] Negri, E., Franceschi, S., Bosetti, C., Levi, F., Conti, E., Parpinel, M., & La Vecchia, C. (2002). Selected micronutrients and oral and pharyngeal cancer., *International Journal of Cancer*, 86(1), 122-127.

[25] (2009). Shariff A.K, Patil S.R, Shukla P.S, Sontakke A.V, Hendre A.S, Gudur A.K. Effect of oral antioxidant supplementation on lipid peroxidation during radiotherapy in head and neck malignancies., *Indian Journal of Clinical Biochemistry*, 24(3), 307-311.

Chapter 11

Role of Oxidative Stress in Calcific Aortic Valve Disease: From Bench to Bedside - The Role of a Stem Cell Niche

Nalini Rajamannan

Additional information is available at the end of the chapter

1. Introduction

Calcific Aortic Stenosis is the most common cause of aortic valve disease in developed countries. This condition increases in prevalence with the advancing age of the U.S. population, afflicting 2-3 % by age 65 [1]. Aortic valve replacement is the number one indication for surgical valve replacement in the United States and in Europe. The natural history of severe symptomatic aortic stenosis is associated with 50% mortality within 5 years [2]. Bicuspid aortic valve disease is the most common congenital heart abnormality and it is the most common phenotype of calcific aortic stenosis. The bicuspid aortic valve (BAV) is the most common congenital cardiac anomaly, having a prevalence of 0.9 to 1.37% in the general population [3]. Understanding the cellular mechanisms of tricuspid versus bicuspid aortic valve lesions will provide further understanding the mechanisms of this disease. Currently, there are three fundamental mechanisms defined in the development of aortic valve disease: 1) oxidative stress via traditional cardiovascular risk factors [4-8,6, 7, 9-12], 2) cellular proliferation [13] and 3) osteoblastogenesis in the end stage disease process [14, 15]. Previously, the Wnt/Lrp5 signaling pathway has been identified as a signaling mechanism for cardiovascular calcification [5, 16, 17]. The corollaries necessary to define a tissue stem cell niche: 1) physical architecture of the endothelial cells signaling to the adjacent subendothelial cells: the valve interstitial cell along the valve fibrosa. 2) defining the oxidative-mechanical stress gradient necessary to activate Wnt3a/Lrp5 in this tissue stem niche to induce disease. Recently, the mechanisms of oxidative stress have been identified in the development of calcific aortic valve disease. This chapter will outline the factors important in the role of calcific aortic valve disease.

2. The role of lipids in vascular and valvular disease

The role of lipids in vascular atherosclerosis has been defined in the literature for years. Atherosclerosis is a complex multifactorial process which produces a lesion composed of lipids [18, 19], macrophages [20], and proliferating smooth muscle cells [21] apoptosis [22] and extracellular bone matrix production [23] in the vascular wall [24, 25]. The activation of these cellular processes is regulated by a number of pathways. Integrins provide an important role in the regulation of cellular adhesion in atherosclerosis [26]. Another critical regulator of vascular endothelial biology is nitric oxide (NO) [27, 28]. Cholesterol-rich LDL also has a critical role in the onset and further progression of the atherosclerotic lesion via an inactivation of endothelial nitric oxide synthase (eNOS) [22, 29-31] contributing to an abnormal oxidation state within the vessel. In this inflammatory environment, growth factors and cytokines are secreted to induce vascular smooth cell proliferation and recruitment of macrophage cells [32-37] which are important in the development of the atherosclerotic plaque lesion.

Recently, similar risk factors for calcific aortic valve disease have recently been described including male gender, hypertension, elevated levels of LDL, and smoking [38, 39] which mimic those that promote the development of vascular atherosclerosis. Surgical pathological studies have demonstrated the presence of LDL and atherosclerosis in calcified valves, demonstrating similarities between the genesis of valvular and vascular disease and suggesting a common cellular mechanism [40, 41]. Patients who have the diagnosis of familial hypercholesterolemia develop aggressive peripheral vascular disease, coronary artery disease, as well as aortic valve lesions which calcify with age [10, 42]. Rajamannan et al, have shown that the development of atherosclerosis occurs in the aortic valve in a patient with Familial Hypercholesterolemia with the Low density lipoprotein receptor mutation [10]. The atherosclerosis develops along the aortic surface of the aortic valve and in the lumen of the left circumflex artery [10]. This provides the first index case of atherosclerotic aortic valve disease in this patient population. Studies have confirmed in experimental hypercholesterolemia that both atherosclerosis and osteoblast markers are present in the aortic valves [4, 6, 13]. This background provides the foundation for studying valve calcification in an experimental atherosclerotic *in vivo* model.

3. Aortic valve calcification

The presence of calcification in the aortic valve is responsible for valve stenosis. Severe aortic stenosis can result in symptomatic chest pain, as well as syncope and congestive heart failure in patients with severe aortic valve stenosis. For years, aortic valve stenosis was thought to be a degenerative process. However, the pathologic lesion of calcified aortic valves demonstrate indicate the presence of complex calcification in these tissues. Furthermore, there are a growing number of descriptive studies delineating the presence of bone formation in the aortic valve [15, 43, 44].

Until recently the etiology of valvular heart disease has been thought to be a degenerative process related to the passive accumulation of calcium binding to the surface of the valve leaflet. Recent descriptive studies have demonstrated the critical features of aortic valve calcification, including osteoblast expression, cell proliferation and atherosclerosis [6, 14, 15, 45] and mitral valve degeneration, glycosaminglycan accumulation, proteoglycan expression, and abnormal collagen expression [46-49]. These studies define the biochemical and histological characterization of these valve lesions. We and others, have also shown that specific bone cell phenotypes are present in calcifying valve specimens in human specimens [16, 50]. These data provide the evidence that the aortic valve calcification follows the spectrum of bone formation in calcifying tissues.

4. The role of Lrp5/beta-catenin activation in cardiovascular calcification and osteoblast bone formation: Connection with the bone axis

Bone and cartilage are major tissues in the vertebrate skeletal system, which is primarily composed of three cell types: osteoblasts, chrondrocytes, and osteoclasts. In the developing embryo, osteoblast and chrondrocytes, both differentiate from common mesenchymal progenitors in situ, where as osteoclasts are of hematopoietic origin and brought in later by invading blood vessels. Osteoblast differentiation and maturation lead to bone formation controlled by two distinct mechanisms: intramembranous and endochondral ossification, both starting from mesenchymal condensations.

To date only two osteoblast-specific transcripts have been identified: 1) Cbfa1 and 2) osteocalcin (OC). The transcription factor Cbfa1 [51] has all the attributes of a 'master gene' differentiation factor for the osteoblast lineage and bone matrix gene expression. During embryonic development, Cbfa1 expression precedes osteoblast differentiation and is restricted to mesenchymal cells destined to become osteoblast. In addition to its critical role in osteoblast commitment and differentiation, Cbfa1 appears to control osteoblast activity, i.e., the rate of bone formation by differentiated osteoblasts [51]. We have shown previously that cholesterol upregulates Cbfa1 gene expression in the aortic valve and atorvastatin decreases the gene expression [6] in an animal model. We have also demonstrated that Sox9 and Cbfa1 are expressed in human degenerative valves removed at the time of surgical valve replacement [16]. The regulatory mechanism of osteoblast differentiation from osteoblast progenitor cells into terminally differentiated cells is via a well orchestrated and well studied pathway which involves initial cellular proliferation events and then synthesis of bone matrix proteins, which requires the actions of specific paracrine/hormonal factors and the activation of the canonical Wntpathway [52].

Genes which code for the bone extracellular matrix proteins in osteoblast cells include alkaline phosphatase (AP), osteopontin (OP), osteocalcin (OC), and bone sialoprotein (BSP). This data supports a potential regulatory mechanism that these matrix proteins play a critical role in the development of biomineralization. To date, many of these markers have been shown to be critical in the extracellular mineralization and bone formation that develops in normal osteoblast differentiation (Fig.5). Dr. Spelsberg and Dr. Rajamannan have extensive experience in osteoblast cell biology and will contribute to the translational studies in the aortic valve involving the differentiation and mineralization [53, 54].

A link between lipids and osteoporosis have been studied extensively [55-60]. These groups have shown in *in vitro* and *in vivo* studies that lipids decrease bone formation and increase vascular calcification. Hurska's group from the University of Washington have studied this important hypothesis in the LDLR[-/-] mice with renal disease [55]. This studied correlated the important understanding of chronic kidney disease with decreased bone formation rates and increase in vascular calcification. This study demonstrates that accelerated vascular calcification found in patients with end stage renal disease may be related to multifactorial mechanisms including traditional atherosclerotic risk factors and elevated serum phosphate levels. Giachelli has also studied extensively the hypothesis of a sodium phosphate abnormality in the vascular smooth muscle cell [61]. Her group has also shown that osteopontin expression by vascular smooth muscle cells may have an inhibitory effect in the development of calcification [62] which further defines the complexity of the matrix synthesis phase of bone formation. Demer's laboratory has also studied extensively the correlation of lipids with vascular calcification and osteoporosis via inhibition of Cbfa1 in osteoblast cells [60, 63]. This paradoxical finding between the calcifying vascular aorta and osteoporosis is an important link in the hypercholesterolemia hypothesis. The development of cardiovascular calcification is a multifactorial process which includes a number of mechanisms. Studies in the different laboratories provide important evidence towards the development of therapies depending on the patient population i.e. end stage renal disease versus treatment of the traditional risk factors for vascular disease.

Our lab (43) and Towler's laboratory (44) have shown that the Lrp5/Wnt/beta-catenin pathway plays an important role in the development of vascular and valvular calcification. Studies have shown that different mutations in Lrp5, an LDL receptor related protein; develop a high bone mass phenotype and an osteoporotic phenotype (45, 46). In the presence of the palmitoylation of Wnt an active beta-catenin accumulates in the cytoplasm, presumably in a signaling capacity, and eventually translocates to the nucleus via binding to nucleoporins [64], where it can interacts with LEF-1/TCFs in an inactive transcription complex [65, 66], The Wnt/Lrp5/frizzled complex turns on downstream components such as Dishevelled (Dvl/Dsh) which leads to repression of the glycogen synthase kinase-3 (GSK3) [67]. Inhibition of GSK3 allows beta-catenin to accumulate in the nucleus, interacting with members of the LEF/TCF class of architectural HMG box of transcription factors including Cbfa1 involved in cell differentiation and osteoblast activation [68, 69, 70-72] and Sox 9, a HMG box transcription factor, is required for chondrocyte cell fate determination and marks early chondrocytic differentiation of mesenchymalprogenitors [73].

To determine a potential signaling pathway for the development of aortic valve disease there are numerous pathways which may be implicated in this disease process [50, 74, 75]. Recent evidence suggests that the Wnt pathway regulates the expression of bone mineral markers in cells responsive to the Wnt pathway. Furthermore the Wnt pathway has been shown to be activated by lipids. Therefore we chose to assess this pathway in our model of experimental hypercholesterolemia to determine how lipids may be regulating Lrp5 in the aortic valve. This background outlines the potential for lipids in the regulation of aortic valve mineralization via the canonical Wnt pathway.

5. Echocardiography and Computerized Tomography (CT) evaluation of the development of calcification and stenosis

Currently the non-invasive "gold standard" for the diagnosis of aortic valve stenosis is 2-Dimensional doppler echocardiography. It is the test of choice to quantify the severity of valve stenosis and pressure differential across the aortic valve. There are a increasing number of studies which have demonstrated the utility of calculating the volume of calcium and the rate of progression of the disease process in the aortic valve [76-80]. Confirmation of hemodynamic valve stenosis by echo will provide the degree of valve stenosis using ultrasound techniques[4]. MicroCT will assess the degree of calcification within the mineralizing tissues.

6. Development of future medical therapies for calcific aortic stenosis

The natural history studies of valvular aortic stenosis as defined by clinical and histopathologic parameters have provided landmark developments towards the understanding of this disease. HMG CoA reductase inhibitors may provide an innovative therapeutic approach by employing both lipid lowering and possibly non-lipid lowering effects to forestall critical stenosis in the aortic valve. Our laboratory has shown that atorvastatin has a number of effects in the aortic valve including: 1) inhibition of foam cell accumulation [6], 2) inhibition of Cbfa1 activation [6], 3) eNOS enzymatic activation [11] and 4) attenuation of Lrp5 receptor activation [81]. Statins have potent LDL lowering effects via inhibition of the rate-limiting step in cholesterol synthesis. There are a number of experimental models which demonstrate the potential for treating the vasculature with statins to inhibit matrix formation [24, 25], cellular proliferation [6] and vascular aneurysm formation [82]. Although valve replacement is the current treatment of choice for severe critical aortic stenosis, future insights into the mechanisms of calcification and its progression may indicate a role for lipid lowering therapy in modifying the rate of progression of stenosis.

There are a growing number of retrospective studies demonstrating that statins may have benefits in slowing the progression of aortic stenosis [83-85]. A recent clinical trial by Cowell et al, demonstrated that high dose atorvastatin did not slow the progression of aortic stenosis in patients [86]. However, the timing of the initiation of the statin therapy was at a later stage of aortic valve disease. A clinical trial in Portugal called RAAVE- Rosuvastatin Affecting Aortic Valve Endothelium demonstrated prospectively that statins slow progression in CAVD in an open label study. In the RAAVE study we found a change in aortic valve area (AVA) in the control group was -0.10 ± 0.09 cm^2 per year versus -0.05 ± 0.12 cm^2 per yearin the

Rosuvastatin group (p=0.041). In addition there was an increase in peak aortic valve velocity was +0.24±0.30 m/sec/yr in the control group as compared to the increase in +0.04±0.38 m/sec/yr in the Rosuvastatin group (p=0.007), indicating that in this prospective hypothesis driven study we found by echocardiography a slowing of progression in the aortic valve disease. SALTIRE initiated atorvastatin in patients who had more advanced aortic stenosis as defined by the mean aortic valve area 1.03 cm^2 as compared to the average aortic valve area in RAAVE of 1.23 cm^2 as the baseline aortic valve area prior to treatment with Rosuvastatin [87]. The investigators of RAAVE hypothesize that the beneficial effect of the statin was secondary to the early initiation of treatment. Furthermore, the SALTIRE investigators recently acknowledged the potential of medical therapies may be found if the treatment of this disease is initiated earlier in the disease process [86]. The studies planned in this application should lead to an important understanding of the molecular and cellular mechanisms of aortic valve disease. Furthermore, the experimental approach will also correlate the development of valve calcium by MicroCT and hemodynamic progression by echocardiography in this important disease process.

The bicuspid aortic valve (BAV) is the most common congenital cardiac anomaly, having a prevalence of 0.9 to 1.37 % in the general population [3]. The natural history of the BAV is progressive stenosis that typically occurs at a faster rate than tricuspid aortic valves requiring earlier surgical intervention in the BAV patients [2, 3]. With the decline of acute rheumatic fever, calcific aortic stenosis has become the most common indication for surgical valve replacement. Despite the high prevalence of aortic stenosis, few studies have investigated the mechanisms responsible for aortic valve disease. The cellular mechanism for the development of this disease is not well known. Previously, we and others have demonstrated that aortic valve calcification is associated with an osteoblast bone-like phenotype [14, 15]. This bone phenotype is regulated by the canonical Wnt pathway in experimental cardiovascular calcification [5, 17]. We have alsoshown that the canonical Wnt/Lrp5 pathway is upregulated in diseased human valves from patients with valvular heart disease [16]. These studies implicate that inhibition of the canonical Wnt pathway provides a therapeutic approach for the treatment of degenerative valvular heart diseases. A recent study [88], discovered that a loss of function mutation in Notch1 was associated with accelerated aortic valve calcification and a number of congenital heart abnormalities. Normal Notch1 receptor functions to inhibit osteoblastogenesis [89, 90]. Evaluation of Notch1 gene and protein expression in human bicuspid calcified valves compared to normal aortic valves removed at the time of surgical valve replacement is shown in Figure 1, Panel A. Notch1 protein expression was decreased in the BAV compared to controls by immunhistochemistry and Western Blot expression Figure 1, Panel B1 and B2, and C. RNA expression by RTPCR indicates a spliced Notch1 receptor in the diseased valves as compared to controls as shown in Figure 1, Panel D. This Notch1 splicing may be the regulatory switch important for the activation of the Wnt pathway and downstream calcification in these diseased valves [5, 17, 90].

Risk factors for the development of calcific aortic valve disease(CAVD) have been elucidated in a number of epidemiologic databases [38]. The risk factors for CAVD are similar to those of vascular atherosclerosis which include: elevated LDL, hypertension, male gender, smoking and increased body mass index [38]. The elucidation of these risk factors have provided the experimental basis for hypercholesterolemia as a method to induce aortic valve disease [4-8]. Furthermore, studies have shown that the eNOS[-/-] mouse is a novel mouse model which develops anatomic bicuspid aortic valves (BAV) [91]. To understand if eNOS[-/-] mice with the BAV phenotype, develops accelerated stenosis earlier than tricuspid aortic valves via the Lrp5 pathway activation, eNOS[-/-] mice were given a cholesterol diet versus cholesterol and atorvastatin. The Visual Sonics mouse echocardiography machine was used to screen for the BAV phenotype. Echocardiography hemodynamics was also performed to determine the timing of stenosis in bicuspid vs. tricuspid aortic valves eNOS-/- mice on different diets.

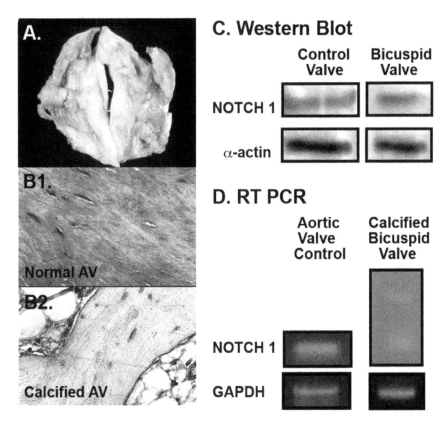

Figure 1. Histology of the aortic valves from human bicuspid calcified valves compared to normal aortic valves removed at the time of surgical valve replacement; **Panel A.** Bicuspid Aortic Valve Removed from patient at the time of surgical valve replacement. **Panel B1.**Notch1 Immunohistochemistry of a Normal Aortic Valve. **Panel B2.**Notch1 Immunohistochemistry of a Bicuspid Aortic Valve. **Panel C.** Notch1 protein expression was decreased in the BAV compared to controls by immunhistochemistry and Western Blot expression. **Panel D.** Notch1 RNA expression was decrease in the BAV as compared to Control aortic Valve.

Figure 2 demonstrates the characterization of the eNOS phenotype as defined by histology, RTPCR and echocardiography. In Figure 2, Panel A is the histology for BAV, Figure 2, Panel B is the semi-quantitative RTPCR from the BAV eNOS[-/-] mice, and echocardiographic data for the bicuspid vs. tricuspid aortic valves Figure 2, Panel C. We measured Notch1, Wnt3a and downstream markers of the canonical Wnt pathway by protein and RNA expression. Notch1 protein was diminished and the RNA expression demonstrates a similar spliced variant with lipid treatments which was not present with the control and atorvastatin treatment. Cholesterol diets increased the members of the canonical Wnt pathway and Atorvastatin diminished these markers significantly (p<0.05).

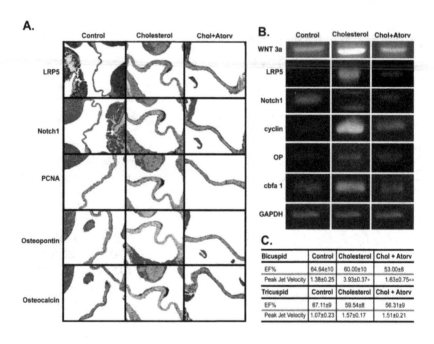

Figure 2. Characterization of the Bicuspid Aortic valve from the eNOS[-/-] bicuspid aortic valves. **Panel A**. Immunohisto-chemistry stain for Lrp5, Notch1, Proliferating Cell Nuclear Antigen, Osteopontin and Osteocalcin from the eNOS[-/-] aortic valves on the control, Left column, control diet; middle column, cholesterol diet; right column, cholesterol diet plus atorvastatin. In each panel, the aortic valve leaflet is in the center. (All frames 20X magnification) **Panel B**. RTPCR for Wnt3a, Lrp5, Notch1, cyclin1, Osteopontin and Cbfa1 from the eNOS[-/-] aortic valves on the control, A1, Cholesterol A2, and the Cholesterol + Atorvastatin diets A3. **Panel C**. Echocardiographic results of the tricuspid versus bicuspid eNOS[-/-] null mice.

BAV is a complex model to study the mechanisms of calcification. The importance of cell-cell communication within a stem cell niche is necessary for the development of valvular heart disease. The two corollaries necessary for an adult stem cell niche is to first define the physical architecture of the stem-cell niche and second is to define the gradient of proliferation to differentiation within the stem-cell niche. The endothelial lining cell located along the aortic surface is responsible for the secretion of a growth factors [92]. These cells interact with the subendothelial cells that are resident below the endothelial layer of cells. These cells have been characterized as myofibroblast cells [75, 93, 94].

To test the hypothesis that BAV disease develops secondary to a stem cell niche process, the physical cell-cell communication needed to be established [95]. In the aortic valve the communication for the stem cell niche would be between the aortic valve endothelial cell and the adjacent myofibroblast cell located below the aortic lining endothelial cell. Conditioned media was produced from untreated aortic valve endothelial cells for the microenvironment that activates signaling in the myofibroblast cell. A mitogenic protein (Wnt3a) was isolated from the conditioned media and then tested directly on the responding mesenchymal cell, the cardiac valve myofibroblast [93, 96,95]. This transfer of isolated protein to the adjacent cell was necessary to determine if the cell would proliferate directly in the presence of this protein. This system is appealing because the responding mesenchymal cell is isolated from the anatomic region adjacent and immediately below that of the endothelial cells producing the growth factor activity along the fibrosa surface. Very little is known regarding the characterization of the endothelial cell conditioned media. These experiments test the corollary that the physical architecture described above is necessary for disease development in the aortic valve.

Figure 3 demonstrates the isolation and characterization of the Wnt3a from the conditioned media microenvironment. Figure 3, Panel A, is light microscopy of aortic valve endothelial cells isolated from the aortic surface of the aortic valve. The results of the mitogen assays for fractions eluting from a DEAE- Sephadex column are shown in Figure 3, Panel B. It can be seen that the mitogenic activity appeared as a single peak eluting at approximately 0.25 M NaCl. The material eluting from DEAE- Sephadex was then applied to Sephadex G-100; the results of mitogen assays on fractions eluting from such a gel filtration column are shown in Figure 3, Panel C. It can be seen that under these native, non-denaturing conditions the bulk of the mitogenic activity eluted as a peak corresponding to standard proteins of 30- 40,000 molecular weight. A SDS denaturing protein gel was run on each sample from the eluted proteins and the bulk of activated protein correlated with the protein peak at 46kd as shown in Figure 3, Panel D. The protein size and charge determination is similar to that previously characterized as Wnt3a [97]. This material lost all activity when heated to 100°C for 5 minutes; disulfide bond reduction with dithiothreitol also abolished all mitogenic activity; and treatment with trypsin destroyed all activity, implicating a protein structure.

The second corollary for identifying a stem cell niche is to define the gradient responsible for the proliferation to differentiation process. The main postulate for this corollary stems from the risk factor hypothesis for the development of aortic valve disease. If traditional atherosclerotic risk factors are necessary for the initiation of disease, then these risk factors are responsible for the gradient necessary for the differentiation of myofibroblast cells to become an osteoblast calcifying phenotype [5, 17, 62, 75, 94, 95, 98, 99]. If traditional risk factors are responsible for the development of valvular heart disease, then an oxidative stress mechanism is important for the development of a gradient in this niche.

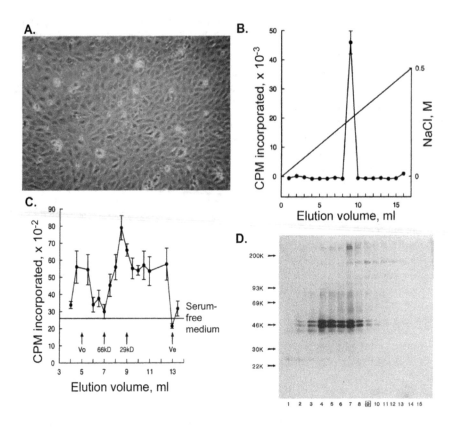

Figure 3. Protein Isolation and Characterization of Aortic Valve Endothelial Cell Conditioned Media; **Panel A**. Light Microscopy for Aortic Valve Endothelial Cells. **Panel B**. Cell Proliferation for fractions eluting from a DEAE- Sephadex column. (p<0.001) **Panel C**. Fractions from DEAE- Sephadex to characterize weight with Sephadex G-100. (p<0.001) **Panel D**. Southern Blot for Protein Expression of Fractions isolated form the DEAE-Sephadex column.

Nitric oxide is important in terms of the mechanism in adult disease processes and also in the developmental abnormalities such as the bicuspid aortic valve phenotype in the eNOS null mouse. To answer this question of the role of oxidative stress and nitric oxide in the aortic valve, I performed *in vitro* experiments to determine eNOS enzymatic and protein regulation in the presence of lipids and attenuation with Atorvastatin. We have previously published that eNOS is regulated in the aortic valve in an experimental hypercholesterolemia model of valvulardisease [11]. Figure 4, demonstrates the eNOS regulation in the endothelial cells in the presence of lipids with and without Atorvastatin. A number of standard assays were performed to measure eNOS functional activity. Figure 4, Panel A1, tests for eNOS enzymatic activity in the aortic valve endothelial cells (AEC) in the presence of LDL with and without Atorvastatin. ENOS enzymatic activity was decreased in the presence of lipids and Atorvastatin improved functional enzyme activity. Figure 4, Panel A2, shows results for tissue nitrites measured in the endothelial cells providing indirect evidence for the enzyme activity. There was an increase is nitrites with lipid treatments and attenuation with Atorvastatin. This increase in nitrite levels correlates with a decrease in the functional activity of the eNOS enzyme in the aortic valve endothelium.

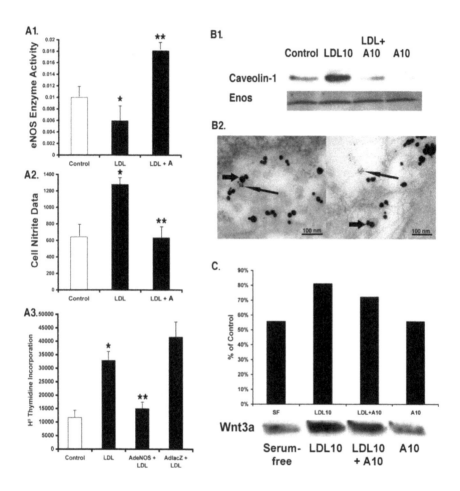

Figure 4. Evidence for eNOS regulation and Wnt3a Secretion from Aortic Valve Endothelial Cells. *p<0.001 for control compared to cholesterol, **p<0.001 for cholesterol compared to cholesterol + Atorvastatin. **Panel A1.** eNOS enzymatic activity in the aortic valve endothelial cells (AEC) in the presence of LDL with and without atorvastatin. **Panel A2.** Cell Nitrite activity in the aortic valve endothelial cells (AEC) in the presence of LDL with and without atorvastatin. **Panel A3.** Thymidine incorporation in LDL treated media, compared to AdeNOS treated myofibroblast cells, versus control LacZ virus. **Panel B1.** Caveolin-1 and eNOS protein expression isolated from the lipid with and without atorvastatin treated cells as shown by Western Blot. **Panel B2.** Electron microscopy immunogold labeling for eNOS and Caveolin-1 localize in the aortic valve endothelial cells in caveolae. **Panel C.** Wnt3a Immunoprecipitate from Conditioned Media treated with LDL with and without Atorvastatin.

The proof of principle experiment to test the importance of eNOS enzymatic activity is an overexpression experiment to determine if eNOS is able to inhibit cell proliferation, an early cellular event in the development of aortic stenosis [13]. Experiments were performed to overexpress eNOS to determine if eNOS overexpression in the aortic valve endothelial cells would regulate cell proliferation. The in vitro myofibroblast cells were directly transduced with an eNOS adenoviral gene construct. Thymidine incorporation was measured to test if overexpressing eNOS can inhibit cellular proliferation. Figure 4, Panel A3, eNOS overexpression inhibits the cell proliferation in the oxidized LDL treated cells induced as compared to the LacZ control treated cells.

A key regulator of eNOS function is caveolin-1 which is expressed in aortic valve endothelial cells [19]. Caveolin-1 upregulation in the presence of lipids inactivates eNOS enzymatic function and further promotes oxidative stress [100, 101]. Experiments were performed to localize the expression of Caveolin1 and eNOS in the aortic valve endothelial cell caveolae. A well defined mechanism to inactivate eNOS enzymatic activity is functional binding of eNOS with caveolin1 in the presence of lipids [29, 95, 102]. Figure 4, Panel B1, demonstrates that Caveolin-1 is upregulated in the lipid treated cells and decreases with atorvastatin treatment with no change in the eNOS protein expression as shown by Western Blot. Figure 4, Panel B2, demonstrates the ultrastructural evidence by immunogold labeling for eNOS and Caveolin-1 present in the aortic valve endothelial cells in caveolae, similar to previously reported data [9, 11]. This caveolin1 upregulation is indirect evidence in addition to the direct data of a decrease in the enzyme activity, that caveolin-1 may play a similar role in AEC found in the aortic valve similar to the vascular endothelium.

Experiments were performed to determine if Wnt3a secretion changes in the microenvironment of the aortic valve endothelial cells with and without lipids. Figure 4, Panel C, demonstrates that Wnt3a protein concentration in the conditioned media in the presence of LDL with and without Atorvastatin. There is a significant increase in the protein with the lipids and attenuation of this protein secretion with the Atorvastatin treatments. This experiment tests the effects of lipids regulating the development of a "Wnt3a" gradient in the microenvironment. If LDL increases Wnt3a secretion into the conditioned media or the microenvironment of the diseased aortic valve, this further contributes to the activation of the canonical Wnt pathway in the subendothelial space of the aortic valve.

The final experiment to test the importance of a stem cell niche to activate the cellular osteoblast gene program in the subendothelial layer cells was to test for the gene expression of the Wnt/Lrp5 pathway in the myofibroblast cells. The stem cell niche is a unique model for the development of an oxidative stress communication within the aortic valve endothelium. As shown in Figure 5, oxidative stress contributes to the release of Wnt3a into the subendothelial space to activate Lrp5/Frizzeled receptor complex on the extracellular membrane of the myofibroblast. This trimeric complex then induces glycogen synthase kinase to be phosphorylated. This phosphorylation event causes β-catenin translocation to the nucleus. β-catenin acts as a coactivator of osteoblast specific transcription factor Cbfa1 to induce mesenchymalosteoblastogenesis in the aortic valve myofibroblast cell.

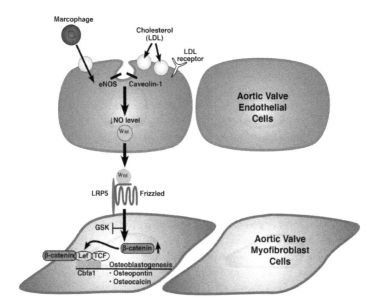

Figure 5. Schematic Modeling for Calcification in the Aortic Valve Stem Cell Niche.

Adult tissues stem cells are a population of functionally undifferentiated cells, capable of (i) homing (ii) proliferation, (iii) producing differentiated progeny, (iv) self-renewing, (v) re-generation, and (vi) reversibility in the use of these options. Within this definition, stem cells are defined by virtue of their functional potential and not by a specific observable character-istic. This data is the first to implicate a cell-cell communication between the aortic valve en-dothelial cell and the myofibroblast cell to activate the canonical Wnt pathway. Lrp5 is important in normal valve development [103], in this stem cell niche, reactivation of latent Lrp5 expression [5, 16], regulates osteoblastogenesis in these mesenchymal cells. The two corollary requirements necessary for an adult stem cell niche is to first define the physical architecture of the stem-cell niche and second is to define the gradient of proliferation to dif-ferentiation within the stem-cell niche. The aortic valve endothelial cell communicates with the myofibroblast cell to activate the myofibroblast to differentiate to form an osteoblast-like phenotype [14]. This concept is similar to the endothelial/mesenchymal transition critical in normal valve development [104]. This data fulfills these main corollaries of the plausibility of a stem cell niche responsible for the development of valvular heart disease. Within a stem cell niche there is a delicate balance between proliferation and differentiation. Cells near the stem-cell zone are more proliferative, and Wnt likely plays a role in directing cell differentia-tion. Stem cell behavior is determined by the number of its stem cell neighbors, which in the valve is the endothelial cell. This assumption is aimed at simply describing the fact that cy-tokines, secreted by cells into the micro-environment are capable of activating quiescent stem cells into differentiation [105].

The important inhibitor in this model is Notch1. Notch1 plays a roll in cellular differentiation decisions. In the osteoblast cell, it serves as an inhibitor of osteoblast differentiation [89, 90]. In the aortic valve, it serves to turn off bone formation via the cell-cell crosstalk between the endothelial and the myofibroblast cells. Normal Notch1 receptor functions to maintain normal valve cellular composition and homeostasis. In the presence of lipids, Notch1 is spliced and therefore activates osteoblastogenesis. In turn, the Wnt3a is secreted and binds to Lrp5 and Frizzled on the extracellular membrane to regulate the osteoblast gene program. This developmental disease process follows a parallel signaling pathway that has been observed in the normal embryonic valve development that has been well delineated by previous investigators [104]. A similar cell-cell communication is necessary for the development of valve disease.

This study provides the correlates described in the mathematical modeling by Agur [106]. This mathematical model has demonstrated the principal that the universal properties of the stem cells can be described in a simple discrete model as derived from hemopoietic stem cell behavior [106]. The transition of hemopoietic stem cells from quiescence into differentiation, is governed by their cell-cycling status, by stimulatory hormones secreted by neighboring cells into the micro-environment and by the level of amplification of stem-cell population [105, 107]. The model of Agur, defines the corollaries necessary to identify a stem cell niche, first the physical architecture of the stem cell niche and second the gradient necessary to regulate the niche. In the BAV the gradient is defined by the niche's microenvironment. The initiation of event of oxidative stress inhibits normal endothelial nitric oxide synthase function, activates notch1 splicing which in turn induces Wnt3a secretion to activate bone formation within the valve [5, 17], [99].

The model proposed in the study as described in Figure 5, provides the cellular architecture for the development of this disease process. This model does not take into account other cytokine/growth factor mediated mechanisms that have been shown to also be important in this disease process [108]. However, understanding CAVD from a development disease perspective will provide a foundation for understanding this and other development disease processes. Clinical trials in the field of CAVD are demonstrating variable results [86, 87]. The possible differences in the published trials are secondary to the timing of therapy and the biological targeting of the lipid levels in these patients. Future medical therapies targeting stem cell niche mediated diseases provides a novel model system to test and to translate clinically for patients in the future.

Author details

Nalini Rajamannan

Molecular Biology and Biochemistry, Mayo Clinic School of Medicine, Rochester MN, USA

References

[1] Lindroos, M., Kupari, M., Heikkila, J., & Tilvis, R. Prevalence of aortic valve abnormalities in the elderly: an echocardiographic study of a random population sample.
Journal of the American College of Cardiology. (1993). , 21(5), 1220-1225.

[2] Ross, J., Jr , , Braunwald, E., Aortic, stenosis., & Circulation, . (1968). Suppl):, 61-67.

[3] Roberts WC, Ko JM. (2005). Frequency by decades of unicuspid, bicuspid, and tricuspid aortic valves in adults having isolated aortic valve replacement for aortic stenosis, with or without associated aortic regurgitation. *Circulation*, 111(7), 920-925.

[4] Drolet, M. C., Arsenault, M., & Couet, J. (2003). Experimental aortic valve stenosis in
rabbits. *Journal of the American College of Cardiology*, 41(7), 1211-1217.

[5] Rajamannan, N. M., Subramaniam, M., Caira, F., Stock, S. R., & Spelsberg, T. C. Atorvastatin inhibits hypercholesterolemia-induced calcification in the aortic valves via
the Lrp5 receptor pathway. Circulation(2005). Suppl):I, 229-234.

[6] Rajamannan, N. M., Subramaniam, M., Springett, M., Sebo, T. C., Niekrasz, M., Mc
Connell, J. P., Singh, R. J., Stone, N. J., Bonow, R. O., & Spelsberg, T. C. (2002). Atorvastatin inhibits hypercholesterolemia-induced cellular proliferation and bone matrix production in the rabbit aortic valve. *Circulation*, 105(22), 2260-2265.

[7] Weiss, R. M., Ohashi, M., Miller, Young. S. G., & Heistad, D. D. (2006). Calcific aortic
valve stenosis in old hypercholesterolemic mice. *Circulation*, 114(19), 2065-2069.

[8] Aikawa, E., Nahrendorf, M., Sosnovik, D., Lok, V. M., Jaffer, F. A., Aikawa, M., &
Weissleder, R. (2007). Multimodality molecular imaging identifies proteolytic and os-
teogenic activities in early aortic valve disease. *Circulation*, 115(3), 377-386.

[9] Rajamannan NM, Springett MJ, Pederson LG, Carmichael SW.Localization of caveolin 1 in aortic valve endothelial cells using antigen retrieval. Journal of Histochemis-
try&Cytochemistry. (2002). , 50(5), 617-628.

[10] Rajamannan NM, Edwards WD, Spelsberg TC. Hypercholesterolemic aortic-valve
disease.New England Journal of Medicine. (2003). , 349(7), 717-718.

[11] Rajamannan, N. M., Subramaniam, M., Stock, S. R., Stone, N. J., Springett, M., Igna-
tiev, K. I., Mc Connell, J. P., Singh, R. J., Bonow, R. O., & Spelsberg, T. C. Atorvastatin
inhibits calcification and enhances nitric oxide synthase production in the hypercho-
lesterolaemic aortic valve. Heart (British Cardiac Society). (2005). , 91(6), 806-810.

[12] Makkena, B., Salti, H., Subramaniam, M., Thennapan, S., Bonow, R. H., Caira, F., Bonow, R. O., Spelsberg, T. C., & Rajamannan, N. M. (2005). Atorvastatin decreases cellular proliferation and bone matrix expression in the hypercholesterolemic mitral
valve. *Journal of the American College of Cardiology*, 45(4), 631-633.

[13] Rajamannan, N. M., Sangiorgi, G., Springett, M., Arnold, K., Mohacsi, T., Spagnoli, L.
G., Edwards, W. D., Tajik, A. J., & Schwartz, R. S. Experimental hypercholesterolemia

induces apoptosis in the aortic valve. Journal of Heart Valve Disease. (2001). , 10(3), 371-374.

[14] Rajamannan, N. M., Subramaniam, M., Rickard, D., Stock, S. R., Donovan, J., Spring-
      ett, M., Orszulak, T., Fullerton, D. A., Tajik, A. J., Bonow, R. O., & Spelsberg, T.
      (2003). Human aortic valve calcification is associated with an osteoblast phenotype.
      Circulation, 107(17), 2181-2184.

[15] Mohler, E. R., 3rd Gannon, F., Reynolds, C., Zimmerman, R., Keane, M. G., & Kaplan,
      F. S. (2001). Bone formation and inflammation in cardiac valves. Circulation, 103(11),
      1522-1528.

[16] Caira, F. C., Stock, S. R., Gleason, T. G., Mc Gee, E. C., Huang, J., Bonow, R. O., Spels-
      berg, T. C., Mc Carthy, P. M., Rahimtoola, S. H., & Rajamannan, N. M. (2006). Human
      degenerative valve disease is associated with up-regulation of low-density lipopro-
      tein receptor-related protein 5 receptor-mediated bone formation. Journal of the Amer-
      ican College of Cardiology, 47(8), 1707-1712.

[17] Shao, J. S., Cheng, S. L., Pingsterhaus, J. M., Charlton-Kachigian, N., Loewy, A. P.,
      Towler, D. A., Msx, promotes., cardiovascular, calcification., by, activating., para-
      crine, Wnt., & signals, . The Journal of clinical investigation. (2005). , 115(5),
      1210-1220.

[18] Desai, M. Y., Rodriguez, A., Wasserman, Gerstenblith. G., Agarwal, S., Kennedy, M.,
      Bluemke, D. A., & Lima, J. A. Association of cholesterol subfractions and carotid lip-
      id core measured by MRI. ArteriosclerThrombVasc Biol. (2005). e, 110-111.

[19] Subbaiah, P. V., Gesquiere, L. R., & Wang, K. Regulation of the selective uptake of
      cholesteryl esters from high density lipoproteins by sphingomyelin. J Lipid Res.
      (2005). , 46(12), 2699-2705.

[20] Kim, W. J., Chereshnev, I., Gazdoiu, M., Fallon, J. T., Rollins, B. J., Taubman, M. B. M.
      C. P., deficiency, is., associated, with., reduced, intimal., hyperplasia, after., arterial,
      injury., & Biochem, . BiochemBiophys Res Commun. (2003). , 310(3), 936-942.

[21] Tanner, F. C., Boehm, M., Akyurek, L. M., San, H., Yang, Z. Y., Tashiro, J., Nabel, G.
      J., & Nabel, E. G. Differential effects of the cyclin-dependent kinase inhibitors Kip1),
      p21(Cip1), and p16(Ink4) on vascular smooth muscle cell proliferation. Circulation.
      (2000). , 27.

[22] Zhang, R., Luo, D., Miao, R., Bai, L., Ge, Q., Sessa, W. C., Min, W., Hsp9, , Akt, phos-
      phorylates. A. S. K., inhibits, A. S., & K1-mediated, apoptosis. (2005). Oncogene,
      24(24), 3954-3963.

[23] (Bostrom K, Watson KE, Horn S, Wortham C, Herman IM, Demer LL. Bone morpho-
      genetic protein expression in human atherosclerotic lesions. The Journal of clinical
      investigation. 1993;91(4):1800-1809). , 91(4), 1800-1809.

[24] Aikawa, M., Rabkin, E., Sugiyama, S., Voglic, S. J., Fukumoto, Y., Furukawa, Y., Shio-
      mi, M., Schoen, F. J., Libby, P., An-Co, H. M. G., reductase, A., inhibitor, cerivastatin.,

suppresses, growth., of, macrophages., expressing, matrix., metalloproteinases, , tissue, factor., in, vivo., & in, vitro. (2001). *Circulation*, 103(2), 276-283.

[25] Williams, J. K., Sukhova, G. K., Herrington, D. M., & Libby, P. (1998). Pravastatin has cholesterol-lowering independent effects on the artery wall of atherosclerotic monkeys. *Journal of the American College of Cardiology*, 31(3), 684-691.

[26] Shyy, J. Y., & Chien, S. Role of integrins in endothelial mechanosensing of shear stress. Circ Res. (2002). , 91(9), 769-775.

[27] Laufs, U., & Liao, J. K. Post-transcriptional regulation of endothelial nitric oxide synthase mRNA stability by Rho GTPase. Journal of Biological Chemistry. (1998). , 273(37), 24266-24271.

[28] Venema RC, Sayegh HS, Kent JD, Harrison DG. Identification, characterization, and comparison of the calmodulin-binding domains of the endothelial and inducible nitric oxide synthases.Journal of Biological Chemistry. (1996). , 271(11), 6435-6440.

[29] Blair, A., Shaul, P. W., Yuhanna, I. S., Conrad, P. A., & Smart, E. J. Oxidized low density lipoprotein displaces endothelial nitric-oxide synthase (eNOS) from plasmalemmalcaveolae and impairs eNOS activation. J Biol Chem. (1999). , 274(45), 32512-32519.

[30] Smart EJ, Anderson RG.Alterations in membrane cholesterol that affect structure and function of caveolae. Methods Enzymol. (2002). , 353, 131-139.

[31] Pritchard, K. A., Ackerman, A. W., Ou, J., Curtis, M., Smalley, D. M., Fontana, J. T., Stemerman, M. B., & Sessa, W. C. Native low-density lipoprotein induces endothelial nitric oxide synthase dysfunction: role of heat shock protein 90 and caveolin-1. Free RadicBiol Med. (2002). , 33(1), 52-62.

[32] Banka CL, Black AS, Dyer CA, Curtiss LK. THP-1 cells form foam cells in response to coculture with lipoproteins but not platelets.J Lipid Res. (1991). , 32(1), 35-43.

[33] Curtiss LK, Dyer CA, Banka CL, Black AS.Platelet-mediated foam cell formation in atherosclerosis. Clin Invest Med. (1990). , 13(4), 189-195.

[34] Brand, K., Banka, C. L., Mackman, N., Terkeltaub, R. A., Fan, S. T., Curtiss, L. K., Oxidized, L. D. L., enhances, lipopolysaccharide-induced., tissue, factor., expression, in., human, adherent., & monocytes, Arterioscler. ArteriosclerThromb. (1994). , 14(5), 790-797.

[35] Leibovich, S. J., Polverini, P. J., Shepard, H. M., Wiseman, D. M., Shively, V., & Nuseir, N. (1987). Macrophage-induced angiogenesis is mediated by tumour necrosis factor-alpha. *Nature*, 329(6140), 630-632.

[36] Leibovich, S. J., Chen, J. F., Pinhal-Enfield, G., Belem, P. C., Elson, G., Rosania, A., Ramanathan, M., Montesinos, C., Jacobson, M., Schwarzschild, Fink. J. S., & Cronstein, B. Synergistic up-regulation of vascular endothelial growth factor expression in murine macrophages by adenosine A(2A) receptor agonists and endotoxin. Am J Pathol. (2002). , 160(6), 2231-2244.

[37] Subramanian SV, Polikandriotis JA, Kelm RJ, Jr., David JJ, Orosz CG, Strauch AR.Induction of vascular smooth muscle alpha-actin gene transcription in transforming growth factor beta1-activated myofibroblasts mediated by dynamic interplay between the Pur repressor proteins and Sp1/Smadcoactivators. MolBiol Cell. (2004). , 15(10), 4532-4543.

[38] Stewart, B. F., Siscovick, D., Lind, B. K., Gardin, J. M., Gottdiener, J. S., Smith, V. E., Kitzman, D. W., & Otto, C. M. Clinical factors associated with calcific aortic valve disease. Cardiovascular Health Study. Journal of the American College of Cardiology(1997). , 29(3), 630-634.

[39] Aronow, W. S., Ahn, C., Kronzon, I., & Goldman, . Association of coronary risk factors and use of statins with progression of mild valvular aortic stenosis in older persons. American Journal of Cardiology. (2001). , 88(6), 693-695.

[40] O'Brien, K. D., Reichenbach, D. D., Marcovina, S. M., Kuusisto, J., Alpers, Otto. C. M., Apolipoproteins, B., (a, , accumulate, E., in, the., morphologically, early., lesion, of., 'degenerative', valvular., & aortic, stenosis. ArteriosclerosisThrombosis & Vascular Biology. (1996). , 16(4), 523-532.

[41] Olsson, M., Thyberg, J., & Nilsson, J. Presence of oxidized low density lipoprotein in nonrheumaticstenotic aortic valves. ArteriosclerThrombVasc Biol. (1999). , 19(5), 1218-1222.

[42] Sprecher, D. L., Schaefer, E. J., Kent, K. M., Gregg, R. E., Zech, L. A., Hoeg, J. M., Mc Manus, B., Roberts, W. C., Brewer, H. B., & Jr , . Cardiovascular features of homozygous familial hypercholesterolemia: analysis of 16 patients. American Journal of Cardiology. (1984). , 54(1), 20-30.

[43] O'Brien, K. D., Kuusisto, J., Reichenbach, D. D., Ferguson, M., Giachelli, C., & Alpers, Otto. C. M. (1995). Osteopontin is expressed in human aortic valvular lesions. *Circulation*, 92(8), 2163-2168.

[44] Mohler, E. R., 3rd Adam, L. P., Mc Clelland, P., Graham, L., & Hathaway, D. R. Detection of osteopontin in calcified human aortic valves. ArteriosclerosisThrombosis & Vascular Biology. (1997). , 17(3), 547-552.

[45] O'Brien, K. D., Kuusisto, J., Reichenbach, D. D., Ferguson, M., Giachelli, C., & Alpers, Otto. C. M. Osteopontin is expressed in human aortic valvular lesions. [comment]. Circulation. (1995). , 92(8), 2163-2168.

[46] Whittaker, P., Boughner, D. R., Perkins, D. G., & Canham, P. B. Quantitative structural analysis of collagen in chordae tendineae and its relation to floppy mitral valves and proteoglycan infiltration. British heart journal. (1987). , 57(3), 264-269.

[47] Wooley, C. F., Baker, P. B., Kolibash, A. J., Kilman, J. W., Sparks, E. A., Boudoulas, H., The, floppy., myxomatous, mitral., valve, mitral., valve, prolapse., mitral, regurgitation., & Prog, Cardiovasc. Dis. (1991). , 33(6), 397-433.

[48] Grande-Allen, K. J., Borowski, A. G., Troughton, R. W., Houghtaling, P. L., Dipaola, N. R., Moravec, C. S., Vesely, I., & Griffin, B. P. Apparently normal mitral valves in patients with heart failure demonstrate biochemical and structural derangements: an extracellular matrix and echocardiographic study. [see comment].(2005). Jan 2004., 54-61.

[49] Grande-Allen, K. J., Calabro, A., Gupta, V., Wight, T. N., Hascall, V. C., & Vesely, I. Glycosaminoglycans and proteoglycans in normal mitral valve leaflets and chordae: association with regions of tensile and compressive loading.(2004). Jul., 621-633.

[50] Jian, B., Jones, P. L., Li, Q., Mohler, E. R., 3rd Schoen, F. J., Levy, R. J., Matrix, metallo-proteinase., is, associated., with-C, tenascin., in, calcific., & aortic, stenosis. (2001). *The American journal of pathology*, 159(1), 321-327.

[51] Ducy, P., Zhang, R., Geoffroy, V., Ridall, A. L., Karsenty, G., Osf, , Cbfa, , transcrip-tional, a., activator, of., osteoblast, differentiation. [see., & comment], . Cell. (1997). , 89(5), 747-754.

[52] Aubin, J. E., Liu, F., Malaval, L., & Gupta, A. K. Osteoblast and chondroblast differ-entiation. Bone(1995). Suppl):, 77S EOF-83S EOF.

[53] Robinson JA, Harris SA, Riggs BL, Spelsberg TC. (1997). Estrogen regulation of hu-man osteoblastic cell proliferation and differentiation. *Endocrinology*, 138(7), 2919-2927.

[54] Spelsberg TC, Harris SA, Riggs BL.Immortalized osteoblast cell systems (new human fetal osteoblast systems). Calcif Tissue Int. (1995). Suppl 1):S, 18-21.

[55] Davies, M. R., Lund, R. J., Mathew, S., & Hruska, K. A. Low turnover osteodystrophy and vascular calcification are amenable to skeletal anabolism in an animal model of chronic kidney disease and the metabolic syndrome. J Am SocNephrol. (2005). , 16(4), 917-928.

[56] Davies MR, Lund RJ, Hruska KA. BMP-7 is an efficacious treatment of vascular calci-fication in a murine model of atherosclerosis and chronic renal failure. J Am Soc-Nephrol. (2003). , 14(6), 1559-1567.

[57] Parhami, F., Garfinkel, A., & Demer, L. L. Role of lipids in osteoporosis. Arterioscler-osisThrombosis & Vascular Biology. (2000). , 20(11), 2346-2348.

[58] Parhami, F., Mody, N., Gharavi, N., Ballard, A. J., Tintut, Y., & Demer, L. L. Role of the cholesterol biosynthetic pathway in osteoblastic differentiation of marrow stro-mal cells. Journal of Bone & Mineral Research. (2002). , 17(11), 1997-2003.

[59] Parhami, F., Morrow, A. D., Balucan, J., Leitinger, N., Watson, A. D., Tintut, Y., Ber-liner, J. A., & Demer, L. L. Lipid oxidation products have opposite effects on calcify-ing vascular cell and bone cell differentiation. A possible explanation for the paradox of arterial calcification in osteoporotic patients. ArteriosclerThrombVasc Biol. (1997). , 17(4), 680-687.

[60] Parhami, F., Tintut, Y., Beamer, W. G., Gharavi, N., Goodman, W., & Demer, L. L. Atherogenic high-fat diet reduces bone mineralization in mice. J Bone Miner Res. (2001). , 16(1), 182-188.

[61] Jono, S., Mc Kee, Murry., Shioi, A., Nishizawa, Y., Mori, K., Morii, H., & Giachelli, C. M. Phosphate regulation of vascular smooth muscle cell calcification. Circ Res. (2000). E, 10-17.

[62] Wada, T., Mc Kee, Steitz. S., & Giachelli, C. M. Calcification of vascular smooth muscle cell cultures: inhibition by osteopontin. CircRes. (1999). , 84(2), 166-178.

[63] Parhami, F., Mody, N., Gharavi, N., Ballard, A. J., Tintut, Y., & Demer, L. L. Role of the cholesterol biosynthetic pathway in osteoblastic differentiation of marrow stromal cells. J Bone Miner Res. (2002). , 17(11), 1997-2003.

[64] Willert, K., Nusse, R., Beta-catenin, a., key, mediator., of, Wnt., & signaling, . (1998). *Current Opinion in Genetics & Development*, 8(1), 95-102.

[65] Behrens, J., von, Kries. J. P., Kuhl, M., Bruhn, L., Wedlich, D., Grosschedl, R., & Birchmeier, W. (1996). Functional interaction of beta-catenin with the transcription factor LEF-1. *Nature*, 382(6592), 638-642.

[66] Huber, O., Korn, R., Mc Laughlin, J., Ohsugi, M., Herrmann, B. G., & Kemler, R. Nuclear localization of beta-catenin by interaction with transcription factor LEF-1. Mech Dev. (1996). , 59(1), 3-10.

[67] Holmen, S. L., Salic, A., Zylstra, C. R., Kirschner, M. W., Williams, B. O. A., novel, set., of-Frizzled, Wnt., fusion, proteins., identifies, receptor., components, that., activate, beta-catenin-dependent., & signaling, . Journal of Biological Chemistry. (2002). , 277(38), 34727-34735.

[68] Caverzasio, J. [., Wnt, L. R. P., new, a., regulation, osteoblastic., pathway, involved., in, reaching., peak, bone., & masses], . Revue Medicale de la Suisse Romande. (2004). , 124(2), 81-82.

[69] Kahler RA, Westendorf JJ. Lymphoid enhancer factor-1 and beta-catenin inhibit Runx2-dependent transcriptional activation of the osteocalcin promoter.Journal of Biological Chemistry. (2003). , 278(14), 11937-11944.

[70] Smith, E., & Frenkel, B. Glucocorticoids Inhibit the Transcriptional Activity of LEF/TCF in Differentiating Osteoblasts in a Glycogen Synthase Kinase-3{beta}-dependent and-independent Manner. J. Biol. Chem. (2005). , 280(3), 2388-2394.

[71] Wang HY, Malbon CC. Wnt signaling, Ca2+, and cyclic GMP: visualizing Frizzled functions. Science (New York,N.Y. (2003). , 300(5625), 1529-1530.

[72] Gregory, Perry., Reyes, E., Conley, A., Gunn, W. G., & Prockop, D. J. Dkk-1-derived Synthetic Peptides and Lithium Chloride for the Control and Recovery of Adult Stem Cells from Bone Marrow. J. Biol. Chem. (2005). , 280(3), 2309-2323.

[73] Yano, F., Kugimiya, F., Ohba, S., Ikeda, T., Chikuda, H., Ogasawara, T., Ogata, N., Takato, T., Nakamura, K., Kawaguchi, H., & Chung, U. I. The canonical Wnt signaling pathway promotes chondrocyte differentiation in a Sox9-dependent manner. BiochemBiophys Res Commun. (2005). , 333(4), 1300-1308.

[74] Jian, B., Narula, N., Li, Q. Y., Mohler, E. R., & 3rd Levy, R. J. Progression of aortic valve stenosis: TGF-beta1 is present in calcified aortic valve cusps and promotes aortic valve interstitial cell calcification via apoptosis. Ann Thorac Surg. (2003). discussion 465-456., 75(2), 457-465.

[75] Mohler, E. R., 3rd Chawla, M. K., Chang, A. W., Vyavahare, N., Levy, R. J., Graham, L., & Gannon, F. H. Identification and characterization of calcifying valve cells from human and canine aortic valves. Journal of Heart Valve Disease. (1999). , 8(3), 254-260.

[76] Alkadhi, H., Wildermuth, S., Plass, A., Bettex, D., Baumert, B., Leschka, S., Desbiolles, L. M., Marincek, B., Boehm, T., Aortic, Stenosis., Comparative, Evaluation., of, 1., Detector, Row. C. T., & Echocardiography, . (2006). Radiology.

[77] Budoff, Takasu. J., Katz, R., Mao, S., Shavelle, D. M., O'Brien, K. D., Blumenthal, R. S., Carr, J. J., Kronmal, R., Reproducibility, of. C. T., measurements, of., aortic, valve., calcification, mitral., annulus, calcification., aortic, wall., calcification, in., the, multiethnic., study, of., & atherosclerosis, Acad. AcadRadiol. (2006). , 13(2), 166-172.

[78] Liu, F., Coursey-Clarke, Grahame., Sciacca, C., Rozenshtein, R. R., Homma, A., Austin, S., & , J. H. Aortic valve calcification as an incidental finding at CT of the elderly: severity and location as predictors of aortic stenosis. AJRAm J Roentgenol. (2006). , 186(2), 342-349.

[79] Boxt LM. CT of valvular heart disease.Int J Cardiovasc Imaging. (2005). , 21(1), 105-113.

[80] Shavelle, D. M., Takasu, J., Budoff, Mao. S., Zhao, X. Q., O'Brien, K. D. H. M. G., Co, A., reductase, inhibitor., (statin, , aortic, valve., & calcium. [comment], . Lancet. (2002). , 359(9312), 1125-1126.

[81] Rajamannan, N. M., Subramaniam, M., Caira, F. C., Stock, S. R., & Spelsberg, T. C. Atorvastatin Inhibits Hypercholesterolemia-Induced Calcification in the Aortic Valves via the Lrp5 Receptor Pathway. Circulation. (2005). In Press.

[82] Steinmetz, E. F., Buckley, C., Shames, M. L., Ennis, T. L., Vanvickle-Chavez, S. J., Mao, D., Goeddel, L. A., Hawkins, C. J., & Thompson, R. W. Treatment with simvastatin suppresses the development of experimental abdominal aortic aneurysms in normal and hypercholesterolemic mice. Ann Surg. (2005). , 241(1), 92-101.

[83] Aronow, W. S., Ahn, C., Kronzon, I., & Goldman, . Association of coronary risk factors and use of statins with progression of mild valvular aortic stenosis in older persons. Am J Cardiol. (2001). , 88(6), 693-695.

[84] Shavelle, D. M., Takasu, J., Budoff, Mao. S., Zhao, X. Q., O'Brien, K. D. H. M. G., Co, A., reductase, inhibitor., (statin, , aortic, valve., & calcium, . (2002). *Lancet*, 359(9312), 1125-1126.

[85] Rosenhek, R., Rader, F., Loho, N., Gabriel, H., Heger, M., Klaar, U., Schemper, M., Binder, T., Maurer, G., & Baumgartner, H. (2004). Statins but not angiotensin-converting enzyme inhibitors delay progression of aortic stenosis. *Circulation*, 110(10), 1291-1295.

[86] Cowell, S. J., Newby, D. E., Prescott, R. J., Bloomfield, P., Reid, J., Northridge, D. B., Boon, N. A. A., randomized, trial., of, intensive., lipid-lowering, therapy., in, calcific., & aortic, stenosis. (2005). *The New England journal of medicine*, 352(23), 2389-2397.

[87] Moura, L. M., Ramos, S. F., Zamorano, J. L., Barros, I. M., Azevedo, L. F., Rocha-Goncalves, F., & Rajamannan, N. M. (2007). Rosuvastatin affecting aortic valve endothelium to slow the progression of aortic stenosis. *Journal of the American College of Cardiology*, 49(5), 554-561.

[88] Garg, V., Muth, A. N., Ransom, J. F., Schluterman, M. K., Barnes, R., King, I. N., Grossfeld, P. D., Srivastava, D., Mutations, in. N. O. T. C. H., cause, aortic., & valve, disease. (2005). *Nature*, 437(7056), 270-274.

[89] Sciaudone, M., Gazzerro, E., Priest, L., Delany, A. M., Canalis, E., Notch, ., impairs, osteoblastic., & cell, differentiation. (2003). *Endocrinology*, 144(12), 5631-5639.

[90] Deregowski, V., Gazzerro, E., Priest, L., Rydziel, S., Canalis, E., Notch, ., overexpression, inhibits., osteoblastogenesis, by., suppressing, Wnt/beta-catenin., but, not., bone, morphogenetic., & protein, signaling. J Biol Chem. (2006). , 281(10), 6203-6210.

[91] Lee TC, Zhao YD, Courtman DW, Stewart DJ. (2000). Abnormal aortic valve development in mice lacking endothelial nitric oxide synthase. *Circulation*, 101(20), 2345-2348.

[92] Rajamannan NM, Helgeson SC, Johnson CM.Anionic growth factor activity from cardiac valve endothelial cells: Partial purification and characterization. Clinical Research. (1988). A.

[93] Johnson CM, Hanson MN, Helgeson SC.Porcine cardiac valvularsubendothelial cells in culture: cell isolation and growth characteristics. J Mol Cell Cardiol. (1987). , 19(12), 1185-1193.

[94] Osman, L., Yacoub, M. H., Latif, N., Amrani, M., & Chester, A. H. Role of human valve interstitial cells in valve calcification and their response to atorvastatin. Circulation(2006). Suppl):I, 547-552.

[95] Rajamannan NM.Oxidative-mechanical stress signals stem cell niche mediated Lrp5 osteogenesis in eNOS(-/-) null mice. Journal of cellular biochemistry. (2012).

[96] Johnson CM, Helgeson SC.Glycoproteins synthesized by cultured cardiac valve endothelial cells: unique absence of fibronectin production. BiochemBiophys Res Commun. (1988). , 153(1), 46-50.

[97] Willert, K., Brown, Danenberg. E., Duncan, A. W., Weissman, I. L., Reya, T., Yates, J. R., & 3rd Nusse, R. (2003). Wnt proteins are lipid-modified and can act as stem cell growth factors. *Nature*, 423(6938), 448-452.

[98] Tintut, Y., Alfonso, Z., Saini, T., Radcliff, K., Watson, K., Bostrom, K., & Demer, L. L. (2003). Multilineage potential of cells from the artery wall. *Circulation*, 108(20), 2505-2510.

[99] Kirton, J. P., Crofts, N. J., George, S. J., Brennan, K., & Canfield, A. E. Wnt/beta-catenin signaling stimulates chondrogenic and inhibits adipogenic differentiation of pericytes: potential relevance to vascular disease? Circ Res. (2007). , 101(6), 581-589.

[100] Garcia-Cardena, G., Martasek, P., Masters, Skidd. P. M., Couet, J., Li, S., Lisanti, M. P., & Sessa, W. C. Dissecting the interaction between nitric oxide synthase (NOS) and caveolin. Functional significance of the noscaveolin binding domain in vivo. J Biol Chem. (1997). , 272(41), 25437-25440.

[101] Garcia-Cardena, G., Oh, P., Liu, J., Schnitzer, J. E., & Sessa, W. C. Targeting of nitric oxide synthase to endothelial cell caveolae via palmitoylation: implications for nitric oxide signaling. ProcNatlAcadSci U S A. (1996). , 93(13), 6448-6453.

[102] Feron, O., Dessy, C., Moniotte, S., Desager, J. P., & Balligand, J. L. Hypercholesterolemia decreases nitric oxide production by promoting the interaction of caveolin and endothelial nitric oxide synthase. J Clin Invest. (1999). , 103(6), 897-905.

[103] Hurlstone, A. F., Haramis, A. P., Wienholds, E., Begthel, H., Korving, J., Van Eeden, F., Cuppen, E., Zivkovic, D., Plasterk, R. H., & Clevers, H. (2003). The Wnt/beta-catenin pathway regulates cardiac valve formation. *Nature*, 425(6958), 633-637.

[104] Paruchuri, S., Yang, J. H., Aikawa, E., Melero-Martin, J. M., Khan, Z. A., Loukogeorgakis, S., Schoen, F. J., & Bischoff, J. Human pulmonary valve progenitor cells exhibit endothelial/mesenchymal plasticity in response to vascular endothelial growth factor-A and transforming growth factor-beta2. Circ Res. (2006). , 99(8), 861-869.

[105] de Haan, G., Dontje, B., & Nijhof, W. Concepts of hemopoietic cell amplification. Synergy, redundancy and pleiotropy of cytokines affecting the regulation of erythropoiesis. Leuk Lymphoma. (1996).

[106] Agur, Z., Daniel, Y., & Ginosar, Y. The universal properties of stem cells as pinpointed by a simple discrete model. J Math Biol. (2002). , 44(1), 79-86.

[107] Veiby OP, Mikhail AA, Snodgrass HR.Growth factors and hematopoietic stem cells. HematolOncolClin North Am. (1997). , 11(6), 1173-1184.

[108] O'Brien KD.Pathogenesis of calcific aortic valve disease: a disease process comes of age (and a good deal more). ArteriosclerThrombVasc Biol. (2006). , 26(8), 1721-1728.

Chapter 12

Menopause Induces Oxidative Stress

Claudia Camelia Calzada Mendoza and
Carlos Alberto Jiménez Zamarripa

Additional information is available at the end of the chapter

1. Introduction

1.1. Menopause: endocrinology and symptoms

Menopause is a physiologic process in women that occurs around 45-55 years old, which is defined as permanent cessation of menstruation by one year in row [1]. The age of menopause depends on multiple factors such as number of ovules from the female at birth, the frequency of loss of these ovules through her life and the number of ovarian follicles required maintaining the menstrual cycle. The diagnosis of menopause is retrospective and is established after a year without menses [2], and their symptoms may have different intensity for each woman [3].

This process is characterized by gradual decrease of estrogen (E) secretion and changes related with sex hormones, so that estradiol levels ranging from 5 to 25 pg/mL, while increasing titers of gonadotrophins, so that the values of follicle stimulating hormone (FSH) between 40 and 250 mU/mL and luteinizing hormone (LH), from 30 to 150 mU/Ml [4, 5].

Irregular uterine bleeding is a characteristic symptom which is due to both depletion and resistance of ovarian receptors to gonadotropins and increased FSH, leading to alterations in the volume and frequency of bleeding (polymenorrhea, hypo-or menorrhagia, oligomenorrhea) [6, 7].

Among symptoms are those related to the genitourinary tract by the common embryological origin of vulva, vagina, bladder, and urethra, consequently alterations as dysuria, urinary urgency and incontinence, epithelial atrophy, decreased production of mucus and vaginal dryness (phenomena that can cause dyspareunia), urethritis, vaginitis or cystitis and local infections [8, 9, 10] (Figure 1).

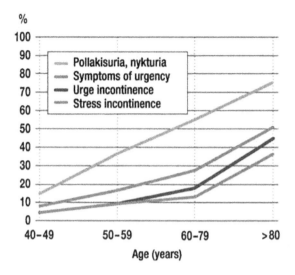

Figure 1. Main genitourinary abnormalities according to age in women [10].

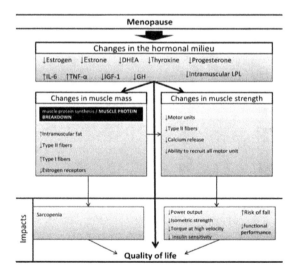

Figure 2. Changes observed in muscle mass and strength after menopause [16].

Hot flushes are one of the main symptoms associated with menopause and occur in more than 75% of menopausal, consisting of intense episodes of heat that begins on chest and spreads to face, sweating, and flushing of face. Hot flushes are associated with headache, anxiety and palpitations, and it usually lasts 2-4 minutes and can vary in frequency, in some women may be daily while others may have one episode per month [11, 12]. The mechanism of hot flushes is not clear, however, it is known that hypothalamus, pituitary gonadotropin releasing hormone and gonadotrophins may be involved in hot flushes [13]. Another frequent symptom is an oral dryness and intense burning sensation that affects mainly the tongue and sometimes lips and gums [14].

On the other hand decreases the content of collagen and elastic fibers of the skin, so that it becomes thinner and brittle losing elasticity and firmness. The epidermis thins, increases water loss and reduces the number of blood vessels, compromising the supply of oxygen and nutrients [15]. Additionally aging is associated with a natural decline in physiological functions, including a loss of muscle mass and strength. Overall, the decline in muscle mass averages 0.4 to 0.8 kg per decade, starting at the age of 20 years, especially around menopause [16] (Figure 2).

Another alteration that occurs is the osteoporosis, which is defined as a skeletal disorder characterized by decreased bone density and an increased risk of fractures [17, 18]. Before reports have confirmed that postmenopausal women have highest incidence of hip fractures [19, 20, 21] (Figure 3).

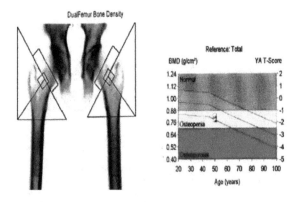

Figure 3. Bone mineral density values by age in women. Bone mineral density decreases around menopause [21].

Menopause is a stage that favors weight gain and development or worsening of obesity, and causes of this problem are many; some are clearly related to hypoestrogenism and other age-dependent, conditioning increased intake and decreased energy expenditure [22, 23] (Figure 4).

During this period there is an abnormal atherogenic lipid profile characterized by increased lipoprotein cholesterol, low density (LDL-C), triglycerides (TG) and small dense LDL particles [24] with reduced HDL-C and elevated serum glucose and insulin, perhaps as a direct result of ovarian failure or indirectly as a result of central redistribution of body fat, and this favors the formation of atheromatous plaques and progression of coronary atherosclerosis and therefore cardiovascular disease incidence increases substantially in postmenopausal women [25, 26]. Other disorders such as obesity and metabolic syndrome also occurs at menopause, suggesting that menopause may be the trigger of the metabolic syndrome at that stage of life [27, 28].

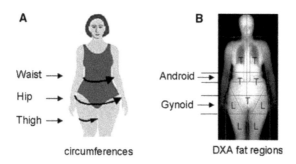

Figure 4. Body fat distribution. Android-type distribution is present in postmenopausal women. DXA= Dual-energy X-ray absorptiometry [23].

Postmenopausal women have higher insulin resistance than premenopausal, which could participate to age, the increase in total body fat, central adiposity, estrogen deficiency, alterations in lipid profile and glucose homeostasis and insulin are more frequent and favor the high cardiovascular morbidity and mortality after menopause. In this sense the transition of menopause is marked by changes in hormonal balance, with increased visceral fat, which are associated with insulin resistance, although it has been found that the change in insulin sensitivity does not alter the lipid profile in early postmenopausal women [24, 26] (Figure 5).

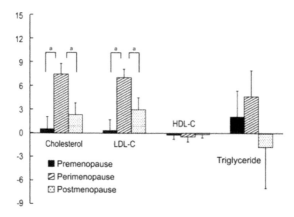

Figure 5. Changes lipid during transition from premenopause to Postmenopause [24].

Depression occurs frequently in postmenopausal women, which is explained by the loss of estrogenic effect in modulating neuronal excitability, synaptic plasticity, neuronal survival induced expression of regenerative responses, regional neurogenesis, regulation of differentiation and neuronal development [29], in the processes of cognition, modulation of mood and other mental states, as well as improving learning and memory [30], regulate the synthesis of tryptophan hydroxylase which is the limiting enzyme in serotonin synthesis so this decline in estrogen at menopause may explain the occurrence of psychological symptoms characteristic of depression (fatigue, irritability, sleep problems, abrupt changes of mood, [31]. With respect depressive symptoms in the Multiethnic Study of Atherosclerosis were analyzed testosterone, estradiol, steroid hormone binding globulin (SHBG) and dehydroepiandrosterone; indicating that in early postmenopausal women, sex hormones were associated with incident depressive symptoms [32].

2. Pro and antioxidants propierties of estrogens

Throughout menopause there are factors that predispose women to the development of oxidative stress, such as estrogen deficiency, as it has been confirmed that they have an antioxidant capacity independently of its binding to receptors, so for example the 17β-estradiol (E2), estriol, estrone, ethinylestradiol and 2-hidroxiestradiol besides reducing neuronal death with antioxidant activity, due to the presence of an intact hydroxyl group on ring A of the molecule [33].

Estrogens are synthesized from different androgen precursors such as androstenedione and testosterone, yielding as products estrone and 17β-estradiol, respectively. The synthesis is catalyzed by aromatase (ARO), the enzyme cytochrome P450 (CYP19) and estrogen synthesizing different tissue-specific manner, and the major estrogen in adipose tissue is estrone, the placenta is estriol and in cells granulosa is 17β-estradiol [34].

The 2-hidroxiestradiol and 2-hydroxyestrone (4-hidroxiestradiol type) (Figure 6) can participate in redox cycling to generate free radicals such as superoxide and chemically reactive estrogen semiquinone/quinone, which can damage DNA and other intracellular constituents.

4-hidroxiestradiol participates in a redox cycle to generate free radicals such as superoxide, and intermediate semiquinone/quinone, these intermediaries may induce cell transformation and initiate tumoral growth [35].

4 - hydroxyestrogens have estrogenic effects and can stimulate the growth of cell lines of breast cancer, with greater intensity than the 4-hydroxyestrone are unstable and can become highly reactive quinone with the formation of semiquinones as intermediary, this reaction produces oxygen free radicals, which can have toxic effects on DNA, such effects include the formation of 8-hydroxy-2-deoxiguanosine a mutagen, resulting from oxidative damage. The toxic effect of 4-hydroxyestrogens probably is prevented under normal conditions intracellular defense mechanisms. Oxygen free radicals can be removed immediately transformed into water by enzymes such as catalase and superoxide dismutase and antioxidant vitamins

such as ascorbic acid and alpha tocopherol, quinone themselves can be inactivated by sulfo compounds, such as glutathione [36].

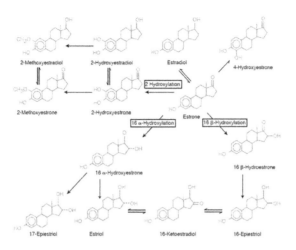

Figure 6. Estrogen metabolism.

Menopause seems to accelerate the development of atherosclerosis and cardiovascular diseases and in order to identify this correlation, was assessed the correlations between intima-media thickness, homocysteine serum levels and oxidative stress both in fertile and postmenopausal women and it was founded that were increased levels of homocysteine, oxidative stress and intima-media tickness (IMT) in postmenopausal women having a positive correlation with IMT, which reinforce the idea that a hyperhomocysteinemia may play a role in the progression of atherosclerosisas a result the lack of estrogens [37].

Vasculo protective effects of estrogen are due in part to the modulation of the balance between nitric oxide (mainly derived from endothelial vasodilator molecule) [38] and superoxide anion (oxygen-free radical highly reactive), promoting the availability of the first such so the lack of protection induces high levels of oxidative stress and low concentrations of NO, these processes are interacting with hypertension, as seen in menopause. In addition, estrogen induces the expression of oxide reductasesthiol / disulfide, such as disulfide isomerase, thioredoxin, thioredoxinreductase and glutaredoxin in the endothelium and inhibits apoptosis mediated by hydrogen peroxide. On the other hand has been described that genetic factors related to dyslipidemia are most important than due to age, for example antioxidant enzymes (SOD, catalase, GR, inflammatory markers CPR, ALT), oxidative stress ($O(2)(-)$, LOO•), hypoxia (HBNO) and all this related to increase vascular resistance, disorders in oxygen supply in tissue and hypoxic competitions of there metabolism may cause, postmenopausal hypertension, hart ischemic disease, impaired hepatic beta-oxidation of fatty acids and hepathosteatosis [39].

17-β-estradiol plays a critical role in neuroprotection through both genomic and non-genomic mechanisms and recently was discovered that a new G-protein-coupled receptor 30 (GPR30) participates in the neuroprotection against oxidative insult, which is agonist G1. E2 attenuated apoptosis induced by H_2O_2 exposure, furthermore, G1 or E2 significantly increased the levels of phosphorylated extracellular signal-regulated kinase 1/2 (p-ERK1/2), Bcl-2 and pro-caspase-3, which is an anti-apoptotic effect [40].

3. Oxidative stress and postmenopause

Actually several oxidative stress biomarkers have been studied in menopause, however, each researcher has used different marker, methodologies and women with dissimilar characteristics (age, ethnic group, postmenopause time), fact does difficult to make a conclusion about the development of oxidative stress during peri, menopause and postmenopause.

Recently has been propose as indicator to γ-glutamyltransferase (GGT) which is an enzyme involved in the transfer of the γ-glutamyl residue from γ-glutamyl peptides to amino acids, H_2O, and other small peptides and can be donated by glutathione [41]. On the other hand, GGT is also involved in the production of glutathione [42], which is limited by cysteine availability. GGT participates in the pathway of extracellular GSH in consequence the biosynthesis of cellular glutathione, the most important cell antioxidant, depends of GGT activity; hence this enzyme may play an important role in the anti-oxidative defense system of the cell [43].

Abdul et al, founded a highly significant reduction in glutathione levels in the post-menopausal-group which could be due to the increase in its free radical scavenging property and increased consumption to counteract the oxidative stress and to inhibit membrane lipid peroxidation which indicates that the increase in serum GGT with enhanced oxidative stress and reduced antioxidant defense system in the post-menopausal women may lead to the speculation that GGT could be considered an index or a oxidative stress marker [43] (Table 1).

Serum level	Premenopausal Group (n=17)	Postmenopausal Group (n=16)	p value
GGT (U/L)	5.96±2.99	9.44±2.89	0.025
GSH (mmole/L)	0.62±0.17	0.47±0.11	0.008
MDA (µmole/L)	1.04±0.06	1.32±0.05	0.035

Table 1. Serum γ-glutamyltransferase, glutathione and malondialdehyde levels in the pre- and postmenopausal women [43].

Supplementary it was found that perimenopausal women have higher total cholesterol values and lower paraoxonase-1 (PON1) activity compared to reference values, 8-oxoG levels were unchanged compared with those of healthy control women, lipoperoxide ranks were significantly increased compared with those of premenopausal women and an indirect correlation between PON1 arylesterase (PON1 A) activity and lipoperoxide levels, between PON1 A activity and atherogenic index, between age and TAS, and between age and 8-oxoG levels. Moreover perimenopausal women had higher total cholesterol levels and PON1 A levels were lower than physiological values (table 2) [44].

Variable	Average±SD or median	Physiological values
TCH	**5.673±0.856 mmol/L**	5.17 mmol/L
TG	1.424±0.66 mmol/L	1.9 mmol/L
LDL	3.103±0.649 mmol/L	3.5 mmol/L
HDL	1.563±0.445 mmol/L	1.4 mmol/L
Atherogenic index (TCH/HDL)	3.853±1.009	5.2
PON1 A	**89.628±14.798 U/mL**	100-200 U/mL
Pon1 L	**12.213±2.956 U/mL**	13-20 U/mL
Homocysteine	8.48±2.97 µmol/L	12 µmol/L
Glycemia	5.43µ0.65 mmol/L	4.2-6.2 mmol/L
Uric acid	246.5 (209.9-296.9) µmol/L	339µmol/L

Table 2. Data showing departures from normality are expressed as median values with the respective lower and upper quartile. The boldfaced entries indicate values beyond the reference range. PON1 A, paraoxonase with arylesterase activity; PON1 L, paroxonase-1 with lactonase activity.Paroxonase-1 levels in perimenonausal women [44].

Another finding is the lipoperoxide level which was significantly increased in perimenopausal women (Table 3). The levels of the marker of oxidative damage to DNA-8-oxoG were not statistically between pre and perimenopausal. In contrast women in perimenopause had repair ability 4 times higher compared with premenopausal women and significantly increased plasma total antioxidant capacity (TAS) [44] (Table 3).

Variable	Perimenopausal women	Controls (premenopausal)
TAS	1.532±0.095 mmol/L[a]	1.230±0.100 mmol/mL
Lipoperoxides	37.995 (32.035-44.849) nMol/mL[a]	28.096 (23.103-30.850) nmol/mL
8-oxoG	0.464 (0.283-0.957) per 10^6 G	0.503 (0.337-0.674) per 10^6 G
Repair ability	36.919% (30.679%-47.046%)[a]	10.539% (8.665%-11.475%)

Table 3. Data showing departures from normality are given as median values with the respective lower and upper quartile. [a]these values are significantly different (P<0.005) compared with controls. Profile oxidant and antioxidant between premenopausal and perimenopausal women [44].

In another study were determined age, body weight, and superoxide dismutase (SOD), cata-lase (CAT) and malondialdehyde (MDA) in disease-free women aged 25-65 years and did found that postmenopausal women had the highest oxidative stress and body weight, also superoxide dismutase, catalase and malondialdehyde were correlated significantly with body weight [45].

Pansini demonstrated that the total body fat mass increases significantly in postmenopause in comparison with premenopause, with specific increases in fat deposition at the level of trunk (abdominal and visceral) and arms. Concomitantly, the antioxidant status adjusted for age showed that antioxidant status was retained. Also both antioxidant status and hydro-peroxide level increased with trunk fat mass [46].

Also has been carried out protocols that analyze the connection between menopause and pe-riodontal conditions, though to compare serum and gingival crevicular fluid (GCF) total an-tioxidant capacity (TAOC) and superoxide dismutase (SOD) concentrations in post-menopausal patients with chronic periodontitis (PMCP) with those of pre-menopausal chronic periodontitis patients (CP). The results showed Serum and GCF TAOC and SOD concentrations were significantly lower in menopause and periodontitis, the lowest values were in the PMCP group, whereas the highest values were in premenopausal. While the ef-fect of menopause was more evident in serum antioxidant analysis, the consequence of pe-riodontitis was observed to be more apparent in GCF and a decrease in systemic and local AO defense was observed owing to both menopause and periodontitis [47].

Risk factor	OR	95% CI	Pa
Menopause (hypoestrogenism)	2.62	1.35-5.11	0.005
Consumption of alcoholic beverages (≥2 glasses/d)	2.49	0.28-22.50	0.417
Smoking (≥2 cigarettes/d)	1.98	0.58-6.82	0.277
Overweight (≥25 kg/m2)	1.43	0.64-3.18	0.383
Insomnia (AIS score ≥8)	1.13	0.58-2.22	0.715
Age, y	1.04	0.94-1.14	0.466
Physical inactivity (≥30 min/d of physical activity)	0.85	0.43-1.68	0.632

Table 4. OR, odds ratio; AIS, Athens Insomnia Scale.aLogistic regression, R^2=0.106, P=0.036.Risk factors for high lipoperoxide levels, as oxidative stress biomarker, in perimenopausal women [49].

Unsaturated fatty acids have a role in the pathogenesis of atherosclerosis. They are very sen-sitive to oxidation caused by excess free oxygen radicals and the consequent oxidative sta-tus, and it is well known that lipid and lipoprotein metabolism is markedly altered in postmenopausal women as it was demonstrated by Signorelli who founded that the oxida-tive stress is involved in the pathophysiology of atherosclerosis. Malonaldehyde (MDA), 4-hydroxynenal (4-HNE), oxidized lipoproteins (ox LDL) were higher in postmenopausal while GSH-PX concentrations were significantly higher in fertile women [48].

Similar findings were found in pre and postmenopausal Mexican women by Sánchez. Lipoperoxides, erythrocyte superoxide dismutase and glutathione peroxidase activities, the total antioxidant status, pro-oxidant factors, body mass index were evaluated. The lipoperoxide levels were significantly higher in the postmenopausal group than in the premenopausal group, which concluded that menopause is the main risk factor for oxidative stress [49].

However, there are other contrasting studies, in example in a report was found than postmenopausal women had lower levels of lipid hydroperoxide oxidation, the MDA levels did not differ between pre- and postmenopausal women, no differences in advanced oxidation protein products (AOPP) and nitrite levels were observed between pre- and postmenopausal women. Postmenopausal women also exhibited a higher total radical antioxidant level [50].

Another study included pre-menopausal, peri-menopausal, and post-menopausal women classified according to the Staging of Reproductive Aging Workshop (STRAW) criteria. No significant correlations between E2 levels and OS markers were detected and consequently, estrogen decline during menopausal transition is not a determinant factor for oxidative stress [51].

4. Associated diseases to oxidative stress

There are several evidences that related to oxidative stress with diseases present in postmenopausal women in example depression, osteoporosis, cardiovascular diseases and leg vasoconstriction.

DEPRESSION

The depression is the most frequent symptom in postmenopausal women, even is a major cause of medical consultation. This disorder has cerebral implications, as showed post-mortem studies in patients with depressive disorder pointed a significant decrease of neuronal and glial cells in cortico-limbic regions which can be seen as a consequence of alterations in neuronal plasticity. This could be triggered by an increase of free radicals which in its turn eventually leads to cell death and consequently atrophy of vulnerable neuronal and glial cell population in these regions [52]. In addition elevated levels of MDA adversely affected the efficiency of visual-spatial and auditory-verbal working memories; short-term declarative memory and the delayed recall declarative memory were founded. 1. Higher concentration of plasma MDA in recurrent depressive disorder (rDD) patients is associated with the severity of depressive symptoms 2. Elevated levels of plasma MDA are related to the impairment of visual-spatial and auditory-verbal working memory and short-term and delayed declarative memory [53]. Actually too is known that estrogen protect neurons against oxidative damage excitotoxins, and beta-amyloid-induced toxicity in cell culture, reduces the serum monoamino oxidase levels and might regulate learning and memory. Nitric oxide (NO) is a messenger and in the central nervous system and acts as neurotransmitter/neuromodulator like serotonin, bradykinin, endothelin, acetylcholine and noradrenaline. Estrogen induces activity of constitutive NO synthase, reduces hyperphosphorylated of Tau and stimulates phosphorylated GSK3b [54]; due to in menopause its reduction induces a depressive disorder [55].

OSTEOPOROSIS

Oxidative stress participates in decreasing bone formation and stimulating bone resorption. Furthermore, antioxidant enzymes have been observed to have low protective activity in women with osteoporosis, also has been determined higher urine deoxypyridinoline, total Peroxide (TPx), MDA, nitric oxide, also lower TAS and glutathione reductase, compared with postmenopausal women whitout osteoporosis [56, 57]. Likewise has been studied polymorphism associated with enzymes involved in oxidative balance such as of the glutathione S-reductase (GSR), superoxide dismutase (SOD1 and SOD2), and catalase (CAT), of which polymorphisms from GSR were associated to bone mineral density [58]. Both oxidative stress and associated polymorphisms are useful tool to predict which patients might develop osteoporosis.

CARDIOVASCULAR DISEASES

Oxidative stress biomarkers have been linked with the presence and severity of the CVD, and to the presence and number of risk factors. It is known that young women during their fertile life are at lower risk of cardiovascular events compared with men, being protected by estrogen action and that oxidative stress is generally higher in men than in premenopausal women. However, after menopause the risk of experiencing cardiovascular events rapidly rises in women, in conjunction with a parallel increase in oxidative stress. Moreover, although oxidative stress results are lower in females compared to males during the first decades of life, this difference decreases until the age range which corresponds to the onset of menopause for women [59].

An analyses of relationship among excess iron, oxidative stress, and centralized fat mass in healthy postmenopausal women showed that almost 14% of the variability in oxLDL was accounted for by centralized fat mass AndGynFM ratio (waistþhip=thigh¼AndGynFM), age, and serum iron. Similarly, 16% the variability in 15-isoprostane $F2_{a\alpha}$ (PGF $F2_{a\alpha}$) was accounted for by the AndGynFM ratio, HOMA, and serum iron. Also it was accounted for 33% of the variability in AndGynFM ratio by high-density lipoprotein cholesterol (HDL-C), ferritin, HOMA, oxLDL, and PGF F2aα, all of before suggests that reducing centralized fat mass and maintaining a favorable lipid profile, antioxidant status, and iron status all may be important in protecting postmenopausal women from atherosclerotic CVD [60]. Similar findings has been observed in diabetic postmenopausal women in whom it has been reported higher levels of total cholesterol (TC), triglyceride (TG), low density lipoprotein cholesterol (LDL-C), very low density lipoprotein cholesterol (VLDL-C), catalase (CAT), and malondialdehyde (MDA) and significantly lower levels of HDL-C, reduced glutathione (GSH), glutathione reductase (GR), glutathione peroxidase (GPx), and superoxide dismutase (SOD) [61]. Fact means a cardiovascular risk.

LEG VASOCONSTRICTION

Leg vasoconstriction has been linked to oxidative stress due to the fact that Intravenous administration of a supraphysiological dose of the antioxidant ascorbic acid increased leg

blood flow in the postmenopausal women as a result of an increase in leg vascular conductance, but it did not affect leg blood flow in premenopausal controls or mean arterial pressure, also changes in leg blood flow and leg vascular conductance with ascorbic acid were
related to high plasma oxidized LDL an low antioxidant status [62] (Table 5, Figure 7).

Variable	Premenopausal	Postmenopausal
Oxidized LDL, U/l	36.9±4.9	55.6±3.3*
TAS, mmol/l	1.4±0.1	1.1±0.1*
ACE, U/l	26.2±3.9	30.3±2.6
Endothelin-1, pg/ml	4.8±0.5	5.3±0.3
Norepinephrine, pg/ml	149±39	343±28*
Epinephrine, pg/ml	20±3	27±2

Table 5. Values are means ±SE, ACE angiotensin-converting enzyme. *P⬛.05 vs premenopausal.Serum biomarkers
associated to leg vasoconstriction [62].

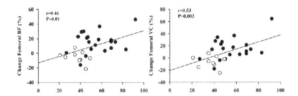

Figure 7. Relationship between plasma oxidized low-density lipoprotein (LDL) and the change in femoral artery BF
(top) and vascular conductance VC (bottom) with ascorbic acid in premenopausal (⬛) and postmenopausal (●) women
[62].

In addition, long-term studies indicate that total cholesterol (TC), LDL cholesterol (LDL-C),
triglycerides (TG), MDA and common carotid artery wall intima-media thickness (IMT) are
higher in women with hormonal depletion over 5 years, reveling a close temporal correlation between plasma oxidative and carotid wall IMT as postmenopause proceeds [63].

Further investigations are needed to examine the roll of oxidative stress as an endogenous
bioactive agent related to disease in post-menopausal women. Since oxidative stress is the
imbalance between total oxidants and antioxidants in the body, any single oxidant/ antioxidant parameter may not reflect oxidative stress. Further studies are needed to understand
the underlying mechanisms of before findings.

5. Hormonal replacement therapy

Hormone replacement therapy (HRT) is defined as treatment that estrogen provides women
to improve the characteristic symptoms of menopause [64], especially osteoporosis, dyslipi-

demias, mood among others, is also important to note that hormone replacement therapy is not without risks.

There are three Hormonal Replacement Therapies (HRT) treatment regimens:

- 1.- Estrogens. They may be natural or synthetic. Estrogens (17β-Estradiol and Estriol) and conjugated equine estrogens, and these are administered orally. Estrogens may administrated by oral, subcutaneous routes, also intravaginal estrogen (tablets, creams, ovules), alone or combined with progestin, are suitable for vaginal symptoms, with no significant increase in endometrial hyperplasia or proliferation.

- 2 - Progestogens. They are administered in combination with estrogen to reduce the risk of endometrial hyperplasia and cancer. Currently most used active ingredients TH are: oral micronized progesterone, medroxyprogesterone and norethisterone. Progestins are mainly used orally, although there are preparations to be administered in combination with estrogen transdermal route [65].

- 3 - Another group of drugs called STEAR (Selective Tissue Estrogenic Activity Regulator) is widely used because it has tissue-specific metabolism, and a main representative is Tibolone, this is a synthetic steroid with weak estrogenic, androgenic and gestagenic activities, which controls vasomotor symptoms, prevents bone demineralization and improves mood [66]. Tibolone improves vaginal symptoms and no significant differences when compared to estrogen, decreases menopausal symptoms, although moderately increases bone density and inhibit bone resorption. In the cardiovascular system there is no evidence of efficacy for the primary or secondary prevention of diseases associated with menopause at this level [67].

6. Effects of hormonal replacement therapy on oxidative stress

As mentioned above, there are different pathologies in the menopause that improve after administrating of hormone replacement therapy, fact that aroused the interest in evaluating their effects on biomarkers of oxidative stress, which has been recognized its participation in illnesses as cancer, atherogenesis, Alzheimer's and aging among others. Below are described the findings on changes in oxidative stress biomarkers after administrating HRT by periods time.

LESS THAN THREE MONTHS

In African American and Caucasian posmenopausal women the HRT reduced plasma levels of free 8-isoprostane after 6 weeks of HT, at the same time nitrite increased, principally in Caucasian women. Both ethnics groups have reduced levels of oxidative stress but the differences were not statistically significant [68]. Even the combined therapy for 3 months had an antioxidant effect in posmenopausal hemodialysis women, who showed reduced levels of MDA although TAC, uric acid and C- reactive protein were not changed [69].

Oxidized low-density lipoprotein (oxLDL)/β2-glycoprotein I (β2GPI) complexes are etiologically important in the development of atherosclerosis. Combined HRT led to a significant increase in TAC and a minor but statistically nonsignificant decrease of oxLDL/β2GPI complexes when compared with the baseline control levels. There was also no significant association between TAC and oxLDL/β2GPI complexes changes related to HRT. This study indicates that, HRT in postmenopausal women leads to an increase in TAC without an equivalent change in serum levels of oxLDL/β2GPI complexes. It is concluded that beneficial effects of HRT could be explained, at least in part, by improving antioxidant status, but may not be directly associated with a change in oxidized lipoprotein production [70] (Figure 8).

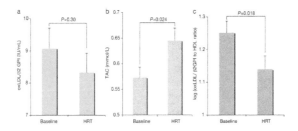

Figure 8. Effects of hormone replacement therapy (HRT) on oxLDL/β2GPI (a), total antioxidant capacity (TAC) (b), oxLDL/β2GPI to HDL-C ratio (c) in the studied subjects. Values are means ± SE. P < 0.05 (paired sample t-test). total serum antioxidant capacity (TAC) as ferric reducing ability of plasma (FRAP) and related these to HRT and oxLDL/β2GPI complexes level [70].

SIX MONTHS

The main studies about effect of HRT has been carried out with combination estrogen plus gestagen, and in them has been founded in example that carbonils groups determined by ELISA showed a reduction of serum levels after six months oral or transdermal treat when compared with control group, and there was not difference between oral and transdermal, which indicates that hormonal therapy reduces the of carbonyl protein, a marker of oxidative stress, suggesting potential protective effect [71]. Similar result were founded with the serum level of malondialdehyde, superoxide dismutase and sulfhydryl groups without changes on plasma total homocysteine (tHcy) (used as atherogenic indicator) [72]. Another study with equal number of months of follow-up showed that carbonyls, MDA and oxLDL were reduced, while erythrocyte glutathione (GSH) were increased, and nitrotyrosine (NT) levels were not changed [73, 74] (Table 6).

On the other hand Tibolone treatment leads to a decrease in concentrations of plasma lipid peroxide, increase plasma concentrations of vitamin E and alpha-tocopherol and significant decrease in lipid peroxide concentrations [75, 76].

ONE YEAR

However, the combined treatment by one year significantly reduced the levels of catecholamines, mean blood pressure and LDL cholesterol while it increased levels of nitrite/nitrate, indicating cardiovascular benefit in healthy recent postmenopausal women. Levels of 8-epi PGF2alpha did not change, suggesting no evident relationship between HRT and oxidative

stress [77]. Although another study reports that conjugated estrogens alone (EHRT) or conjugated estrogen with medroxyprogesterone acetate can reduce lipoprotein lipase (LPL),
hepatic lipase (HL), oxidized apolipoprotein B in LDL [78] also platelet MDA, glutathione-S-
transferase (GST) and SOD levels were lower and total thiol (t-SH) content was higher than
pre-treatment levels. These results indicate that hormone replacement therapy may affect
platelet membrane fatty acid content and oxidant-antioxidant balance in postmenopausal
women [79].

	MPA n=25	NETA n=20	Total n=45
Total-C	204.1±30.0/	225.7±30.1/	213.7±31.7/
Pre/postreatment (mg/100ml)	178.6±10.7*	191.6±21.1*	184.4±17.2*
HDL-C	45.7±10.6/	49.1±11.0/	47.2±10.8/
Pre/postreatment (mg/100ml)	53.2±7.6*	56.6±8.2	54.8±8.0*
LDL-C	131.7±24.9/	147.5±25.2/	138.71±26.08/
Pre/postreatment (mg/100ml)	102.8±13.6*	112.7±21.2*	107.2±17.9*
Triglycerides	133.5±62.0/	138.4±74.4/	135.67±67.1/
Pre/postreatment (mg/100ml)	115.2±35.2	113.5±37.0	114.4±35.6
MDA	4.7±0.4/	4.9±0.4/	4.82±0.4/
Pre/postreatment (mg/100ml)	4.0±0.4*	3.6±0.3*	3.8±0.4*
OxLDL	54.9±8.5/	53.1±8.3/	54.13±8.34/
Pre/postreatment (mg/100ml)	47.9±4.0*	44.8±4.4*	46.6±4.4*
PON 1	51.5±6.7/	51.4±9.0/	51.47±7.7/
Pre/postreatment (mg/100ml)	73.9±11.2*	67.4±10,0*	71.0±11.1*

Table 6. Comparinson of parameters before and after HRT in postmenopausal women p⬚.005Serum lipid parameters, MDA, oxLDL and PON1 levels in postmenopausal women before and after HRT [74].

Similarly the effect of DNA damage by oxidative stress has been evaluated. The 8-hydroxy-
deoxyguanosine (8-OHdG) is widely used for determination of DNA damage since it is ex-
cised from oxidative damaged DNA with endonuclease repair enzymes coded (OGG1).
After HT, mean blood 8-OHdG (DNA damage marker) level significantly decreased compared to those before HT, while urinary 8-OHdG level did not show any difference, this
without relation with S326C polymorphism [80].

Tibolone acts as an antioxidant upon increase the concentration of reduced sulfhydryl [81],
however, the exact mechanism has not been elucidated, but it could participate in a direct
mechanism, ie through the structure tibolone and its metabolites as it is similar to the structure of 17β-estradiol, considered as an antioxidant for its phenol ring, which can act neutralizing to free radical [82, 83]. Moreover tibolone reduces the concentration of
malondialdehyde compared to those who had no treatment [84, 85].

Although there are reports that indicate the antioxidant effect of HRT, there are also studies that indicate otherwise; in example in another study combined HRT led to decreased plasma total and LDL cholesterol, but did not affect oxidizability and oxidation of LDL. Circulating levels of antioxidant vitamins (beta-carotene, vitamin C, vitamin E/triglycerides) and total antioxidant capacity of plasma and lipid peroxidation, assessed by plasma TBARs, were not different from controls in postmenopausal women receiving HRT, which indicates that combined HRT modifies the blood lipid profile, however it does not appear to influence oxidative status [86]. Additionaly DNA damage, GPx activity and nitrite level as well as a decreased GSH level were observed after oral administrating of estrogens alone or combinated [87].

With respect to hot flushes, they have been associated to smaller level of total antioxidant activity in plasma, without differences in nitrite-nitrate concentrations, and after HRT there is an increase in total antioxidant activity level and nitrite-nitrate concentrations in menopausal women, with and without hot flushes [88].

On the other side estrogen increases vasodilatation and inhibits the response of blood vessels to injury and the development of atherosclerosis, it has been related to hormone´s effect on serum lipid concentration, that is reducing MDA and oxLDL levels and increasing activity of paroxonase PON1, which a calcium-dependent enzyme and in serum is exclusively located on HDL. PON is synthesized and secreted by liver and tightly binds to HDL subfractions that also contain apoA-1 and apoJ or clusterin and it has the capacity to protect LDL against oxidation [89, 90, 91].

7. Effect of nutrition and exercise on oxidative stress biomarkers

Adequate nutrition and physical exercise are two factors of health promotion and its effect on oxidative stress has been investigated in postmenopausal women, which has given controversial data. With respect to foods, they contain large amount of antioxidant molecules from there arouse the interest to check if their use can reduce the oxidative stress observed in postmenopause.

For example it was reported that the intake of fresh, greenhouse-grown vegetables for 3-wk did not induced changes in the urine concentrations of 8-isoprostane F2α, hexanoyl lysine, and serum high sensitivity C-reactive protein despite that plasma carotenoids were elevated in overweight postmenopausal women [92]. Something similar was established with a 2-month supplementation period with the Klamath algae extract, which is an extract naturally rich in powerful algal antioxidant molecules (AFA-phycocyanins) and concentrated with Klamath algae's natural neuromodulators (phenylethylamine as well as natural selective MAO-B inhibitors), whose effect was to increase in the plasma levels of carotenoids, tocopherols and retinol, however in this study oxidative stress was not measured [93, 94].

Otherwise is soy milk consumption for four weeks, which did not reduced markers of inflammation and oxidative stress as (tumor necrosis factor alpha [TNF-alpha], interleukin

[IL]-1beta, IL-6) and oxidative stress (superoxide dismutase [SOD], glutathione peroxidase [GPx], cyclooxygenase-2 [COX-2]) [95]. In this sense has also been found that de-alcoholised wine (DAW) with different polyphenol content by one month does not exert a protective activity towards oxidative DNA damage by comet assay, nor modifies significantly the gene expression profile of peripheral lymphocytes, whereas it shows blood-fluidifying actions, expressed as a significant decrease in blood viscosity. However, this effect does not correlate with the dosage of polyphenols from (DAW) [96].

In contrast, the intake for 8 weeks of soya protein diet and soya nut diet decreased MDA and increased the total antioxidant capacity in postmenopausal women with metabolic syndrome [97]. Extra-virgin olive oils (EVOO) is another example of food which reduces DNA damage by oxidative stress when is administrated for 8 weeks in healthy postmenopausal, which was evaluated by the comet assay in peripheral blood lymphocytes [98].

Nevertheless there are also reports of foods that raise oxidative stress as with American ginseng (AG) and wine. The AG supplementation at 500 mg every day for 4 months causes oxidative stress in postmenopausal women, due to reduced total antioxidant capacity, elevated plasma malondialdehyde and urine 8-hydroxydeoxyguanosine concentrations and increased erythrocyte antioxidant enzyme activity such as erythrocyte superoxide dismutase and GSH reductase [99].

Previous studies do not allow a conclusion on the effect of foods rich in antioxidant compounds, because were used different markers and administration time, and still more the age range of the postmenopausal differs considerably. This shows the need for more studies. Before shows the importance of further research with other foods with antioxidant properties known such as:

Vitamin C - Citrus fruits and their juices, berries, dark green vegetables (spinach, asparagus, green peppers, brussel sprouts, broccoli, watercress, other greens), red and yellow peppers, tomatoes and tomato juice, pineapple, cantaloupe, mangos, papaya and guava.

Vitamin E - Vegetable oils such as olive, soybean, corn, cottonseed and safflower, nuts and nut butters, seeds, whole grains, wheat, wheat germ, brown rice, oatmeal, soybeans, sweet potatoes, legumes (beans, lentils, split peas) and dark leafy green vegetables. Selenium - Brazil nuts, brewer's yeast, oatmeal, brown rice, chicken, eggs, dairy products, garlic, molasses, onions, salmon, seafood, tuna, wheat germ, whole grains and most vegetables.

Beta Carotene - Variety of dark orange, red, yellow and green vegetables and fruits such as broccoli, kale, spinach, sweet potatoes, carrots, red and yellow peppers, apricots, cantaloupe and mangos [100].

With respect to the lycopene, the following mechanism of the role of lycopene in chronic diseases has been mentioned by Agarwal and Rao [101] and Waliszewski and Blasco [102].

This highlights the importance of promote healthy lifestyles (balanced diet and moderate intensity exercise) in vulnerable populations, such as menopausal women, in order to prevent aging induced oxidative stress-related diseases.

Whit respect to the exercise has been reported that who practice yoga or tai chi (TC) for a year have same BMI, and even more no effects were shown on erythrocyte superoxide dismutase activity, plasma lipid peroxidation (TBARS) or total homocysteine concentrations, but the activity of erythrocyte glutathione peroxidase - an aerobic training-responsive enzyme - was higher in TC practitioners [103]. Although short term aerobic physical activity program (8 or 15 weeks) exhibited similar results, namely a reduction of serum glucose, LDL-cholesterol (LDL-C), plasma TBARS concentrations, decreasing of HOMA(IR) as well as an increased of total antioxidant status (TAS) of plasma and reduced glutathione (GSH) concentrations in red blood cells (RBC) increased significantly [104, 105].

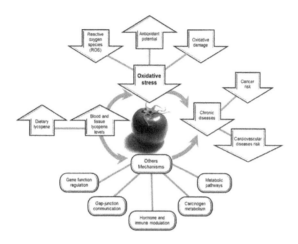

Figure 9. Lycopene and its mechanism in preventing of chronic diseases (Adapted from 101 and 102).

Regular physical training has been shown to upregulate antioxidant enzymatic systems, which may slow down the usual increase of oxidative stress in postmenopausal women, since it has been identified significant negative associations between oxidative stress and indices of physical fitness-activity (malondialdehyde, 8-iso-prostaglandin F2alpha, 8-hydroxy-2'-deoxyguanosine). Conversely, glutathione peroxidase is positively correlated with fitness level, furthermore mean arterial blood pressure (MABP) and cerebrovascular conductance (CVC) are directly associated with 8-hydroxy-2'-deoxyguanosine, nitrotyrosine and nitric oxide (NO) These findings demonstrate that, after menopause, fitness level and regular physical activity mediate against oxidative stress by maintaining antioxidant enzyme efficiency. Furthermore, these results suggest that oxidative stress and NO production modulate MABP and CVC [106].

Contrary to the above has also been reported, that the exercise does not modify the antioxidant status (although this is lower in metabolic healthy obese postmenopausal women than non-metabolic healthy obese postmenopausal women) and worse increases serum levels of thiobarbituric acid-reactive substances [107].

This highlights the importance of promote healthy lifestyles (balanced diet and moderate intensity exercise) in vulnerable populations, such as menopausal women, in order to prevent aging induced oxidative stress-related diseases.

8. Conclusion

The studies presented here were performed with different number of patients, methodologies and biomarkers, but most of them indicate that estrogen depletion induces oxidative stress and hormone replacement therapy seems to reduce it. With respect to the modification of biomarkers of oxidative stress damage by food and exercise needs more research because so far no conclusive data have been obtained.

Author details

Claudia Camelia Calzada Mendoza[1*] and Carlos Alberto Jiménez Zamarripa[2]

*Address all correspondence to: cccalzadam@yahoo.com.mx

1 Section of Post graduate Studies and Research of Escuela Superior de Medicina- Instituto Politécnico Nacional. Street Salvador Díaz Mirón S/N, Colony Casco de Santo Tomás, Delegation Miguel Hidalgo, C.P. 11340, México D.F.

2 Hospital psychiatry "Dr. Samuel Ramírez Moreno"-psychiatric careservices- Secretaria de Salud, highway México-Puebla Km 5.5, Colony Santa Catarina, Tláhuac, C.P. 13100, México D.F.

References

[1] Reddish, S. (2011). Menopausal transition- assessment in general practice. *Australian Family Physician*, 40(5), 266-72.

[2] Burger, H. G., Hale, G. E., Robertson, D. M., & Dennerstein, L. (2007). Review of hormonal changes during the menopausal transition: focus on findings from the Melbourne Women's Midlife Health Project. *Human Reproduction Update*, 13(6), 559-65.

[3] de Cetina , Canto T. (2006). Los síntomas en la menopausia. *Revista de Endocrinología y Nutrición*, 14(3), 141-148.

[4] Guthrie, J. R., Dennerstein, L., Hopper, J. L., & Burger, H. G. (1996). Hot flushes, menstrual status and hormone levels in a population-based sample of midlife women. *Obstetrics and Gynecology*, 88, 437-441.

[5] Zárate, A., Hernández-Valencia, M., Austria, E., Saucedo, R., & Hernánde, M. (2011). Diagnosis of premature menopause measuring circulating anti-Müllerian hormone. *Ginecologia Obstetricia Mexicana*, 79(5), 303-7.

[6] Lockwood, C.J. (2011). Mechanisms of normal and abnormal endometrial bleeding. *Menopause*, 18(4), 408-411.

[7] Goldstein, S.R. (2011). Significance of incidentally thick endometrial echo on transvaginal ultrasound in postmenopausal women. *Menopause*, 18(4), 434-436.

[8] Chapple, C. R., Wein, A. J., Abrams, P., Dmochowski, R. R., Giuliano, F., Kaplan, S. A., Mc Vary, K. T., & Roehrborn, C. G. (2008). Lower urinary tract symptoms revisited: a broader clinical perspective. *European Urology*, 54(3), 563-569.

[9] Labrie, F., Archer, D., Bouchard, C., Fortier, M., Cusan, L., Gomez, J. L., Girard, G., Baron, M., Ayotte, N., Moreau, M., Dubé, R., Côté, I., Labrie, C., Lavoie, L., Berger, L., Gilbert, L., Martel, C., & Balser, J. (2009). Intravaginaldehydroepiandrosterone (Prasterone), a physiological and highly efficient treatment of vaginal atrophy. *Menopause*, 16(5), 907-22.

[10] Goepel, M., Kirschner-Hermanns, R., Welz-Barth, A., Steinwachs, K. C., & Rübben, H. (2010). Urinary incontinence in the elderly: part 3 of a series of articles on incontinence. *Dtsch ArzteblInt*, 107(30), 531-6.

[11] Morrison, L. A., Sievert, L. L., Brown, D. E., Rahberg, N., & Reza, A. (2010). Relationships between menstrual and menopausal attitudes and associated demographic and health characteristics: the Hilo Women's Health Study Women Health,, 50(5), 397-413.

[12] Pachman, D. R., Jones, J. M., & Loprinzi, C. L. (2010). Management of menopause-associated vasomotor symptoms: Current treatment options, challenges and future directions. *International Journal Womens Health*, 9(2), 231-35.

[13] Archer, D. F., Sturdee, D. W., Baber, R., de Villiers, T. J., Pines, A., Freedman, R. R., Gompel, A., Hickey, M., Hunter, M. S., Lobo, R. A., Lumsden, M. A., Mac, Lennan. A. H., Maki, P., Palacios, S., Shah, D., Villaseca, P., & Warren, M. (2011). Menopausal hot flushes and night sweats: where are we now ? *Climacteric*, 14(5), 155-128.

[14] Frutos, R., Rodríguez, S., Miralles-Jorda, L., & Machuca, G. (2002). Oral manifestations and dental treatment in menopause. *Medicine Oral*, 7(1), 26-30.

[15] Safoury, O., Rashid, L., & Ibrahim, M. A. (2010). Study of androgen and estrogen receptors alpha, beta in skin tags. *Indian Journal Dermatology*, 55(1), 20-4.

[16] Maltais, M.L., Desroches, J., & Dionne, I.J. (2009). Changes in muscle mass and strength after menopause. *Journal Musculo skeletand Neuronal Interactions*, 9(4), 186-197.

[17] Sosa, M., Saavedra, P., Jódar, E., Lozano, T. C., Quesada, J. M., Torrijos, A., Pérez, C. R., Nogués, X., Díaz, C. M., Moro, M. J., Gómez, C., Mosquera, J., Alegre, J., Olmos, J.,

Muñoz, T. M., Guañabens, N., Del Pino, J., & Hawkins, F. (2009). Bone mineral density and risk of fractures in aging, obese post-menopausal women with type 2 diabetes. The GIUMO Study. *Aging Clinical Experimental Research*, 21(1), 27-32.

[18] Koroglu, B. K., Kiris, F., Ersoy, I. H., Sutcu, R., Yildiz, M., Aksu, O., Ermis, F., Ersoy, S., & Tamer, M. N. (2011). Relation of leptin, adiponectin and insulin resistance to bone mineral density in type 2 diabetic postmenopausal women. *Endokrynologia Polska*, 62(5), 429-35.

[19] Nelson, H. D., Helfand, M., Woolf, S. H., & Allan, J. D. (2002). Screening for Postmenopausal Osteoporosis: A Review of the Evidence for the U.S. Preventive Services Task Force. *Annals of Internal Medicine*, 137(6), 529-41.

[20] del Rio, B. L., Romera, B. M., Pavia, S. J., Setoain, Q. J., Serra, M. L., Garces, R. P., & Lafuente, N. C. (1992). Bone mineral density in two different socio-economic population groups. *Bone and Mineral*, 18(2), 159-68.

[21] Suresh, S., Kumar, T. S., Saraswathy, P. K., & Pani, S. K. H. (2010). Periodontitis and bone mineral density among pre and post menopausal women: A comparative study. *Journal of Indian Society of Periodontology*, 14(1), 30-34.

[22] Pavón, P., Alameda, H. C., & Olivar, R. J. (2006). Obesidad y menopausia. *Nutrición Hospitalaria*, 21, 633-637.

[23] Zillikens, M. C., Uitterlinden, A. G., van Leeuwen, J. P., Berends, A. L., Henneman, P., van Dijk, K. W., Oostra, B. A., van Duijn, C. M., Pols, H. A., & Rivadeneira, F. (2010). The role of body mass index, insulin, and adiponectin in the relation between fat distribution and bone mineral density. *Calcified Tissue International*, 86(2), 116-25.

[24] Cho, E. J., Min, Y. J., Oh, M. S., Kwon, J. E., Kim, J. E., Lee, W. S., Lee, K. J., Kim, S. W., Kim, T. H., Kim, M. A., Kim, C. J., & Ryu, W. S. (2011). Effects of the transition from premenopause to postmenopause on lipids and lipoproteins: quantification and related parameters. *Korean Journal Internal Medicine*, 26(1), 47-53.

[25] Kallikazaros, I., Tsioufis, C., Zambaras, P., Skiadas, I., Toutouza, M., Tousoulis, D., Stefanadis, C., & Toutouzas, P. (2008). Estrogen-induced improvement in coronary flow responses during atrial pacing in relation to endothelin-1 levels in postmenopausal women without coronary disease. *Vascular Health Risk and Management*, 4(3), 705-14.

[26] Zavala, A. G., & Grover, F. (2007). Perfil lipídico y cambio en sensibilidad a la insulina en posmenopáusicas. *Revista Médica deChile* 135:, 613-619.

[27] Garay, S., & Arellano, S. (2006). Diabetes mellitus (DM), menopausia y reemplazo hormonal. *Revista de Endocrinología y Nutrición*, 14(3), 191-195.

[28] Whitcroft, S., & Herriot, A. (2011). Insulin resistance and management of the menopause: a clinical hypothesis in practice. *Menopause International*, 17(1), 24-8.

[29] Ryan, J., Scali, J., Carrière, I., Peres, K., Rouaud, O., Scarabin, P. Y., Ritchie, K., & An-
      celin, M. L. (2011). Estrogen receptor alpha gene variants and major depressive epi-
      sodes. *Jornal of Affective Disorders*, 136(3), 1222-6.

[30] Parry, B.L. (2010). Optimal management of perimenopausal depression. *International
      Joornal of Women´s Health*, 9(2), 143-151.

[31] Zender, R., & Olshansky, E. (2009). Women's mental health: depression and anxiety.
      The Nursing Clinics of North America, 44(3), 250-256.

[32] Colangelo, L. A., Craft, L. L., Ouyang, P., Liu, K., Schreiner, P. J., Michos, E. D., &
      Gapstur, S. M. (2012). Association of sex hormones and sex hormone-binding globu-
      lin with depressive symptoms in postmenopausal women: the Multiethnic Study of
      Atherosclerosis. *Menopause.* Mar 12.PMID: 22415566 [PubMed- as supplied by pub-
      lisher] PMCID: PMC3376685.

[33] Escalante, G. C., Quesada, M. S., & Zeledón, S. F. (2009). Oxidative Profile of the
      Menopausal Woman: Estrogens´ Rol in the Prevention and Treatment of Diseases.
      Acta Médica Costarricense, 51(4), 206-2011.

[34] Ghosh, D., Griswold, J., Erman, M., & Pangborn, W. (2009). Structural basis for an-
      drogen specificity and oestrogen synthesis in human aromatase. *Nature*, 457(7226),
      219-223.

[35] Liehr J.G. (2000). Is Estradiol a Genotoxic Mutagenic Carcinogen?. *Endocrine Reviews*,
      21(1), 40-54.

[36] Lippert, T. H., Seeger, H., & Mueck, A. O. (2000). The impact of endogenous estradiol
      metabolites on carcinogenesis. *Steroids*, 65, 357-369.

[37] Pulvirenti, D., Signorelli, S., Sciacchitano, S., Di Pino, L., Tsami, A., Ignaccolo, L., &
      Neri, S. (2007). Hyperhomocysteinemia, oxidative stress, endothelial dysfunction in
      postmenopausal women. *La Clinica Terapeutica*, 158(3), 213-7.

[38] Nascimento, G. R., Barros, Y. V., Wells, A. K., & Khalil, R. A. (2009). Research into
      Specific Modulators of Vascular Sex Hormone Receptors in the Management of Post-
      menopausal Cardiovascular Disease. *Current Hypertension Reviews*, 5(4), 283-306.

[39] Ratiani, L., Parkosadze, G., Cheishvili, M., Ormotsadze, G., Sulakvelidze, M., & Sani-
      kidze, T. (2012). Role of estrogens in pathogenesis of age-related disease in women of
      menopausal age. *Georgian Medical News*, 203, 11-6.

[40] Liu, S. B., Han, J., Zhang, N., Tian, Z., Li, X. B., & Zhao, M. G. (2011). Neuroprotective
      effects of oestrogen against oxidative toxicity through activation of G-protein-cou-
      pled receptor 30 receptor. *Clinical and Experimental Pharmacology and Physiology*, 38(9),
      577-85.

[41] Johnson, D. K., Mc Millin, G. A., Bishop, M. L., Fody, E. P., & Schoeff, L. E. (2010).
      Enzymes. *In: Clinical Chemistry Techniques, Principles, Correlations* 6th ed. Philadel-
      phia: Lippincott Williams and Wilkins., 300.

[42] Vasudevan, D. M., & Sreekumari, S. (2005). Iso-enzymes and clinical enzymology. *In: Vasudevan DM, Sreekumari S, editors. Textbook of Biochemistry (for medical students).* 4th ed. New Delhi: Jaypeebrothers medical publishers (P) Ltd., 57.

[43] Abdul, R. O. F., Al, S. G. A., & Bushra, H. Z. B. H. (2010). Serum γ-glutamyltransferase as Oxidative Stress Marker in Pre-and Postmenopausal Iraqi Women. *Oman Medical Journal,* 25(4), 286-8.

[44] Zitňanová, I., Rakovan, M., Paduchová, Z., Dvořáková, M., Andrezálová, L., Muchová, J., Simko, M., Waczulíková, I., & Duračková, Z. (2011). Menopause Oxidative stress in women with perimenopausal symptoms. *Menopause,* 18(11), 1249-55.

[45] Mittal, P. C., & Kant, R. (2009). Correlation of increased oxidative stress to body weight in disease-free post menopausal women. *Clinical Biochemistry* 42(10-11) 1007-11.

[46] Pansini, F., Cervellati, C., Guariento, A., Stacchini, M. A., Castaldini, C., Bernardi, A., Pascale, G., Bonaccorsi, G., Patella, A., Bagni, B., Mollica, G., & Bergamini, C. M. (2008). Oxidative stress, body fat composition, and endocrine status in pre- and postmenopausal women. *Menopause,* 15(1), 112-8.

[47] Baltacioğlu, E., Akalin, F. A., Alver, A., Balaban, F., Unsal, M., & Karabulut, E. (2006). Total antioxidant capacity and superoxide dismutase activity levels in serum and gingival crevicular fluid in post-menopausal women with chronic periodontitis. *Journal of Clininal Periodontology,* 33(6), 385-92.

[48] Signorelli, S. S., Neri, S., Sciacchitano, S., Pino, L. D., Costa, M. P., Marchese, G., Celotta, G., Cassibba, N., Pennisi, G., & Caschetto, S. (2006). Behaviour of some indicators of oxidative stress in postmenopausal and fertile women. *Maturitas,* 53(1), 77-82.

[49] Sánchez, R.M.A., Zacarías, F.M., Arronte, R.A., Correa, M.E., & Mendoza, Núñez V.M. (2012). Menopause as risk factor for oxidative stress. *Menopause,* 19(3), 361-7.

[50] Victorino, V. J., Panis, C., Campos, F. C., Cayres, R. C., Colado-Simão, A. N., Oliveira, S. R., Herrera, A. C., Cecchini, A. L., & Cecchini, R. (2012). Decreased oxidant profile and increased antioxidant capacity in naturally postmenopausal women. *Studio controversial.* Age (Dordrecht Netherlands) May 28. PMID: 22645022.

[51] Cervellati, C., Pansini, F. S., Bonaccorsi, G., Bergamini, C. M., Patella, A., Casali, F., Fantini, G. F., Pascale, G., Castaldini, C., Ferrazzini, S., Ridolfi, F., Cervellati, G., Cremonini, E., Christodoulou, P., & Bagni, B. (2011). Estradiol levels and oxidative balance in a population of pre-, peri-, and post-menopausal women. *Gynecological Endocrinology,* 27(12), 1028-32.

[52] Michel, T. M., Pülschen, D., & Thome, J. (2012). The role of oxidative stress in depressive disorder. *Current Pharmaceutical Design* Jun 6 PMID: 22681168.

[53] Talarowska, M., Gałecki, P., Maes, M., Bobińska, K., & Kowalczyk, E. (2012). Total antioxidant status correlates with cognitive impairment in patients with recurrent depressive disorder. *Neurochemical Research,* 37(8), 1761-7.

[54] Pinto, A. R., Calzada, M. C. C., Campos, L. M. G., & Guerra, A. C. (2012). Effect of Chronic Administration of Estradiol, Progesterone, and Tibolone on the Expression and Phosphorylation of Glycogen Synthase Kinase-3b and the Microtubule-Associated Protein Tau in the Hippocampus and Cerebellum of Female Rat. *Journal of Neuroscience Research*, 90, 878-886.

[55] Wagner, J. A., Tennen, H., Finan, P. H., White, W. B., Burg, M. M., & Ghuman, N. (2011). Lifetime History of Depression, Type 2 Diabetes, and Endothelial Reactivity to Acute Stress in Postmenopausal Women. *International Jornal of Behavioral Medicine* Oct 2. PMID: 21964983.

[56] Yilmaz, N., & Eren, E. (2009). Homocysteine oxidative stress and relation to bone mineral density in post-menopausal osteoporosis. *Aging Clinical and Experimental Research* 21(4-5)353-7.

[57] Sendur, O. F., Turan, Y., Tastaban, E., & Serter, M. (2009). Antioxidant status in patients with osteoporosis: a controlled study. *Joint Bone Spine*, 76(5), 514-8.

[58] Mlakar, S. J., Osredkar, J., Prezelj, J., & Marc, J. (2012). Antioxidant enzymes GSR, SOD1, SOD2, and CAT gene variants and bone mineral density values in postmenopausal women: a genetic association analysis. *Menopause*, 19(3), 368-76.

[59] Vassalle, C., Mercuri, A., & Maffei, S. (2009). Oxidative status and cardiovascular risk in women: Keeping pink at heart. *World Journal of Cardiology*, 1(1), 26-30.

[60] Crist, B. L., Alekel, D. L., Ritland, L. M., Hanson, L. N., Genschel, U., & Reddy, M. B. (2009). Association of oxidative stress, iron, and centralized fat mass in healthy postmenopausal women. *Journal Women´s Health (Larchmt)*, 18(6), 795-801.

[61] Kumawat, M., Sharma, T. K., Singh, N., Ghalaut, V. S., Vardey, S. K., Sinha, M., & Kaushik, G. G. (2012). Study of changes in antioxidant enzymes status in diabetic post menopausal group of women suffering from cardiovascular complications. *Clinical Laboratory*58(3-4) 203-7.

[62] Moreau, K. L., De Paulis, A. R., Gavin, K. M., & Seals, D. R. (2007). Oxidative stress contributes to chronic leg vasoconstriction in estrogen-deficient postmenopausal women. *Journal of Applied Physiology*, 102, 890-895.

[63] Signorelli, S. S., Neri, S., Sciacchitano, S., Di Pino, L., Costa, M. P., Pennisi, G., Ierna, D., & Caschetto, S. (2001). Duration of menopause and behavior of malondialdehyde, lipids, lipoproteins and carotid wall artery intima-media thickness. *Maturitas*, 39(1), 39-42.

[64] Smith, C. C., Vedder, L. C., Nelson, A. R., Bredemann, T. M., & Mc Mahon, L. L. (2010). Duration of estrogen deprivation, not chronological age, prevents estrogen's ability to enhance hippocampal synaptic physiology. *Proceedings of National Academy of Science of United States of America*, 107(45), 19543-19548.

[65] Griffiths, F., & Convery, B. (1995). Women's use of hormone replacement therapy for relief of menopausal symptoms, for prevention of osteoporosis, and after hysterectomy. *British Journal of General Practice*, 45(396), 355-358.

[66] Modelska, K., & Cummings, S. (2002). Tibolone for postmenopausal women: systematic review of randomized trials. *The Journal of Clinical Endocrinology and Metabolism*, 87(1), 16-23.

[67] Huang, K. E., & Baber, R. (2010). Updated clinical recommendations for the use of tibolone in Asian women Climateric,13:, 317-327.

[68] Ke, R. W., Todd, Pace. D., & Ahokas, R. A. (2003). Effect of short-term hormone therapy on oxidative stress and endothelial function in African American and Caucasian postmenopausal women. *Fertiland Steril*, 79(5), 1118-22.

[69] Chang, S. P., Yang, W. S., Lee, S. K., Min, W. K., Park, J. S., & Kim, S. B. (2003). Effects of hormonal replacement therapy on oxidative stress and total antioxidant capacity in postmenopausal hemodialysis patients. *Renal Failure 2002*, 24(1), 49-57.

[70] Darabi, M., Ani, M., Movahedian, A., Zarean, E., Panjehpour, M., & Rabbani, M. (2010). Effect of hormone replacement therapy on total serum anti-oxidant potential and oxidized LDL/ß2-glycoprotein I complexes in postmenopausal women. *Endocrine Journal*, 57(12), 1029-1034.

[71] Polac, I., Borowiecka, M., Wilamowska, A., & Nowak, P. (2011). Oxidative stress measured by carbonyl groups level in postmenopausal women after oral and transdermal hormone therapy. *Journal ofObstetric andGynaecology Research 2012 Apr 30.* doi:j. , 1447-0756.

[72] Gökkuşu, C., Özbek, Z., & Tata, G. (2012). Hormone replacement therapy: relation to homocysteine and prooxidant-antioxidant status in healthy postmenopausal women Archives of Gynecology and Obstetretics,, 285(3), 733-9.

[73] Telci, A., Cakatay, U., Akhan, S. E., Bilgin, M. E., Turfanda, A., & Sivas, A. (2002). Postmenopausal hormone replacement therapy use decreases oxidative protein damage. *Gynecolicand Obstetric Investigation*, 54(2), 88-93.

[74] Topçuoʻlu, A., Uzun, H., Aydin, S., Kahraman, N., Vehid, S., Zeybek, G., & Topçuoʻlu, D. (2005). The Effect of Hormone Replaceent Therapy on Oxidized Low Density Lipoprotein Levels and Paroxonase Activity in Postmenopausal women. *Tohoku Journal of Experimental Medicine*, 205(1), 79-86.

[75] Vural, P., Akgül, C., & Canbaz, M. (2005). Effects of menopause and tibolone on antioxidants in postmenopausal women. *Annals of Clinical Biochemistry*, 42(3), 220-3.

[76] Bednarek, T. G., Tworowska, U., Jedrychowska, I., Radomska, B., Tupikowski, K., Bidzinska, S. B., & Milewicz, A. (2006). Effects of oestradiol and oestroprogestin on erythrocyte antioxidative enzyme system activity in postmenopausal women. *ClinicalEndocrinolology (Oxf)*, 64(4), 463-8.

[77] Maffei, S., Mercuri, A., Prontera, C., Zucchelli, G. C., & Vassalle, C. (2006). Vasoactive biomarkers and oxidative stress in healthy recently postmenopausal women treated with hormone replacement therapy. *Climacteric x*, 9(6), 452-8.

[78] Castanho, V. S., Gidlund, M., Nakamura, R., & de Faria, E. C. (2011). Post-menopausal hormone therapy reduces autoantibodies to oxidized apolipoprotein B100. *Gynecological Endocrinology*, 27(10), 800-6.

[79] Gökkuşu, C., Tata, G., Ademoğlu, E., & Tamer, S. (2010). The benefits of hormone replacement therapy on plasma and platelet antioxidant status and fatty acid composition in healthy postmenopausal women. *Platelets*, 21(6), 439-44, PMID: 20459351.

[80] Kim, H., Ku, S. Y., Kang, J. W., Kim, H., Kim, Y. D., Kim, S. H., Choi, Y. M., Kim, J. G., & Moon, S. Y. (2011). The 8-hydroxydeoxyguanosine concentrations according to hormone therapy and S326C polymorphism of OGG1 gene in postmenopausal women. *Molecular Genetics and Metabolism*, 104(4), 644-7.

[81] Dlugosz, A., Roszkowska, A., & Zimmer, M. (2009). Oestradiol protects against the harmful effects of fluoride more by increasing thiol group levels than scavenging hydroxyl radicals. *Basic and ClinicalPharmacolology and Toxicology*, 105(6), 366-373.

[82] Wen, Y., Doyle, M., Cooke, T., & Feely, J. (2000). Effect of menopause on low density lipoprotein oxidation: is estrogen an important determinant? *Maturitas*, 34, 233-238.

[83] Subbiah, M. T., Kessel, B., Agrawal, M., Rajan, R., Abplanalp, W., & Rymaszewski, Z. (1993). Antioxidant potential of specific estrogens on lipid peroxidation. *The Journal of Clinical Endocrinology and Metabolism*, 77, 1095-1097.

[84] Kassia, E., Dalamagaa, H., Hroussalasa, G., Kazanisa, K., Merantz, A., & Zacharic, E. J. (2010). Giamarellos-Bourbouli, A. Dionyssiou-Asterioua. Adipocyte factors, high-sensitive C-reactive protein levels and lipoxidative stress products in overweight postmenopausal women with normal and impaired OGTT. *Maturitas*, 67, 72-77.

[85] Arteaga, E., Rojas, A., Villaseca, P., Bianchi, M., Arteaga, A., & Duran, D. (1998). In vitro effect of oestradiol, progesterone, testosterone, and of combined estradiol/progestins on low density lipoprotein (LDL) oxidation in postmenopausal women. *Menopause*, 5(1), 16-23.

[86] Bureau, I., Laporte, F., Favier, M., Faure, H., Fields, M., Favier, A. E., & Roussel, A. M. (2002). No antioxidant effect of combined HRT on LDL oxidizability and oxidative stress biomarkers in treated post-menopausal women. *Journal of the American College of Nutrition*, 21(4), 333-8.

[87] Akcay, T., Dincer, Y., Saygili, E. I., Seyisoğlu, H., & Ertunğalp, E. (2010). Assessment of DNA nucleo base oxidation and antioxidant defense in postmenopausal women under hormone replacement therapy. *Indian Journal of Medical Sciences*, 64(1), 17-25.

[88] Leal, H.M., Abellán, A.J., Carbonell, M.L.F., Díaz, F.J., García, S.F.A., & Martínez, S.J.M. Influence of the presence of hot flashes during menopause on the metabolism

of nitric oxide. Effects of hormonal replacement treatment. *Medicina Clínica (Barc) 200022*, 114(2), 41-5.

[89] Itabe, H. (2003). Oxidized low-density lipoprapteins: What is understood and what remains to be clarified. *Biological and Pharmaceutical Bulletin*, 26(1), 1-9.

[90] Mackness, M. I., Mackness, B., Durrington, P. N., Connelly, P. W., & Hegele, R. A. (1996). Paraoxonase: biochemistry, genetics and relationship to plasma lipoproteins: Current Opinionin Lipidology,, 7, 69-76.

[91] Aviram, M., Rosenblat, M., Bisgaier, C. L., Newton, R. S., Primo, P. S. L., & La Du, B. N. (1998). Paraoxonaseinhibitis high-density lipoprotein oxidation and preserves its function. *The Journal of ClinicalInvestigaction*, 101, 1581-1590.

[92] Crane, T. E., Kubota, C., West, J. L., Kroggel, M. A., Wertheim, B. C., & Thomson, C. A. (2011). Increasing the vegetable intake dose is associated with a rise in plasma carotenoids without modifying oxidative stress or inflammation in overweight or obese postmenopausal women. *Journal of Nutrition*, 141(10), 1827-33.

[93] Scoglio, S., Benedetti, S., Canino, C., Santagni, S., Rattighieri, E., Chierchia, E., Canestrari, F., & Genazzani, A. D. (2009). Effect of a 2-month treatment with Klamin, a Klamath algae extract, on the general well-being, antioxidant profile and oxidative status of postmenopausal women. *GynecologicalEndocrinology*, 25(4), 235-40.

[94] Miquel, J., Ramírez, B. A., Ramírez, B. J., & Diaz, A. J. (2006). Menopause: A review on the role of oxygen stress and favorable effects of dietary antioxidants. *Archives of Gerontology and Geriatrics*, 42, 289-306.

[95] Beavers, K. M., Serra, M. C., Beavers, D. P., Cooke, M. B., & Willoughby, D. S. (2009). Soymilk supplementation does not alter plasma markers of inflammation and oxidative stress in postmenopausal women. *Nutrition Research*, 29(9), 616-22.

[96] Giovannelli, L., Pitozzi, V., Luceri, C., Giannini, L., Toti, S., Salvini, S., Sera, F., Souquet, J. M., Cheynier, V., Sofi, F., Mannini, L., Gori, A. M., Abbate, R., Palli, D., & Dolara, P. (2011). Effects of de-alcoholised wines with different polyphenol content on DNA oxidative damage, gene expression of peripheral lymphocytes, and haemorheology: an intervention study in post-menopausal women. *European Journal of Nutrition*, 50(1), 19-29.

[97] Azadbakht, L., Kimiagar, M., Mehrabi, Y., Esmaillzadeh, A., Hu, F. B., & Willett, W. C. (2007). Dietary soya intake alters plasma antioxidant status and lipid peroxidation in postmenopausal women with the metabolic syndrome. *British Journal of Nutrition*, 98(4), 807-13.

[98] Salvini, S., Sera, F., Caruso, D., Giovannelli, L., Visioli, F., Saieva, C., Masala, G., Ceroti, M., Giovacchini, V., Pitozzi, V., Galli, C., Romani, A., Mulinacci, N., Bortolomeazzi, R., Dolara, P., & Palli, D. (2006). Daily consumption of a high-phenol extra-virgin olive oil reduces oxidative DNA damage in postmenopausal women. *British Journal of Nutrition*, 95(4), 742-51.

[99] Dickman, J. R., Koenig, R. T., & Ji, L. L. (2009). American ginseng supplementation induces an oxidative stress in postmenopausal women. *Journal of the American of College of Nutrition*, 28(2), 219-28.

[100] Garcia P.M.C. (2008). Antioxidantes en la dietamediterranea. *Nutrición Clínica en Medicina*, 2(3), 129-140.

[101] Agarwal, S., & Rao, A. V. (2000). Tomato lycopene and its role in human health and chronic diseases. *Canadian Medical Association or its licensors*, 163(6), 739-44.

[102] Waliszewski, K. N., & Blasco, G. (2010). Propiedadesnutraceúticasdellicopeno. *Salud Publica Mexicana*, 52, 254-265.

[103] Palasuwan, A., Margaritis, I., Soogarun, S., & Rousseau, A. S. (2011). Dietary intakes and antioxidant status in mind-body exercising pre- and postmenopausal women. *The Journal of Nutrition Health and Aging*, 15(7), 577-84.

[104] Karolkiewicz, J., Michalak, E., Pospieszna, B., Deskur-Smielecka, E., Nowak, A., & Pilaczyńska-Szcześniak, Ł. (2009). Response of oxidative stress markers and antioxidant parameters to an 8week aerobic physical activity program in healthy, postmenopausal women. *Archives of Gerontology and Geriatrics* 49(1)e67-71.

[105] Schmitz, K. H., Warren, M., Rundle, A. G., Williams, N. I., Gross, M. D., & Kurzer, M. S. (2008). Exercise effect on oxidative stress is independent of change in estrogen metabolism. *Cancer Epidemiology Biomarkers and Prevention*, 17(1), 220-3.

[106] Pialoux, V., Brown, A. D., Leigh, R., Friedenreich, C. M., & Poulin, M. J. (2009). Effect of cardiorespiratory fitness on vascular regulation and oxidative stress in postmenopausal women. *Hypertension*, 54(5), 1014-20.

[107] Lwow, F., Dunajska, K., Milewicz, A., Jedrzejuk, D., Kik, K., & Szmigiero, L. (2011). Effect of moderate-intensity exercise on oxidative stress indices in metabolically healthy obese and metabolically unhealthy obese phenotypes in postmenopausal women: a pilot study. *Menopause*, 18(6), 646-53.

Permissions

The contributors of this book come from diverse backgrounds, making this book a truly international effort. This book will bring forth new frontiers with its revolutionizing research information and detailed analysis of the nascent developments around the world.

We would like to thank Prof. Dr. José Antonio Morales-González, for lending his expertise to make the book truly unique. He has played a crucial role in the development of this book. Without his invaluable contribution this book wouldn't have been possible. He has made vital efforts to compile up to date information on the varied aspects of this subject to make this book a valuable addition to the collection of many professionals and students.

This book was conceptualized with the vision of imparting up-to-date information and advanced data in this field. To ensure the same, a matchless editorial board was set up. Every individual on the board went through rigorous rounds of assessment to prove their worth. After which they invested a large part of their time researching and compiling the most relevant data for our readers. Conferences and sessions were held from time to time between the editorial board and the contributing authors to present the data in the most comprehensible form. The editorial team has worked tirelessly to provide valuable and valid information to help people across the globe.

Every chapter published in this book has been scrutinized by our experts. Their significance has been extensively debated. The topics covered herein carry significant findings which will fuel the growth of the discipline. They may even be implemented as practical applications or may be referred to as a beginning point for another development. Chapters in this book were first published by InTech; hereby published with permission under the Creative Commons Attribution License or equivalent.

The editorial board has been involved in producing this book since its inception. They have spent rigorous hours researching and exploring the diverse topics which have resulted in the successful publishing of this book. They have passed on their knowledge of decades through this book. To expedite this challenging task, the publisher supported the team at every step. A small team of assistant editors was also appointed to further simplify the editing procedure and attain best results for the readers.

Our editorial team has been hand-picked from every corner of the world. Their multi-ethnicity adds dynamic inputs to the discussions which result in innovative outcomes. These outcomes are then further discussed with the researchers and contributors who give their valuable feedback and opinion regarding the same. The feedback is then collaborated with the researches and they are edited in a comprehensive manner to aid the understanding of the subject.

Apart from the editorial board, the designing team has also invested a significant amount of their time in understanding the subject and creating the most relevant covers. They scrutinized every image to scout for the most suitable representation of the subject and create an appropriate cover for the book.

The publishing team has been involved in this book since its early stages. They were actively engaged in every process, be it collecting the data, connecting with the contributors or procuring relevant information. The team has been an ardent support to the editorial, designing and production team. Their endless efforts to recruit the best for this project, has resulted in the accomplishment of this book. They are a veteran in the field of academics and their pool of knowledge is as vast as their experience in printing. Their expertise and guidance has proved useful at every step. Their uncompromising quality standards have made this book an exceptional effort. Their encouragement from time to time has been an inspiration for everyone.

The publisher and the editorial board hope that this book will prove to be a valuable piece of knowledge for researchers, students, practitioners and scholars across the globe.

List of Contributors

Alejandro Chehue Romero, Elena G. Olvera Hernández, Telma Flores Cerón and Angelina Álvarez Chávez
Autonomous University of Hidalgo State, Mexico

María de Lourdes Segura-Valdez, Lourdes T. Agredano-Moreno, Alma Zamora-Cura, Reyna Lara-Martínez and Luis F. Jiménez-García
Laboratory of Cell Nanobiology and Electron Microscopy Laboratory (Tlahuizcalpan), Department of Cell Biology, Faculty of Sciences, National Autonomous University of Mexico (UNAM), Circuito Exterior, Ciudad Universitaria, Coyoacán 04510, México D.F, México

Tomás Nepomuceno-Mejía, Rogelio Fragoso-Soriano and Georgina Álvarez-Fernández
Visiting from the Department of Biochemistry, Faculty of Medicine, UNAM, México

Jorge Alberto Mendoza Pérez
Department of Environmental Systems Engineering at National School of Biological Sciences- National Polytechnic Institute, Mexico, D.F., Mexico

Tomás Alejandro Fregoso Aguilar
Department of Physiology at National School of Biological Sciences-National Polytechnic Institute, Mexico, D.F., Mexico

Antonio Cilla, Amparo Alegría, Reyes Barberá and María Jesús Lagarda
Nutrition and Food Science Area, Faculty of Pharmacy, University of Valencia, Avda. Vicente Andrés Estellés s/n, 46100 - Burjassot, Valencia, Spain

Mirandeli Bautista Ávila, Juan Antonio Gayosso de Lúcio, Nancy Vargas Mendoza, Claudia Velázquez González, Minarda De la O Arciniega and Georgina Almaguer Vargas
Universidad Autónoma del Estado de Hidalgo, Mexico

César Esquivel-Chirino
Facultad de Odontología, Universidad Nacional Autónoma de México, México
Departamento de Medicina y Toxicología Ambiental, Instituto de Investigaciones Biomédicas, Universidad Nacional Autónoma de México, México
Departamento de Bioquímica. (INNCMSZ) Instituto Nacional de Ciencias Médicas y Nutrición "Salvador Zubirán" México D.F., México
Facultad de Odontología, Universidad Intercontinental, México

Jose Luis Ventura-Gallegos and Alejandro Zentella-Dehesa
Departamento de Medicina y Toxicología Ambiental, Instituto de Investigaciones Biomédicas, Universidad Nacional Autónoma de México, México
Departamento de Bioquímica. (INNCMSZ) Instituto Nacional de Ciencias Médicas y Nutrición "Salvador Zubirán" México D.F., México

Luis Enrique Hernández-Mora
Facultad de Odontología, Universidad Nacional Autónoma de México, México

José Antonio Morales-González
Instituto de Ciencias de la Salud, Universidad Autónoma del Estado de Hidalgo (UAEH), México

Delina Montes Sánchez
Programa de Genómica Funcional de Procariotes, Centro de Ciencias Genómicas, Universidad Nacional Autónoma de México, Campus Morelos, México
Departamento de Medicina y Toxicología Ambiental, Instituto de Investigaciones Biomédicas, Universidad Nacional Autónoma de México, México
Departamento de Bioquímica. (INNCMSZ) Instituto Nacional de Ciencias Médicas y Nutrición "Salvador Zubirán" México D.F., México

Jaime Esquivel-Soto
Facultad de Odontología, Universidad Nacional Autónoma de México, México
Facultad de Odontología, Universidad Intercontinental, México

Eduardo Madrigal-Santillán, Sandra Cruz-Jaime, María del Carmen Valadez-Vega, Karla Guadalupe Pérez-Ávila and José Antonio Morales-González
Instituto de Ciencias de la Salud, UAEH, México

Eduardo Madrigal-Bujaidar
Escuela Nacional de Ciencias Biológicas, IPN, México

María Teresa Sumaya-Martínez
Universidad Autónoma de Nayarit, Tepic, México

David M. Small and Glenda C. Gobe
Centre for Kidney Disease Research, School of Medicine, The University of Queensland, Brisbane, Australia

Maria-Luisa Lazo-de-la-Vega-Monroy and Cristina Fernández-Mejía
Unidad de Genética de la Nutrición, Departamento de Medicina Genómica y Toxicología Ambiental, Instituto de Investigaciones Biomédicas, Universidad Nacional Autónoma de México/ Instituto Nacional de Pediatría, México

Mario Nava-Villalba
Dentistry Department, School of Medicine.Autonomous University of Querétaro and Dentistry Department, Health Science Division, University of the Valley of México, Campus Querétaro, Querétaro, México

German González-Pérez
Dentistry Department, School of Medicine, Autonomous University of Querétaro, Querétaro, México

Maribel Liñan-Fernández
Dentistry Department, School of Medicine, Autonomous University of Querétaro, Querétaro, México

Torres-Carmona Marco
Dentistry Department, School of Medicine, Autonomous University of Querétaro and Genetics Department, Comprehensive Rehabilitation Center of Querétaro, Querétaro, México

Nalini Rajamannan
Molecular Biology and Biochemistry, Mayo Clinic School of Medicine, Rochester MN, USA

Claudia Camelia Calzada Mendoza
Section of Post graduate Studies and Research of Escuela Superior de Medicina- Instituto Politécnico Nacional, Street Salvador Díaz Mirón S/N, Colony Casco de Santo Tomás, Delegation Miguel Hidalgo, C.P. 11340, México D.F.

Carlos Alberto Jiménez Zamarripa
Hospital psychiatry "Dr. Samuel Ramírez Moreno"-psychiatric careservices- Secretaria de Salud, highway México-Puebla Km 5.5, Colony Santa Catarina, Tláhuac, C.P. 13100, México D.F.

Printed in the USA
CPSIA information can be obtained
at www.ICGtesting.com
JSHW011503221024
72173JS00005B/1186